CONSTRUCTION
ESTIMATING TECHNIQUES

Glenn M. Hardie
*British Columbia
Institute of Technology*

PRENTICE-HALL, INC., Englewood Cliffs, New Jersey 07632

Library of Congress Cataloging-in-Publication Data

Hardie, Glenn M.
  Construction estimating techniques.

  Bibliography: p.
  Includes index.
    1. Building—Estimates. I. Title.
  TH435.H326 1986   692′.5   86-9415
  ISBN 0-13-168741-7

Editorial/production supervision and
  interior design: Eileen M. O'Sullivan
Cover design: Photo Plus Art
Manufacturing buyer: John Hall

> "Man is the measure of all things"
> Protagoras (481–411 BC)

© 1987 by Prentice-Hall, Inc.
A Division of Simon & Schuster
Englewood Cliffs, New Jersey 07632

All rights reserved. No part of this book may be
reproduced, in any form or by any means,
without permission in writing from the publisher.

Printed in the United States of America
10 9 8 7 6 5 4 3 2 1

ISBN 0-13-168741-7 025

Prentice-Hall International (UK) Limited, *London*
Prentice-Hall of Australia Pty. Limited, *Sydney*
Prentice-Hall of Canada Inc., *Toronto*
Prentice-Hall Hispanoamericana, s.a., *Mexico*
Prentice-Hall of India Private Limited, *New Delhi*
Prentice-Hall of Japan, Inc., *Tokyo*
Prentice-Hall of Southeast Asia Pte. Ltd., *Singapore*
Editora Prentice-Hall do Brasil, Ltda., *Rio de Janeiro*

CONTENTS

PREFACE	*xii*
ACKNOWLEDGMENTS	*xiii*

PART I
Estimating — 1

CHAPTER 1 **INTRODUCTION TO ESTIMATING** — *1*

    1.1 General Objectives 1

        1.1.1 Objectives of the Book, 1
        1.1.2 Objectives of Estimating, 1

    1.2 Scope of the Book 2

        1.2.1 Preamble, 2
        1.2.2 Principles of Estimating, 2
        1.2.3 Construction Classifications, 2
        1.2.4 Construction Disciplines, 4

    1.3 Structure of the Book 4

        1.3.1 Preamble, 4
        1.3.2 Part I: Estimating, 4
        1.3.3 Part II: Measurement, 5
        1.3.4 Part III: Pricing, 5
        1.3.5 Part IV: Applications, 5
        1.3.6 Part V: Appendices, 5
        1.3.7 Part VI: The Project Drawings, 5
        1.3.8 Questions and Assignments, 5

    1.4 Guidelines for Research Assignments 6

        Questions 7

CHAPTER 2   ESTIMATING   9

- 2.1 Function 9
  - *2.1.1 Preamble, 9*
  - *2.1.2 Importance of Estimating, 9*
- 2.2 Process 10
  - *2.2.1 Preamble, 10*
  - *2.2.2 Process Components, 10*
- 2.3 Risk 10
  - *2.3.1 Preamble, 10*
  - *2.3.2 Prioritization, 11*
- Questions 11

CHAPTER 3   ESTIMATORS   13

- 3.1 Introduction 13
- 3.2 Knowledge 13
- 3.3 Skills 14
- 3.4 Abilities 15
- 3.5 Duties 15
- 3.6 Gender 16
- Questions 16

CHAPTER 4   ESTIMATES   19

- 4.1 Classes of Estimates 19
  - *4.1.1 Terminology, 19*
  - *4.1.2 Objectives of Classification, 19*
  - *4.1.3 Approximate Estimates, 20*
  - *4.1.4 Detailed Estimates, 20*
- 4.2 Contents of Estimates 21
  - *4.2.1 Preamble, 21*
  - *4.2.2 Approximate Estimates, 22*
  - *4.2.3 Detailed Estimates, 23*
- Questions 24

CHAPTER 5   CASE STUDY   25

- 5.1 Introduction 25
  - *5.1.1 Preamble, 25*
  - *5.1.2 Objectives, 26*
- 5.2 The Figures 26
  - *5.2.1 Instructions to Bidders, 26*
  - *5.2.2 Blank Bid Form, 26*
  - *5.2.3 Blank Checklist Form, 26*
  - *5.2.4 Blank Measurement Form, 27*
  - *5.2.5 Blank Pricing Form, 27*
  - *5.2.6 Blank Combination Form, 28*
  - *5.2.7 Outline Specification, 29*
  - *5.2.8 Completed Checklist, 30*
  - *5.2.9 Completed Measurement, 30*

Contents   v

        5.2.10   *Completed Pricing, 31*
        5.2.11   *Alternative Procedure, 31*
        5.2.12   *Estimate Summary, 32*
        5.2.13   *Completed Bid Form, 32*
        5.2.14   *Review and Conclusion, 41*

    Questions   41

PART II
Measurement   45

CHAPTER 6   **INTRODUCTION TO MEASUREMENT**   **45**

    6.1   General Objectives   45

        6.1.1   *Preamble, 45*
        6.1.2   *Analysis of Measurement, 45*

    6.2   Measurement in General   46

        6.2.1   *Preamble, 46*
        6.2.2   *Accuracy, 46*
        6.2.3   *Economy, 46*
        6.2.4   *Standards, 46*
        6.2.5   *Confidence, 47*
        6.2.6   *Flexibility, 48*
        6.2.7   *Simplicity, 48*
        6.2.8   *Repetition, 49*
        6.2.9   *Style, 49*

    Questions   50

CHAPTER 7   **METHODS OF MEASUREMENT**   **51**

    7.1   Introduction   51

    7.2   Methods for Contractors   51

        7.2.1   *Preamble, 51*
        7.2.2   *Net Measurement, 52*
        7.2.3   *Materials Take-off, 52*

    7.3   Methods for Designers   53

        7.3.1   *Preamble, 53*
        7.3.2   *Parameter Methods, 54*
        7.3.3   *Elemental Methods, 57*

    Questions   59

CHAPTER 8   **TECHNIQUES OF MEASUREMENT**   **61**

    8.1   Introduction   61

    8.2   Techniques for Detailed Estimates   61

        8.2.1   *Preamble, 61*
        8.2.2   *Checklist, 61*
        8.2.3   *Format, 62*
        8.2.4   *Identification, 62*
        8.2.5   *Recording Data, 62*
        8.2.6   *Preparation, 64*

          8.2.7   Classification, 64
          8.2.8   Description, 65
          8.2.9   Abbreviation, 65
          8.2.10 Dimensions, 65
          8.2.11 Extension, 67
          8.2.12 Units Symbols, 67

8.3   Techniques for Approximate Estimates 67

          8.3.1   Preamble, 67
          8.3.2   Checklist, 67
          8.3.3   Format, 68
          8.3.4   Classification, 68
          8.3.5   Description, 69
          8.3.6   Remainder, 69
          8.3.7   Parameter Estimates, 69

Questions 69

CHAPTER 9   SYSTEMS OF MEASUREMENT   71

9.1   Introduction 71

9.2   The Metric System 71

          9.2.1   Preamble, 71
          9.2.2   Development, 72
          9.2.3   System Components, 72
          9.2.4   Other Aspects, 74

9.3   The Imperial System 75

          9.3.1   Preamble, 75
          9.3.2   Development, 75
          9.3.3   Tables of Imperial Units, 77

Questions 77

CHAPTER 10   CALCULATION   79

10.1   Introduction 79

10.2   Review of Numbers 80

          10.2.1 Types of Numbers, 80
          10.2.2 Bases of Numbers, 80
          10.2.3 Conventions, 80

10.3   Arithmetical Functions 80

          10.3.1 Addition, 80
          10.3.2 Subtraction, 81
          10.3.3 Multiplication, 81
          10.3.4 Division, 81
          10.3.5 Explanatory Notes, 82
          10.3.6 Squaring, 83
          10.3.7 Percentages, 83
          10.3.8 Averaging, 84
          10.3.9 Perimeters, 85

Contents vii

    10.4   Other Functions 88

          10.4.1   Preamble, 88
          10.4.2   Algebraic Functions, 88
          10.4.3   Geometric Functions, 89
          10.4.4   Trigonometric Functions, 89

    10.5   Mental Calculations 90

    10.6   Calculators and Their Use 92

    10.7   Signs and Symbols 92

          Questions 93

CHAPTER 11   MENSURATION   95

    11.1   Introduction 95

          11.1.1   Preamble, 95
          11.1.2   Lengths and Areas, 95
          11.1.3   Volumes and Capacities, 95

    11.2   Definition of Terms 96

    11.3   Formulas 99

          Questions 99

PART III
Pricing   *101*

CHAPTER 12   INTRODUCTION TO PRICING   101

    12.1   General Objectives 101

          12.1.1   Preamble, 101
          12.1.2   Terminology, 101

    12.2   Factors of Cost 102

          12.2.1   Preamble, 102
          12.2.2   Direct Factors, 102
          12.2.3   Indirect Factors, 103

    12.3   Sources of Cost Data 103

          12.3.1   Preamble, 103
          12.3.2   Actualities, 103
          12.3.3   Probabilities, 104
          12.3.4   Other Issues, 106

          Questions 108

CHAPTER 13   ELEMENTS OF PRICING   109

    13.1   Introduction 109

          13.1.1   Preamble, 109
          13.1.2   Terminology, 109

13.2 Establishment of Data 112
- *13.2.1 Preamble, 112*
- *13.2.2 Factual Data, 112*
- *13.2.3 Productivity Data, 114*

13.3 Related Issues 118
- *13.3.1 Preamble, 118*
- *13.3.2 Components of Cost, 118*
- *13.3.3 Discounting, 119*

Questions 120

CHAPTER 14 TECHNIQUES OF PRICING 121

14.1 Introduction 121
- *14.1.1 Preamble, 121*
- *14.1.2 Basic Technique, 121*

14.2 Components of Cost 122
- *14.2.1 Preamble, 122*
- *14.2.2 Materials, 122*
- *14.2.3 Labor, 123*
- *14.2.4 Equipment, 124*
- *14.2.5 Overhead Costs, 127*
- *14.2.6 Profit, 130*

14.3 Application of Prices 130
- *14.3.1 Preamble, 130*
- *14.3.2 Quantity, 130*
- *14.3.3 Process, 131*
- *14.3.4 Technique, 131*
- *14.3.5 Conclusion, 131*

Questions 132

CHAPTER 15 BIDDING AND COST CONTROL 133

15.1 Introduction 133
- *15.1.1 Preamble, 133*
- *15.1.2 Objectives, 133*

15.2 Bidding Procedures 133
- *15.2.1 Preamble, 133*
- *15.2.2 Primary Elements, 134*
- *15.2.3 Secondary Elements, 134*

15.3 Cost Control 135
- *15.3.1 Preamble, 135*
- *15.3.2 Terminology, 136*
- *15.3.3 Process, 138*
- *15.3.4 Engineering, 139*

15.4 Contents of Contracts 140
- *15.4.1 Preamble, 140*
- *15.4.2 Contents, 141*
- *15.4.3 Elements, 143*

Contents ix

	15.5	Parties to Contracts 143
		15.5.1 Preamble, 143
		15.5.2 Configurations, 143
		Questions 145

PART IV
Applications 147

CHAPTER 16	INTRODUCTION	147

16.1 General Objectives 147

16.1.1 Preamble, 147
16.1.2 Materials and Methods, 147

16.2 Discussion of Selected Trades 148

16.2.1 Preamble, 148
16.2.2 02200 Earthworks, 148
16.2.3 03100 Concrete Formwork, 149
16.2.4 03300 Cast-in-Place Concrete, 150
16.2.5 06100 Rough Carpentry, 152
16.2.6 09250 Gypsum Drywall, 156

Questions 157

CHAPTER 17	MEASUREMENT FOR CONTRACTORS	159

17.1 General Objectives 159

17.1.1 Preamble, 159
17.1.2 Objectives, 159

17.2 Methods for Contracting 159

17.2.1 Preamble, 159
17.2.2 02200 Earthworks, 160
17.2.3 03100 Concrete Formwork, 163
17.2.4 03300 Cast-in-Place Concrete, 164
17.2.5 06100 Rough Carpentry, 165
17.2.6 09250 Gypsum Drywall, 165

17.3 Worked Examples 166

17.3.1 Preamble, 166
17.3.2 Trades Measurement 167

17.4 Additional Trades 202

17.4.1 Preamble, 202
17.4.2 Selected Trades, 203

Questions 206

CHAPTER 18	MEASUREMENT FOR DESIGNERS	207

18.1 General Objectives 207

18.1.1 Preamble, 207
18.1.2 Objectives, 207
18.1.3 List of Elements, 207

18.2 Methods for Designers 209

- 18.2.1 Substructure, 209
- 18.2.2 Structure, 209
- 18.2.3 Exterior Cladding, 210
- 18.2.4 Interior Partitions and Doors, 211
- 18.2.5 Vertical Movement, 211
- 18.2.6 Interior Finishes, 212
- 18.2.7 Fittings and Equipment, 212
- 18.2.8 Services, 213
- 18.2.9 Site Development, 213
- 18.2.10 Overhead and Profit, 214
- 18.2.11 Contingencies, 214

18.3 Worked Examples 214

- 18.3.1 Preamble, 214
- 18.3.2 Elements Measured, 215

18.4 Selected Parameters 215

- 18.4.1 Preamble, 215
- 18.4.2 Area Method, 216
- 18.4.3 Volume Method, 216
- 18.4.4 Additional Items, 216

Questions 216

CHAPTER 19 PRICING EXAMPLES—SELECTED TRADES 219

19.1 General Objectives 219

- 19.1.1 Preamble, 219
- 19.1.2 Objectives, 219

19.2 Methods for Contractors 219

- 19.2.1 Preamble, 219
- 19.2.2 02200 Earthworks, 220
- 19.2.3 03100 Concrete Formwork, 225
- 19.2.4 03300 Cast-in-Place Concrete, 230
- 19.2.5 06100 Rough Carpentry, 235
- 19.2.6 09250 Gypsum Drywall, 241

19.3 Conclusion 242

Questions 242

CHAPTER 20 PRICING EXAMPLES—SELECTED ELEMENTS 245

20.1 General Objectives 245

- 20.1.1 Preamble, 245
- 20.1.2 Objectives, 245

20.2 Methods for Designers 245

- 20.2.1 Preamble, 245
- 20.2.2 Element Identification, 246
- 20.2.3 Element Pricing, 250
- 20.2.4 Element Summary, 253

20.3 Worked Example 254

20.4 Practice Example 258

Questions 260

Contents

| CHAPTER 21 | COMPUTER APPLICATIONS | 261 |

21.1 Introduction 261
21.2 Main Elements 261

21.2.1 Computer Principles, 261
21.2.2 Hardware and Software, 262
21.2.3 Programming, 264
21.2.4 Applications, 265
21.2.5 Program Selection, 267

21.3 Case Studies 268

21.3.1 Preamble, 268
21.3.2 Case 1: General Contractor, 270
21.3.3 Case 2: Cost Consultant, 278
21.3.4 Case 3: Estimating Service, 284
21.3.5 Case 4: Design Consultant, 291
21.3.6 Case 5: Specialty Contractor, 295

21.4 Conclusion 302

21.4.1 Inferences, 302
21.4.2 A Final Word, 303

Questions 303

PART V
Appendices 305

APPENDIX A	ABBREVIATIONS	305
APPENDIX B	MASTERFORMAT	309
APPENDIX C	SELECTED BIBLIOGRAPHY	315
APPENDIX D	CONSTRUCTION ASSOCIATIONS	319
APPENDIX E	CHECKLISTS	323

PART VI
Project Drawings 327

1. Residential Foundation (imperial) IMP 1   331
   Residential Foundation (metric) MET 1   332
2. Retaining Wall Project (imperial) IMP 2   333
   Retaining Wall Project (metric) MET 2   334
3. Fireplace Construction (imperial) IMP 3   335
4. Wood-frame House (metric) MET 3   336
5. Post-and-beam House (imperial) IMP 4   339
6. Retail Store Building (metric) MET 4   341

INDEX 349

PREFACE

A TRILOGY

This book on "Construction Estimating Techniques" is the second in a series of three books, prepared by this author. The first book dealt with "Construction Contracts and Specifications," and the third one, still in preparation at this date of publication, will deal with "Construction Materials and Methods." Although these books have been prepared primarily for use as text books in an instructional setting, they may also find their place as reference texts in design and construction offices, in field settings, and of course, in public and private libraries.

All three books have been written so as to fit the information in each to the others in such a manner that all three may be studied in sequence or simultaneously, in whole or in part, in individual or group settings. Together, they form a fairly comprehensive body of reliable information of practical use and current value to many of the participants in the construction industry, such as architects, contractors, draftspersons, engineers, estimators, inspectors, lawyers, project managers, specifications writers, students, superintendents, suppliers, technologists, and others.

ACKNOWLEDGMENTS

In general, acknowledgment is made of the contributions and influences of the many authors, clients, colleagues, colleges, employees, employers, friends, instructors, organizations, peers, professors, students, universities and the rest from whom I have learned many of the valuable lessons which I have been able to set down in this book for the benefit of others.

In particular, acknowledgment is made of the excellent cooperation which I received from all of the various government agencies, professional associations, publishers, and others from whom I requested permission to refer to or reproduce materials of one sort or another for use in this book. Their names are respectively and gratefully noted in various places throughout the book, wherever appropriate.

Finally, I would like to thank the following individuals for their more specific contributions to the production of this book, and to alphabetically acknowledge their cooperation:

1. Gerry Berkenpas, Senior Instructor, Building Department, British Columbia Institute of Technology, for his fine work in preparing the working drawings for all of the projects included in this book.
2. Keith Collier, Chartered Surveyor, Lecturer in Construction Management, Douglas College, for sharing his knowledge of construction, his experience in instruction, and his personal friendship over the last quarter century.
3. Derek Hale, John Lancaster, and Anna Maharajh who teach construction estimating at the British Columbia Institute of Technology, and each of whom contributed directly and indirectly with suggestions and ideas for the book.
4. My wife Lorraine, my son Lindsay, and my daughter Karen, without whose collective cooperation the book would have taken twice as long to write, with half as much satisfaction and enthusiasm.

Glenn M. Hardie
Vancouver, 1986

PART I ESTIMATING

CHAPTER 1
INTRODUCTION TO ESTIMATING

1.1 GENERAL OBJECTIVES

1.1.1 Objectives of the Book

The skills of measurement and pricing of construction work can readily be acquired by any person of average intelligence and competence who has reason and motivation to do so. The general objective of this book is to assist the reader to acquire these skills, by presenting the necessary knowledge and the opportunity to apply such knowledge in a graduated series of structured steps, consisting of some reading, some research, and some application.

It should be understood at the outset, however, that no amount of skill in either measurement or pricing will be of much use to those who do not have a good knowledge of construction materials and methods. One has to be able to visualize and understand building construction systems and processes before a realistic attempt can be made to estimate the probable costs of such systems or processes.

In this introductory portion of the book, the specific knowledge and skills that the competent estimator must acquire are listed in some detail, preliminary to discussion of measurement and pricing techniques. The inherent abilities and some of the assigned duties of the estimator are also listed for review. At the risk of oversimplification, it may be said that to derive the fullest benefit from this book, the reader or student of estimating should bring to this task most of the background knowledge and abilities described in the lists; the skills and duties of the estimator are, in large part, the subject matter of this book.

1.1.2 Objectives of Estimating

In general, the objectives of estimating in construction are to establish the following criteria about any project under consideration:

1. *The project.* The first objective is to discover the location, nature, style, magnitude, quality, and timing of the work.

2. *The documentation.* The second objective is to discover the amount, quality, and type of the information being presented on which the estimate will be based.
3. *The responsibilities.* A third objective is to discover and distribute legal and contractual responsibilities among the various parties who may become involved in the project.
4. *The probable costs.* The fourth objective is to establish probable costs with as much certainty as possible in the given circumstances of time and place.
5. *The estimating system.* A final objective is to achieve the foregoing objectives by the development and application of a reliable and practical system of investigation, measurement, pricing, and summarization which can be integrated into the other operations of a design or contracting organization, whether in the public or private sector of the construction industry.

1.2 SCOPE OF THE BOOK

1.2.1 Preamble

Many books on estimating consist primarily of information on construction materials and methods, with only a small overlay of measurement and pricing scattered throughout the rest of the material. In such books, the estimating elements are often repetitively and superficially presented in general terms, and rely on the memorization of a collection of rules of thumb, often based on the personal experiences, opinions, and biases of the author.

In this book, a deliberate policy has been adopted to focus explicitly on estimating matters, and only incidentally to support explanations or illustrations of these matters with examples of construction materials or methods to the minimum extent necessary to make specific estimating issues clear. In this manner, the attention of the reader will not be diverted from the estimating issues with digressions into construction issues.

1.2.2 Principles of Estimating

A second policy has been adopted to explain estimating issues on the basis of principle first, and only incidentally to make some reference to related rules of thumb, appropriate to the context, and only where such rules are consistent with the principles being established. It is considered important that novices understand the measurement and pricing principles underlying such rules of thumb, for two reasons: first, to apply such rules correctly, and second, to be able to modify existing rules or indeed develop new rules for specific uses.

A third and more difficult policy has been adopted to present practices and techniques which are widely used in the construction industry, as distinct from presenting the narrower views of one author or the practices of one company. The reader should be able to adapt and apply the general principles and precepts recommended and illustrated in this book to suit virtually any construction company configuration or policy.

1.2.3 Construction Classifications

The construction industry embraces many types of construction work, all of which can be classified under two broad headings: **heavy construction,** which includes

Residential construction.

bridges, dams, roads, tunnels, mills, and the like; and **building construction,** which includes residential and nonresidential construction. Residential construction, which comprises private dwellings and small apartment blocks, is so specialized that it can almost be considered as a separate industry. Nonresidential construction includes all commercial and institutional construction, such as auditoria, banks, churches, colleges, courthouses, department stores, garages, hospitals, jails, libraries, office buildings, schools, shopping centers, and the like. Buildings such as large apartment blocks and multiunit condominia, although built for residential occupancy and use, generally also fall into the commercial class of buildings.

Commercial construction. [From S.W. Nunnally, *Construction Methods and Management* (Englewood Cliffs, N.J.: Prentice-Hall, Inc., 1980).]

This book does not deal with heavy construction as defined above. With respect to building construction, the main emphasis in this book is placed on commercial and institutional construction, although the principles and practices have equal application to residential work, for which the actual form of estimating may be modified by simplification to some extent. Some examples of estimating applied to residential construction are included in the book for illustration and discussion. Furthermore, while it is possible to make a number of distinctions between commercial and institutional building types and uses, it is not necessary to do so for the purposes of estimating probable construction costs; the processes are essentially identical.

1.2.4 Construction Disciplines

Construction can be also be subdivided into a number of major disciplines, such as structural work, architectural work, mechanical work, electrical work, among others. This book focuses primarily on *structural and architectural work,* with only coincidental applications in mechanical, electrical, and other work. The main reason for this selection of disciplines arises out of the fact that the cost estimation of structural and architectural work embraces consideration of work done on a numerical, lineal, superficial, and volume basis. If measurement and pricing of structural and architectural work is understood, there will be little difficulty encountered in transferring the principles and practices of such estimating techniques to these other disciplines. A secondary reason for making this selection among these disciplines is to shorten the text (and thus reduce its cost while improving its utility) by eliminating unnecessary repetition of statements of principle and practice.

The context of the book is North American, excluding Mexico. The principles and procedures described herein are based on and specifically suited to work in both Canada and the United States. Approximately equal weight is given to Canadian and American practices, and an equal amount of illustration is given in both the metric and imperial systems of measurement. With minor modifications to suit local practice, the main elements of the book may be used in most other English-speaking countries.

Although most illustrations included in the book exemplify good practice, a number of deliberate errors have been introduced into the text, to bring out certain points for consideration or debate. All such errors are clearly indicated. In some cases, modification of the exemplified material will be necessary before it is included in an actual estimate or contract or office procedure. Cost data shown in worked examples are not necessarily representative of actual prices, and will *in all cases* need to be modified to suit local conditions and current timing.

1.3 STRUCTURE OF THE BOOK

1.3.1 Preamble

The book is divided into six major parts. This arrangement permits study of the subject matter contained in these parts in any sequence appropriate to the needs of the reader. It also permits more rapid reference to specific parts of the text.

1.3.2 Part I: Estimating

The first part introduces the general subject matter of the book: *estimating, estimators, and estimates.*

1.3.3 Part II: Measurement

Part II deals with aspects of *measurement,* regarding both principle and application, of interest to people on both sides of the contractual "fence," such as designers and contractors. Several different approaches to measurement of interest to architects and engineers, as well as to contractors, subcontractors, and suppliers, are presented in a logical sequence for consideration and adoption for trial. Wherever practical, measurement examples have been based on the project drawings included in Part VI.

1.3.4 Part III: Pricing

Part III deals with aspects of *pricing,* embracing the establishment of cost data, the inclusion of such data into unit prices, and the application of such prices to measured quantities of construction work. The principles of pricing are shown by their application to actual examples, based on items of work selected from the project drawings or measurement illustrations.

1.3.5 Part IV: Applications

Part IV consists of an organized sequence of *worked examples,* to show the principles of measurement and pricing applied to selected portions of the projects represented by the drawings in Part VI. While these worked examples cover all of the principles discussed in Parts I, II, and III, they do not, and are not intended to, cover every possible measurement and pricing problem that an estimator might encounter in trying to determine the probable cost of any portion of work or trade responsibility that might arise on any construction project.

It is intended that the estimator will make the necessary adaptations of the material as presented, to make it suit the specific problems or exigencies of his or her particular interests or responsibilities. Furthermore, the student or reader will find a number of measurement and pricing exercises and assignments included in Part IV, to present an opportunity to practice application of the principles described in the text, using the worked examples contained in the text as models on which to base the practice.

1.3.6 Part V: Appendixes

Part V comprises appendixes appropriate to the subject matter of the book, including lists of abbreviations, an excerpt from the Masterformat, a bibliography, a list of trade associations, and checklists for use by estimators.

1.3.7 Part VI: The Project Drawings

Part VI contains drawings for some small buildings, for use as necessary as vehicles to provide worked measurement and pricing examples throughout the rest of the book. These drawings were selected because of their simplicity, clarity, and universal application. Although two of these buildings are residential houses, the styles of construction are fairly typical of common commercial and institutional work.

1.3.8 Questions and Assignments

At the end of each chapter, a series of questions has been prepared and inserted, designed to test the reader's reaction to and comprehension of the various topics presented in the chapters. The answer to each question is to be found in the preceding chapter. At the end of each part, a series of research assignments has been included, each of which is intended to cause the reader to investigate issues related to the content of the book. Practical measurement and pricing exercises are also included on a graduated basis from simple to difficult.

As a general policy, worked examples throughout the book have been based on the trade work to be found in excavation work, concrete and formwork, carpentry work, and the work of one finishing trade. These trades were selected for the following reasons:

1. They are relatively common on most construction projects and are fairly uniform in process and procedure throughout Canada and the United States.
2. They permit presentation of problems ranging from very simple to relatively difficult, in both measurement and pricing.
3. They permit study and discussion of the four basic modes of unit measurement: numerical, lineal, superficial, and volumetric.
4. It is thought better to make a detailed study of a limited number of typical trade types than to try to cover every aspect of every trade, which is, of course, virtually impossible to do in any case.
5. If the reader or student can master the estimating problems presented by these four basic disciplines, the adaptations necessary to handle comparable problems in other disciplines will be minimal and relatively trouble-free.

Finally, it might be said that the subject matter of this book generally covers the period of time between two points, as follows:

1. In the *designer's* office, from the time a commission to design a project is negotiated with a client up to the time when the working contract documents are ready for bidding
2. In the *contractor's* office, from the time the decision is made to bid on a construction project up to the time when the bid has to be submitted to the designer or owner

The construction industry is rapidly moving toward more widespread use of microcomputers in design and construction offices. The measurement and pricing principles and practices enunciated in this book have equal application to both manual and electronic processes in estimating. However, the techniques necessary to operate microcomputers are excluded from the book, although a few examples of their results are included for illustration purposes.

1.4 GUIDELINES FOR RESEARCH ASSIGNMENTS

The research projects included at the end of each major part of this book are intended to provide an opportunity for the reader to investigate some issues that affect the estimating process in a broader sense than the somewhat restricted parameters of the book. These guidelines are intended to assist students and their instructors to develop the full benefits that will accrue from a conscientious effort to develop the issues under a controlled situation, along the following lines:

Read the section or chapter in the book dealing with the assigned topic. Prepare a written outline (between 3000 and 4000 words) of those topics considered to be of interest and importance relative to the topic under review. Discuss the draft outline, first with the instructor or mentor for the course (if any), and then with experienced personnel in the local construction industry. Revise the draft as necessary, if possible in consultation with the instructor. Submit the final report (of approximately 5000 words) for grading and/or group discussion.

Reports should be properly organized, as recommended for any formal term or thesis paper, and be accompanied by appropriate appendixes and acknowledgments. Reports can be graded on the following basis:

1. Accuracy of content   40%
2. Proper use of language 30%
3. Organization of report 20%
4. Bonus for excellence   10%

Topics other than the ones listed can, of course, be added or substituted to reflect the interests of a particular group, region, or development in the industry.

QUESTIONS

1.1. Identify and briefly describe any two of the five objectives of estimating as described in this chapter.
1.2. Distinguish between the terms "heavy construction" and "building construction," by briefly defining these two terms.
1.3. Give one good reason why cost data appearing in a book such as this must be modified before use on an actual project.

CHAPTER 2
ESTIMATING

2.1 FUNCTION

2.1.1 Preamble

Every construction and design company must perform a number of management functions in order to get into and to stay in business. The primary functions of any business include production, financial control, and selling. To perform these functions, a number of activities must be undertaken, some of which include accounting, financing, office and plant management, procurement, purchasing, and public relations, and some others, one of which is estimating.

In turn, estimating is one of the few activities in every construction or design firm that affects and is affected by elements of all three primary functions. The ability of firms in the business of construction design and implementation to bring their projects in on time and budget is a prime factor in the establishment and maintenance of their reputations and their economic lives.

2.1.2 Importance of Estimating

It is important to recognize right at the beginning of this study the importance of estimating in the affairs of a construction or design company. The success of the business will be directly proportional to the quality of the estimating process and the reliability of its results. The construction industry is inherently risky; construction and design firms go bankrupt at a clearly higher rate than the average for all types of business. If it is in the nature of construction to be risky, it logically follows that no more risks should be taken than are absolutely necessary. If all construction activities are divided into two categories, those that occur *outside* the company and therefore beyond its significant control, and those that occur *within* the company and therefore within its direct control, it seems only prudent to remove as much of the risk as possible from all internal activities, not least of which is the activity of estimating, as has been established above.

2.2 PROCESS

2.2.1 Preamble

What, then, is estimating, and what is the estimating process? **Estimating** is, by definition,[1] the formation of an approximate judgment or opinion regarding the value, amount, size, or weight of something. In construction parlance, an estimate is an expression of opinion or the prediction of the probable future costs of certain construction activities, usually based on some data having an acceptable degree of reliability.

One often hears of estimators or contractors using terms like "guestimating" or "throwing in a bid," but as will be shown, such casual attitudes have no defendable place in estimating, or elsewhere in construction, for that matter. The basic principles of estimating are not difficult to absorb, and they are easy to apply in practice. Risk in estimating will diminish as the principles are understood and the skills developed and applied.

2.2.2 Process Components

The estimating process can be divided into two major components: measurement and pricing. There are some other procedures very closely allied to the estimating process—indeed, some say they are part of the process, such as bidding and cost control. A number of these related topics will be addressed in this book, to show their connection to the estimating process, but for the time being, attention will be focused on measurement and pricing, both of which can be conveniently subdivided into three main parts for study.

In *measurement,* the three parts are **descriptions** of work, establishment of **dimensions,** and the calculation of **quantities** of work. In *pricing,* the three parts are the **establishment of data** related to cost and productivity, the **computation of prices** based on the data, and the **application of prices** to the measured quantities of work. Each of these six elements is examined in detail in subsequent chapters.

2.3 RISK

2.3.1 Preamble

As has been said, construction work involves a certain amount of risk taking. It can also be said that risk is directly proportional to the amount and quality of available information. One objective of the estimating process is therefore to maximize information about any given set of proposed construction circumstances, to minimize the unknowns and therefore the risks, and to permit reasonably reliable predictions to be made with respect to construction costs.

Furthermore, it can be said that not every part of every construction project is of equal importance or complexity; the various component parts that go to make up the whole each have varying degrees of economic or other importance. A second objective of the estimating process is therefore to enable varying degrees of attention to be focused on those parts of the construction project that are going to generate more or less cost. In this regard, it is useful to know that there is a precept, articulated by an Italian economist named Parado, which suggests that *a minority* of the items of work in any project will generate the *majority* of costs for that project.

[1]*The American College Dictionary* (New York: Random House, Inc., 1960).

Others have expressed the same idea by making reference to the "significant few" as distinct from the "insignificant many."

2.3.2 Prioritization

Estimators have developed this useful precept into what is now referred to as the **20/80 rule,** which is to say that 20% of the work will generate 80% of the cost, and vice versa. The significance of this precept is that particular care must be taken to clearly identify and accurately estimate the cost of those few items that will contribute most of the cost, while relatively less attention may be paid to the remaining more abundant but less significant items of work and cost. The necessity of making this distinction comes about from the pressures that are imposed on every estimator; there is simply never enough time in the estimating process to give equal consideration to every element of design and construction and to every aspect of cost on any given project. Risks also arise in proportion to the amount of time and pressure on the estimator to do his or her work.

QUESTIONS

2.1. Identify and briefly comment on any one of the three primary functions of any business.

2.2. Explain why the secondary business function of estimating is of such importance to a construction company.

2.3. Define the term "estimating" as it is used in construction.

2.4. Identify and briefly describe each of the two primary or major components of the estimating process.

2.5. Identify the three primary constituent parts of either of the two components identified above.

2.6. Briefly explain the 20/80 rule and show its application to construction estimating.

CHAPTER 3
ESTIMATORS

3.1 INTRODUCTION

Competent estimators possess certain bodies of *knowledge,* demonstrate certain *skills,* and display certain *abilities* in the execution of their duties. Some of the more important of these attributes are identified in the following sections as a guide to those who aspire to become competent in this field.

3.2 KNOWLEDGE

The construction estimator must have reliable knowledge of the main elements in each of the following categories of information, which have been listed in alphabetical order for ease of reference and to avoid prioritizing of importance.

1. *Bidding.* The procedures used for bidding and bid depository in the local region should be fully understood, as well as the nature of bid strategies used by the estimator's company.
2. *Contracts.* The types and contents of most of the common forms of construction contract should be familiar to the estimator, together with knowledge of significant legal decisions and legislation affecting the construction industry.
3. *Costs.* The estimator must have knowledge of the main factors affecting the costs of labor, materials, equipment, and other elements, together with sources of reliable data and methods of converting such data to other more useful forms.
4. *Cost accounting.* Knowledge of the principles and practices of construction cost accounting are essential for the competent estimator to do his or her work.

5. *Documentation.* The estimator must know how technical information is organized in the construction industry, and be familiar with the terminology used in construction specifications and contracts, as well as with current graphical symbols used to depict different types of work in construction drawings.
6. *Materials.* It is essential that the estimator be fully conversant with the physical and chemical properties of common construction materials, products, and systems, particulary those relevant to the special interests of the company for which the estimator works. A special study should be made to discover materials which are incompatible with others or which have inherent difficulties in handling, preparation, application, or removal.
7. *Methods.* Estimators require detailed knowledge of the way in which specific parts of buildings are actually assembled in the field or in the factory, with special emphasis on the work of their own trade.
8. *Productivity.* The estimator should become and remain familiar with labor productivity and equipment output factors used to gauge the amount of effort required for various types of work.
9. *Principles.* Estimators should have a general knowledge of the principles of the common forms of construction, such as concrete, masonry, steel, and wood-frame, as well as knowledge of the common laws of physics and building science having to do with corrosion, gravity, moisture control, movement, and the principles of leverage, to mention a few of the more useful items.
10. *Regulations.* There is a vast array of codes, standards, regulations, and legislation which apply to construction processes and cause either opportunities or restraints in the use of specific materials, methods, or systems. A general knowledge of this field of interest is invaluable to the estimator, coupled with detailed knowledge of particular limitations specifically applicable to the estimator's company and its sphere of activities.

3.3 SKILLS

The estimator must either possess or acquire a reasonable degree of skill in each of the following categories.

1. *Calculation.* Although it is not necessary to have a profound knowledge of higher mathematics, every estimator should be skillful in the application of the basic arithmetical functions of addition, subtraction, multiplication, and division, and be able to apply some simple algebraic and trigonometric functions from time to time. Skill is required in both mental and machine calculation.
2. *Communication.* The estimator must be able to communicate ideas to others using a variety of verbal, graphical, and numerical representations. He or she should therefore be reasonably skillful in writing, in drafting or drawing, and in dealing with numbers.
3. *Interpretation.* The estimator must be able to correctly interpret meanings of verbal, graphical, and numerical data, by recognizing overt statements of fact, and by understanding the significance of implied obligations that may be inferable from given data.

Chap. 3 / Estimators   15

4. *Measurement.* Skill in techniques of measuring quantities of materials, labor, or other items, and skill in recording and using such measured data is clearly a requirement in estimating.
5. *Mensuration.* The estimator must be able to correctly apply the appropriate formula for a number of basic geometric plane shapes and solid figures, to determine perimeters, areas, and volumes of whole or part figures.
6. *Negotiation.* Skill in the art and technique of negotiation is useful, with respect to adjusting conditions of contract, making changes to work in progress, and manipulating cost and other data to bring about results favorable to oneself or one's clients.

3.4 ABILITIES

To the extent that the estimator has the following attributes of character, his or her work will be more valuable and enjoyable.

1. *Clarity of thought.* The ability to think clearly and in an organized manner helps to identify problems and to discover their solutions, which is in essence the primary task of the estimator.
2. *Creativity in ideas.* The ability to see creative or imaginative solutions to new or everyday problems is clearly helpful to the estimator; estimating is as much art as science.
3. *Independence of action.* Although the estimator may follow some common and widely used procedures in his or her work, the ability and strength of character to make independent decisions and to follow separate courses of action with conviction is a desirable trait in an estimator.
4. *Punctuality.* Time and timing are of great importance in estimating; therefore, the ability to be punctual and dependable in keeping track of the time and appointments will be to the advantage of both the estimator and the people with whom he or she has to deal.
5. *Problem solving.* The ability to apply sound problem-solving methodology to the many difficulties and uncertainties that confront the estimator is a most useful attribute.

3.5 DUTIES

In most construction and design organizations, the estimator will perform a number of duties, some of which are listed below, again in approximate alphabetical order. More specific tasks will be added later in the book, in connection with particular activities involved in the measurement and pricing processes.

1. *Bidding.* The estimator will be involved in the assembly of cost and other data for insertion into the bidding process.
2. *Documents.* The estimator will acquire temporary possession of the documents proposed for use in construction of the project under consideration.
3. *Control.* The estimator will usually make field tours to inspect construction work under way and to gather cost and other data for purposes of accounting and control.

4. *Conversion.* The estimator will often be involved in activity to convert data from one form or source to another more appropriate form for use in the estimating/accounting cycle.
5. *Measurement.* The estimator will extract the quantities of work to be done from the construction documents, using standard techniques and stationery.
6. *Negotiation.* The estimator is usually involved in negotiations that lead to a construction contract satisfactory to the principal parties involved.
7. *Pricing.* The estimator will prepare unit prices from cost data and apply these prices to the totals of work previously measured.
8. *Professional development.* The estimator has a duty to keep abreast of new building and office technology and developments in contract procedure. This activity can to some extent be accomplished by becoming a member of appropriate professional organizations, attending meetings of interest to estimators, and by reading related elements of the local and national technical press.
9. *Site visit.* The estimator will almost always make a field trip to examine the site of any proposed new construction in which his or her company may be interested. One reason for this duty is that, in the event of a subsequent dispute concerning some aspect of the site that such a visit might have disclosed, a person who has made the visit will probably be in a stronger legal, technical, and psychological position.

3.6 GENDER

Terminology. For the sake of simplicity and consistency, the use of gender will be avoided wherever possible throughout this book. In any instance where male gender occurs, the female gender can be substituted without difficulty. Some words, such as "journeyman" (meaning a qualified tradesperson), have traditional meanings and have not yet been modified to reflect the new realities. Such words will occasionally be found in context in this text. As more women enter the field of construction design and management, changes in terminology can be anticipated.

Women in Construction. Although employment in the construction industry has not traditionally been a goal for women, in recent years, increasing numbers of women have found acceptable management positions in the construction industry as cost accountants, architects, engineers, estimators, plan checkers, and quantity surveyors. There are also many being employed in various construction trades, such as carpentry, masonry, painting, and welding.

Associations. In both the United States and Canada, there are associations that cater especially to the needs of women in construction. The addresses of several of these organizations are given in Appendix D, together with the addresses of other pertinent associations of interest, use, and value to all estimators, male or female.

QUESTIONS

3.1. Ten areas of knowledge specific to the needs of the estimator have been identified in this chapter. Select any two of these and comment briefly on their particular significance.

3.2. Identify two situations where the skills of negotiation are of use to the estimator. Briefly justify your selections.

3.3. Why should the estimator develop habits of punctuality?

3.4. Give two reasons why estimators should visit the sites of projects on which their companies are about to bid.

3.5. Give two reasons why estimators should visit the sites of projects which their companies are in the process of building.

CHAPTER 4
ESTIMATES

4.1 CLASSES OF ESTIMATES

4.1.1 Terminology

Estimates of materials, personnel, equipment, time, money, or other resources can be classified under two broad titles: **approximate estimates** and **detailed estimates.** Approximate estimates are normally prepared for or by owners, developers, and designers, usually for budget purposes; detailed estimates are normally prepared for or by contractors, subcontractors, and suppliers, usually for contract purposes.

As with so much terminology in building technology, these two terms are somewhat imprecise, because many approximate estimates are quite detailed, while many detailed estimates require considerable approximations to be made. The word "approximate" used in this context certainly does not imply that there is anything casual or offhand about this approach to estimating. The estimating process is always applied with as much care as the available data will permit, to produce a price that will be as close as possible to the probable ultimate reality.

4.1.2 Objectives of Classification

There are many distinctions that can be made between approximate and detailed estimates, but one which is worthy of note at this point is that whereas in a *detailed* estimate, the objective is usually to determine the smallest quantity and the keenest price for the various items of work in the project, in an *approximate* estimate a more liberal view of quantities and prices may be taken. The estimator who is preparing such approximate estimates is not bidding to do the work at a specific price, but is instead offering advice to a client as to how much the building portion of the overall project should cost. It almost goes without saying that care must be taken with all such estimates, to avoid accusations or action with respect to professional negligence if the estimate is later found to be substantially in error.

Both types of estimates may be prepared internally by paid in-house employees of designers or contractors, or they may be prepared externally by cost consultants who specialize in doing this type of work for a fee. It might be noted in passing that whereas approximate estimates are done in several stages, each to confirm the previous stage, detailed estimates are usually done only once, before the award of a contract to do the work based on that estimate. All estimates should be thoroughly checked for accuracy and freedom from significant error.

4.1.3 Approximate Estimates

As stated, the approximate estimate is usually prepared by or for owners, developers, or designers, to establish the probable costs of the construction portion of the overall development budget. Such costs are part of the so-called **hard costs,** as they deal with the actual construction component of the development. Other costs, such as for design fees, financing, marketing, and the like, are referred to as the **soft costs.**

Large variations in the type, amount, and quality of available data can be expected at various stages in the preparation of the design for a construction project. At the earliest stage, there may be very little information of any kind, because the owner's need and the designer's concepts may be only vaguely identified or formed. At the intermediate stage, the needs of the owner will have been more closely analyzed and the design crystallized to produce a program having space requirements defined and construction style and quality determined. At the final stage, most of the design concepts have been translated into working drawings, detailed specifications, and other forms of relatively accurate data.

Approximate estimates are often prepared at some point during each of the foregoing three major stages: first, to establish the general economic feasibility of the proposals; second, to modify either the proposals or the design based on the proposals, to bring one into line with the other; and third, to confirm or further modify the predictions or earlier expectations before the proposed construction documents become part of an actual construction contract. Although two or three approximate estimates are usually sufficient for most designs, on large or complex projects as many as five or six estimates may be made. On many government or public works projects, such budget control is mandatory, to minimize or justify cost overruns.

4.1.4 Detailed Estimates

As stated, the detailed estimate is usually prepared by or for contractors, subcontractors, and suppliers, prior to bidding on or negotiating to do the work required by a proposed construction project. As the words imply, the detailed estimate involves careful and correct measurement of quantities of work required by the proposed contract documents, and the subsequent calculation and application of accurate unit prices to these quantities, to produce the probable total costs of the work to be done.

In detailed estimates, variations in the type, amount, and quality of **detail** may occur. Such variation is normal and necessary, and if properly handled by designers and contractors, variations need not cause much problem for either one.

Type. The detail necessary to estimate one kind of work may not be suitable for another kind. For example, the details required to estimate costs of concrete work are different from those required to estimate costs of reinforcing steel work to concrete, because the nature of the work is different in these two trades. In concrete work, we need to know the length, width, depth, and composition of each component, whereas in reinforcing steel work, we need to know the length, the spacing, the quality, and the weight or size of each bar in each concrete component. In other work, such as in electrical or mechanical systems, the work may be represented in a schematic or symbolic form, whereas in excavation work, the work is really only implied on the drawings and the specific detail is to be found in verbal description in specifications and data tabulated in soils reports.

Amount. The key issue has largely to do with the placement of responsibility for the success of the outcome or result of the design/build process. For those aspects of the project for which the designer will assume responsibility, an adequate amount of detail must be made available to the contractor; for the remainder, less detail may suffice. For example, the amount of detail necessary to estimate the cost of concrete work is far greater than the detail required to estimate the cost of formwork for that concrete, notwithstanding the fact that the formwork is more complicated to design and build, and therefore to measure and price, than the concrete work. The designer must tell the contractor all that the contractor needs to know about the concrete work, but the contractor will then decide for himself how to make and assemble the necessary formwork for the detailed concrete design; he usually does not require or expect the same amount of detail to be presented in the contract documents. This particular example can be proven by reference to the drawings and specifications for almost any concrete framed building, where the documents will show the concrete work in considerable detail, but there may be no mention of the formwork system that the contractor is expected to provide to work the concrete into the shapes required by the design.

Quality. The quality of detailing in the drawings and specification data will indicate the standard of professional competence which the designer brings to this task. Some designers are very conscientious about providing an adequate amount of correctly drawn or specified details of the right type for use by estimators, inspectors, and others involved, whereas other designers seem to presume that the details will somehow work themselves out as the work proceeds. The details will probably work themselves out all right, but not necessarily in a manner that the designer would like to see. In some types of work, such as architectural woodwork or ornamental metalwork in a commercial or institutional building, good-quality detailing is essential, in contrast, say, to standard wood framing in a simple residential building, where the quality of detailing may be of a much lower order, because of the probable familiarity of the builder with the intended end result.

Specific instances of type, amount, and quality of detail necessary for estimating the costs of construction work will be illustrated in many of the worked examples presented later in the book.

4.2 CONTENTS OF ESTIMATES

4.2.1 Preamble

The contents of all estimates, whether approximate or detailed, include the following four basic parts:

1. The **standards of quality** to establish the requirements of the design and the client
2. The **measurement of quantities** of commodities of one type or another

3. The **development of unit prices** to be applied to the measured quantities
4. The **summarization of costs** of the commodities, together with allowances or other contingencies to arrive at the probable total cost

As can be imagined, the techniques for developing the constituents for each of the foregoing parts for both classes of estimates will be different, because of the differing types of data available and the differing needs of the designers and owners on the one hand, and the contractors and suppliers on the other hand. Each of these four basic parts is explained in greater detail in the following sections.

4.2.2 Approximate Estimates

As mentioned above, the contents of an approximate estimate consist of four parts, each identified below with a short title and a brief description.

1. *Criteria.* The designer or owner must establish some criteria or standards of quality to be expected in the proposed development. These standards may encompass building use or occupancy, floor-space requirements, general building type, and quality of anticipated construction.
2. *Elements.* The designer must analyze the proposed building to determine its probable component parts and then select some appropriate units by which the various parts of the building can be adequately measured.
3. *Pricing.* The designer must gather, assemble, and apply appropriate cost data to price the measured components.
4. *Summary.* The designer must tabulate all of the probable quantities and related costs and include all likely allowances to develop the estimated probable total construction cost of the proposed building.

In most cases, the owner is usually satisfied with an estimated cost, either stated as a single figure or stated as a price within a given range of high and low figures. On occasion, an owner may wish to know the costs associated with some major or minor feature in whole or in part; the summary should be arranged to show such detail and to permit some modification upon instruction by the owner.

Designers need to know the approximate amounts of money allocated to the various components of the proposed building: first, to be able to give their own consultants pertinent cost information about the budget for their share of the total design, and second, to keep track of the cost of the designer's own share of the design as it is further refined and developed into the working document stages. Essentially, this can be done in one of two ways: either the designers establish the amount of money to be made available for each major portion of the building and then insist on the cost of the design of that portion being kept within that budget or target figure, or the designers and their consultants prepare whatever design they feel is appropriate for the various portions of the project and then determine what that design will cost in whole and in part.

To conclude, the scope of approximate estimates should cover the costs of the design developed by the primary designer, the costs of the designs prepared by the designer's consultants, and the additional allowances made for overhead, profit, inflation, and any other allowances or contingencies that might legitimately or judiciously be included in the overall budget figure presented to the owner for approval. For a small project, such as a private dwelling or a small commercial building, the designers may prepare an approximate estimate completely with their own forces within their own organization. On larger projects, the designer may enlist the help of consultants with parts of the cost determination, or indeed may even retain a specialist cost consultant, such as a building economist or a quantity surveyor. On many government projects, such as schools, hospitals, and housing developments funded with federal tax dollars, such specialist help is mandatory, as experience has shown that many otherwise capable designers show little competence or interest in establishing realistic budgets. The reasons for such behavior would make an interesting study.

4.2.3 Detailed Estimates

Parallel to the approximate estimate, the contents of a detailed estimate also consist of four parts, identified below by their common or colloquial titles, each followed by a short explanation.

1. *The spec notes.* The estimator must review the proposed conditions of contract, the drawings, the specifications for the work of the various trades, addenda, and all other information issued from the designer's office or otherwise made available for examination in connection with the project in question. Careful notes must be made of all issues likely to affect cost, time, or quality. Such notes, although brief, should be self-explanatory and should be cross-referenced to the particular contract conditions that caused them to be made in the first place. The estimator should always visit the site of the proposed construction work; not to do so is to take an unnecessary chance and risk, as mentioned previously.
2. *The take-off.* Having prepared the spec notes, the estimator can now begin to measure the work to be done. This process is called the take-off, and it consists of the following four parts, about each of which more will be said in later sections of the book.
    a. The **classification of work** to be done at a number of levels in the estimate. For example, concrete work is classified separately from carpentry work, and within concrete work, concrete items are classified separately from formwork or other types of work.
    b. The **description of items** of work to be done at each level. For example, in concrete work, one expects to find foundations, walls, columns, beams, slabs, and so on; suitable words or numbers are chosen to accurately and briefly describe all significant features of each item of work to be measured.
    c. The **dimensions of items** of work are extracted from the drawings or specifications, either by reading figured dimensions directly, by calculation of inferred dimensions, or by judicious and limited scaling of indicated dimensions from drawings, using an architectural or engineering scale rule or some other manual or electronic device.
    d. The **extension of dimensions**; that is, lengths are summarized, or they are multiplied by widths to produce areas; and areas are multiplied by depths, heights, or thicknesses to produce volumes. The totals of lengths, areas, and volumes are then established by addition.
3. *The pricing.* The unit prices used to price measured quantities of work are usually, in the first instance, worked up from first principles by the estimator. This is done by researching costs, taxes, labor productivity, equipment output, and related items, and then making some calculation of the probable price, either on the basis of total cost or unit cost. Such prices are then compared with previous data, known to be reliable, to confirm their appropriateness in the present context. The results are then

usually filed, either manually in loose-leaf folders or electronically on computer storage disks, for later reuse or study.

4. *The summary.* After the quantities and prices have all been established, the final step is to prepare a summary of the estimate. This summary will generally show the work of the various trades arranged into some suitable sequence, often showing the work proposed to be done by the general contractor separate from the work intended to be done by subcontractors. Allowances for overhead and profit will be added at the end of the overall summary.

The purpose of this short explanation is to present to the reader a quick overall view of the estimating process. Detailed examples of all of these aspects of estimates are presented in other parts of the book. The case study described in Chapter 5 shows examples of the elements described in this chapter.

QUESTIONS

4.1. Distinguish between a detailed estimate and an approximate estimate. Identify one major feature of each type.

4.2. Define the term soft costs. Give two clear examples of such costs in connection with the development of a building project.

4.3. Identify the key issue with respect to the amount of detail available for the preparation of detailed estimates.

4.4. Measurement and pricing are two of the four basic parts of any estimate. Identify and briefly comment on the other two parts.

4.5. Name the four parts of the take-off process, and briefly discuss the significance of each part to the whole process.

CHAPTER 5
CASE STUDY

5.1 INTRODUCTION

5.1.1 Preamble

Although everyone has met children who seem born to become doctors, lawyers, or politicians, it is safe to say that few young people grow up with the stated objective of becoming a construction estimator as their life's goal. Many people become involved in construction estimating as a result of circumstances largely beyond their control. They may have been affected by staff changes such as promotions or reductions in their companies, they may have suffered some injury which prevents them from continuing to do construction work of a more physical nature, they may have started a small contracting company and then discovered that estimating is a necessary though possibly unpopular part of the business, they may have been required to take a course in the subject to fulfill other academic objectives, and so on.

As a result, it is the experience of the author that the majority of novices in the construction estimating field understandably have little knowledge of, feeling for, or skill in the processes involved in estimating and bidding, and therefore hold incomplete or incorrect views of the place, function, and importance of the various elements that make up the process. Indeed, in many minds, there appears to be an attitude which sometimes amounts to positive resistance to acquiring facility in estimating method and technique. To some extent, this attitude might be likened to the attitude of people who have to learn to drive a car. They can probably readily acquire the relatively simple mechanical skills; it is their driving habits that will soon disclose the strengths and weaknesses of their character. So it is with estimating. It takes more than just knowledge of technique to be a competent estimator, just as it takes more than technique to be a competent driver. It also requires discipline, judgment, and the ability to assess risk on the run.

5.1.2 Objectives

Because of the various backgrounds that people bring to estimating, the primary objective of this chapter is to present a case study, intended to act as a vehicle or medium to permit discussion and reflection on the whole of the estimating process as it would be applied to a typical and realistic situation, and to put the various parts of the process into a logical order and reasonably sharp focus to permit useful examination of each part, both in this chapter and in succeeding parts of the book.

The case study is based on the Post-and-Beam House project, for which design and working drawings are included in Part VI, numbered IMP-3 and IMP-4. The study is presented in the form of a series of figures contained in the following pages, identified and described in the notes contained in the following sections, which are intended to be read in close conjunction with their respective figures. These figures are realistic, though simplified representations of actual documents that one might encounter during preliminary examinations, estimating, and bidding on such a project. The student is encouraged to acquire and review samples of actual documentation and formats suggested by these figures or models, taken from current local projects in the region in which he or she lives. Readers are also cautioned not to use these models in actual cases without appropriate modification.

5.2 THE FIGURES

5.2.1 Instructions to Bidders

Contractors discover projects that are available for bidding in several ways. They may read a formal notice in a paper or newsletter, they may be invited directly by the owner or designer, they may be referred by a friend, they may come across the bidding documents at builder's exchanges while they are examining other sets of project documents, or they may discover the project in other ways. After an available project has been identified, the contractor's estimator will first quickly glance over all the documents, but especially the drawings, to get a general idea of the type and magnitude of the project. If interest is maintained, attention may then be focused on the Instructions to Bidders, such as the example shown in Figure 5.1, to note specific obligations regarding bidding on the job in question.

5.2.2 Blank Bid Form

Next, attention will be directed to the Bid Form, because it is on the basis of this piece of paper that the contract will subsequently be established. The Bid Form usually consists of a number of paragraphs, dealing separately with the basic offer, any special provisions regarding unit prices, separate or alternate prices, dates or timing, subcontractor lists, and signatures and titles of the bidding party. Even in the simple Bid Form shown in Figure 5.2, all of the primary elements of a valid and enforceable contract are implicit. For further information on contents of contracts, see Section 15.4.

5.2.3 Blank Checklist Form

The next step for the estimator is to start to make notes about the project. This information is extracted from the drawings and specifications, and is best entered onto a form prepared for this purpose. The page heading of a typical Checklist Form intended for use in any construction company is shown in Figure 5.3; the company name and address or logo can be printed at the top of the page, consistent

```
                    THE POST AND BEAM HOUSE
                    ─────────────────────
                     INSTRUCTIONS TO BIDDERS
                     ───────────────────────

  1. Submit bids on the form provided.

  2. Fill in all blank spaces, either with words
     and figures, or crossed out and initialed.

  3. Bids will be received by the Architect on behalf
     of the Owner at the Architect's Office at 4:00 pm
     local time on Thursday, 11 April 1985.

  4. Late bids shall be returned unopened to the bidders.

  5. Attach a list of subcontractors proposed for the
     various parts of the work.

  6. Bids shall be held to be firm for 14 days from the
     date fixed for the close of bidding.

  7. Bid depository shall not be in effect for this job.

  8. It is not necessary to submit copies of bonds nor
     insurance policies with the bid.

  9. A bid bond in the amount of ten (10) percent is
     required to be submitted with each bid.

     END
     ───

     Job No: IMP-3     Waterfront Designers Inc.   Page 1 of 1
```

Figure 5.1 Instructions to Bidders.

with all other stationery used by the firm. Notes on each section of work should be entered on separate forms.

5.2.4 Blank Measurement Form

When the estimator is ready to start measurement, a take-off form of some kind should be selected. The example shown in Figure 5.4 is one that is widely used by construction companies and is typical of most types, in that there should be an identification panel at the top of the page, followed by columns arranged to carry descriptions, dimensions, extensions, and totals. Specialist contractors can arrange the columns in other ways; for example, in mechanical work, the description columns usually occur in the center of the page, with columns for lengths of piping on one side and columns for numbered fittings on the other. Forms for metric dimensions will vary slightly from forms for imperial dimensions, because of the way in which dimensions are expressed in each of these two measurement systems; examples of such minor differences are given in Part IV.

5.2.5 Blank Pricing Form

Once the measurement is complete, the totals of quantities of work for each section are then extracted from the take-off sheets and entered onto summary or recapitulation sheets, shown in Figure 5.5. The word "recapitulation" stems from a Latin

```
                    THE POST AND BEAM HOUSE
                         BID FORM
    To the Owners:
        Having examined the Contract Documents for the above
    project, we hereby offer to furnish all labor, materials,
    equipment, and other services necessary for the proper
    completion of all construction work for the sum of:
    .............................................($.........)
        We enclose herewith a bid bond in the amount of ten
    percent of our bid price. The following init prices shall
    be used to price additions to or deletions from the work.
         ITEM (as specified)       UNIT    ADDITIONS  DELETIONS
    1. Concrete in foundations:     CY      ........  ........
    2. Formwork to foundations:     SF      ........  ........
        Upon acceptance of this offer, we will execute the Form
    of Contract, furnish specified securities, and substantiall
    complete all work within .... calendar days from the date
    of the award of the contract.
    FIRM NAME:..................................................
    SIGNATURE:..........................(authorized officer)
    ADRESS:..............................PHONE:..............

    Job No: IMP-3     Waterfront Designers Inc.    Page 1 of 1
```

Figure 5.2 Blank Bid Form.

```
                         COMPANY NAME
    SPECIFICATION CHECKLIST FOR ESTIMATING      PAGE    OF
    Project:                        No:         Estimator:
    Section:                        No:         Date:
```

Figure 5.3 Blank Checklist Form.

word meaning "heading." The objective is to bring together the summaries of quantities of work with the summaries of the price analysis of such items of work for each separate section of the total project.

5.2.6 Blank Combination Form

On larger or complex projects, it is usually more convenient to use separate measurement and pricing sheets, because the measurement may run on for several pages, whereas the pricing may be consolidated onto one or two pages. For example, the original rough carpentry take-off for the Wood-Framed House (for which drawings

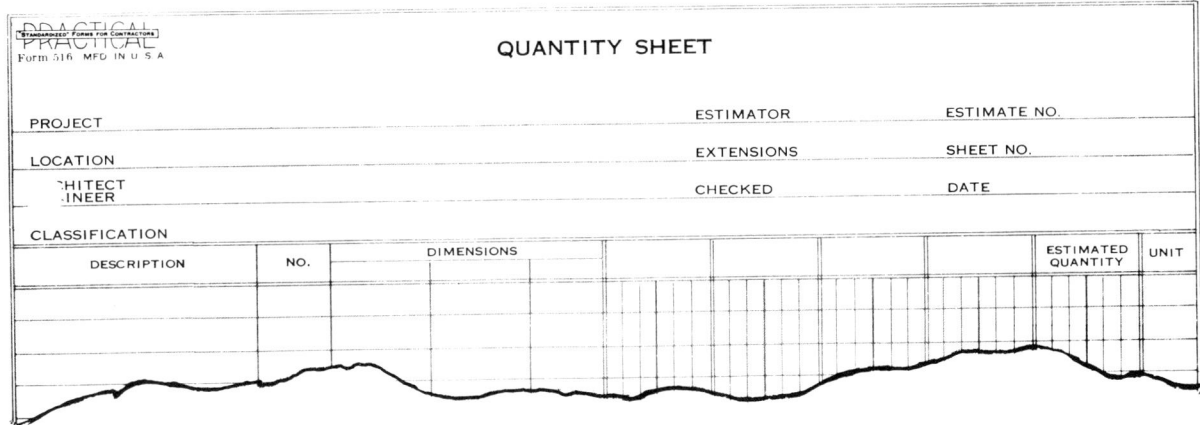

Figure 5.4 Blank Measurement Form. (Form reproduced by permission of Frank R. Walker Company, Chicago.)

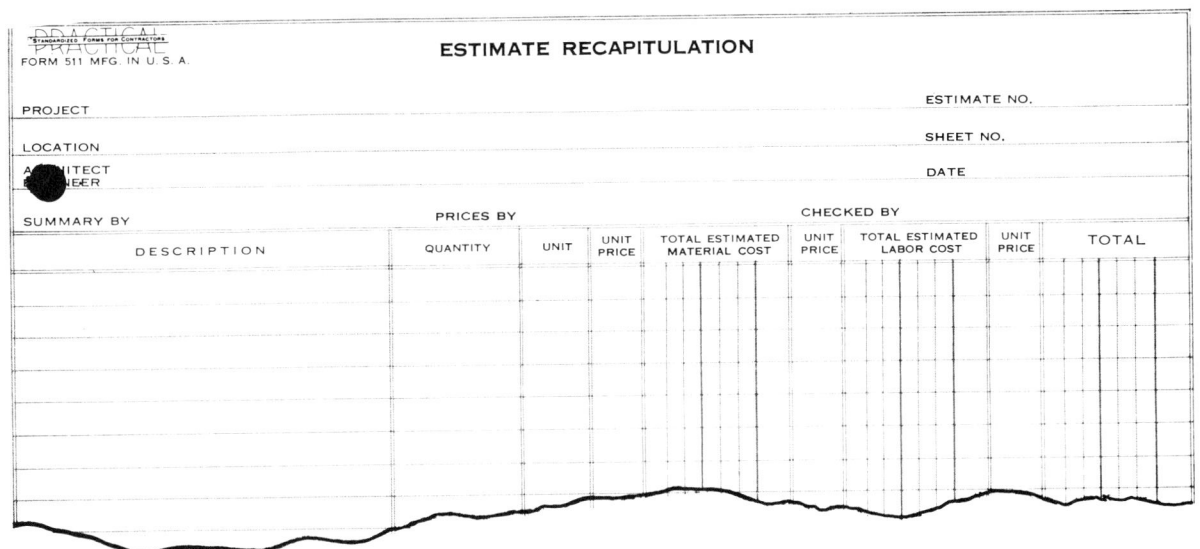

Figure 5.5 Blank Pricing Form. (Form reproduced by permission of Frank R. Walker Company, Chicago.)

are included in Part VI and which is not really a large job) consisted of 13 pages, while the pricing of that work consisted of one page. However, on small or simple sections of work, it is sometimes convenient to do the measurement and the pricing on a combined sheet. Figure 5.6 shows an example of such a form. Selection of forms is a matter of experience and choice, and most construction companies should keep a supply of various types of estimating sheets in stock for use by their estimators. All such documentation should be integrated with related paperwork within each company, such as cost accounting, purchasing, payroll, and the like.

5.2.7 Outline Specification

To illustrate measurement, pricing, and bidding technique in this short case study, the trade section for masonry work has been selected, because of its simplicity and clarity in this particular project. Figure 5.7 is intended to represent some aspects of a specification section for masonry work which might be found in the project manual for a proposed new building. This figure is not a direct extract of actual doc-

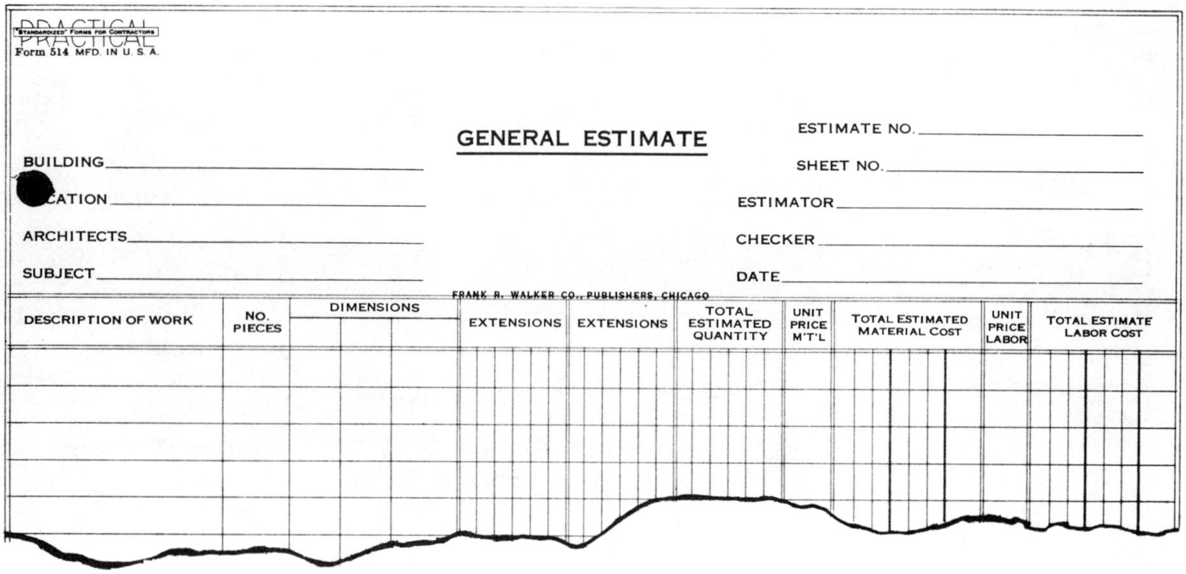

PRACTICAL "STANDARDIZED" FORMS FOR CONTRACTORS
Form 514 MFD. IN U. S. A.

GENERAL ESTIMATE

BUILDING ______
LOCATION ______
ARCHITECTS ______
SUBJECT ______

ESTIMATE NO. ______
SHEET NO. ______
ESTIMATOR ______
CHECKER ______
DATE ______

FRANK R. WALKER CO., PUBLISHERS, CHICAGO

DESCRIPTION OF WORK	NO. PIECES	DIMENSIONS	EXTENSIONS	EXTENSIONS	TOTAL ESTIMATED QUANTITY	UNIT PRICE M'T'L	TOTAL ESTIMATED MATERIAL COST	UNIT PRICE LABOR	TOTAL ESTIMATE LABOR COST

Figure 5.6 Blank Combined Form. (Form reproduced by permission of Frank R. Walker Company, Chicago.)

umentation; it is included simply to raise some points for discussion about the possible contents of such a section. The main reason why an actual extract has not been included in this book is explained in the figure itself, and involves the large extent to which such work is governed by codes and regulations, with only minimal optional features left for specification. These notes are intended to explain the detailed checklist and take-off that follow. The masonry estimator, of course, would read the appropriate specification section and regulations affecting this part of the work.

5.2.8 Completed Checklist

The estimator then prepares the checklist for the particular section about to be measured and priced. Figure 5.8 shows a typical completed list for work of the type stated. Note the details filled in at the page heading. The items in the list primarily concern things that are going to cost money, and they are organized under appropriate headings, usually with the most expensive elements coming first, followed by secondary or less important items. The objective of preparing the checklist is to assist the estimator to get a more precise feeling for the scope and nature of the work, and to consider the sequence in which the items of work should be measured. The sequence is usually the actual chronological sequence of work, although not always. It may be seen that the checklist makes no mention of specific regulations, notwithstanding the major significance of such regulations in this type of work. Note that this particular example has been typed for legibility. Normally such checklists are manually or electronically prepared; they are seldom retyped in practice.

5.2.9 Completed Measurement

Figure 5.9 shows the completed measurement for the masonry work. The items follow the sequence suggested in the checklist, and the various columns of the estimating form are put to their proper use. Although much more detailed discussion of every aspect of measurement is covered in Part II and therefore will not be introduced at this point, it may be observed that in the layout of these two pages, particular care has been taken to make every part readable and understandable.

Chap. 5 / Case Study

> The Post-and-Beam House:
> Outline Specification for Masonry Work
>
> The masonry work in this project is minimal; it is for this reason that this section of the total project has been selected as a vehicle to illustrate aspects of estimating in this case study. The masonry work involves the construction of one chimney base and fireplace. The following notes describe some topics appropriate to the measurement and pricing of this section of the work.
>
> There are some fundamental aspects of construction and the laws of physics relative to chimneys and fireplaces which do not permit much latitude for individual design concepts. Furthermore, such work, whether factory-made or site-built, is usually carefully and extensively regulated by building codes.
>
> Building codes relating to chimneys usually dictate typical attributes, such as overall heights, cross-sectional areas at different elevations, angles, minimum and maximum flue sizes, linings, leakages, clearances, foundations, cappings, cleanouts, and basic material standards for concrete, mortar, metal components, and the like.
>
> Regulations governing fireplaces cover such aspects as wall thicknesses, sizes, and supports of openings, and details of the construction of the hearth, smoke chambers, dampers, and facings and mantel units.
>
> Specifications normally incorporate all such codes and regulations, in addition to describing optional features beyond the minimal requirements, such as the nature and quality of exterior facings and interior finishes associated with the chimney and fireplace construction, degree of cleanup required, and so on.
>
> In this particular project, standard red clay facing bricks are proposed to be used to form an ornamental chimney breast, 48 in. wide and rising full-height from floor to ceiling, across the front of the fireplace inside the living room. Matching bricks shall be used to face the exterior surfaces of the chimney base. A factory-fabricated sectional steel flue unit is also required to be provided, together with standard-quality dampers, cleanouts, and ashpit accessories.
>
> For more comprehensive and detailed information on specification format, style, and content, refer to Part 2 of *Construction Contracts and Specifications,* by Glenn M. Hardie, published by Reston Publishing Co., Inc., Reston, Va., 1981.

Figure 5.7 Specification Notes.

5.2.10 Completed Pricing

Figure 5.10 shows the recapitulation of totals of the measured quantities onto the pricing form. The actual development of the unit prices for each item of work is done on separate sheets or is extracted for separate sources in the contractor's office and is not shown in this case study. Details of how the pricing is actually done are given in Part III. The developed unit prices are entered opposite their appropriate items, and the cost is calculated for materials, labor, and totals.

5.2.11 Alternative Procedure

Figure 5.11 shows the combination sheet, where the measurement on a small construction job, such as the work on one simple trade, is done on the same sheet as the pricing. The items, prices, and totals are identical to those shown in Figures 5.9 and 5.10. On the basis of this estimate, the masonry subcontractor would then make a bid or offer, probably in the amount of $1000 in this case, to the general contractor bidding the whole project to the owner. This subcontractor's bid might be made orally in person or by phone, but it should be confirmed in writing at the

```
                - COMPANY NAME -
SPECIFICATION CHECKLIST FOR ESTIMATING        PAGE 1 OF 1
Project: Post & Beam House        No: IMP-3    Estimator: GMH
Section: Masonry - Fireplace      No: 04200    Date: April

MASONRY ITEMS:

Fire Bricks in Cement Mortar ($400/M)
Clay Facing Bricks in Ditto  ($200/M)
Precast Concrete Cap (3" thick)
Cement Mortar and Parging

PREFABRICATED ITEMS:

Steel Damper Unit (standard)
Sectional Flue Units (8" x 12")
Ashpit and Cleanout Doors

MISCELLANEOUS ITEMS:

Steel Angle Lintel
Cleaning Masonry

END
```

Figure 5.8 Checklist.

earliest moment in a reasonably formal manner, by mail or telegram, more or less along the same lines as the general contractor has to make his bid submission to the owner.

5.2.12 Estimate Summary

Figure 5.12 shows the use of an estimate recapitulation form for the general summary that would be prepared by the general contractor about to bid on the entire project. This form would be prepared by the general contractor's estimator at the start of the estimating process, to assist in determining the various parts of the work that are going to be done by the contractor's own forces, and that which will be done by subcontractors. Later, as the total prices for the various parts become available, they are inserted into their proper places in the general summary, and so the whole bid price is developed. For further limited discussion of aspects of bidding, see Part III.

5.2.13 Completed Bid Form

In Figure 5.13, the blank Bid Form has been completed by the general contractor, and it is now ready for submission to the owner in accordance with the Instructions to Bidders. Notice that the amount of the bid is inserted in both words and figures, and that the contract documents should make it clear which will prevail in the event of a discrepancy. The fact that a specific number of days has been stated makes time of the essence in this contract. Time is an important aspect of any good construction contract. On a larger and more important project, the signature of the signing officer may have to be witnessed, and the seal of the company may have to be affixed in the proper manner; these are discretionary affairs for the owner and the designer to decide. The contractor really only has the option of following all the instructions or not bidding on the job in the first place.

Chap. 5 / Case Study

QUANTITY SHEET

PROJECT: Post and Beam House
LOCATION: Blaine, WA.
ARCHITECT/ENGINEER: Byles/Wesson
CLASSIFICATION: Masonry - Chimney & Fireplace. 04200.
ESTIMATOR: GMW
EXTENSIONS: LFH
CHECKED: GMW
ESTIMATE NO. IMP-3
SHEET NO. 1 of 2
DATE: April

MASONRY

1. Standard Firebricks in cmt. mortar.
   (nom. size: 9" x 4½" x 2½") incl. angle cutting.
   Allow 10% waste; prime cost: $400/M.

Description	No.	l	w	d		Est. Qty	Unit
3" thick hearth	1	3·4	1·10		6 SF (x 3·5 units)	20	ea
4" thick lining	1	6·0	2·6		15 — (x 6·5 units)	100	ea

   2/2·0 = 4·0   2·0
           2·0   0·6
           6·0   2·6

2. Standard Red Clay Facings, in cmt. mortar.
   incl. angle cutting and ties.
   Allow 10% waste; prime cost: $200/M.

   4" to stack, wi. tooled joints.

   2/2·3 = 4·6   5·0   1   7·2   4·9   35 SF
         2·8 -0·3   1   3·4   2·1   7
         7·2   4·9   42 SF (x 6·5 units)   275 ea

   4" interior f.place front, wi. struck joints & ties.

   7·6
   1·0
   0·4 = 1·10   1   4·0   x   8·10   35 SF

   deduct: opening   1   2·8   x   2·0   5
         hearth.   1   3·4   x   0·8   2
                              7
                                        -7
                                28 SF (x 6·5)   185 ea

3. 3" P.C. cap, 46" x 26", wi. flue hole, set in cmt. mortar on brkwrk.   1 ea   1 ea

4. Cement mortar (for 580 units)   allow   1 cy

Figure 5.9 Measurement (IMP-3).

QUANTITY SHEET

PROJECT: P & B House (cont.)
ESTIMATOR: Gunn
ESTIMATE NO.: IMP-3
LOCATION:
EXTENSIONS:
SHEET NO.: 2 of 2
ARCHITECT/ENGINEER:
CHECKED:
DATE:
CLASSIFICATION: 04200 (continued)

DESCRIPTION	NO.	DIMENSIONS	ESTIMATED QUANTITY	UNIT
PREFAB. ITEMS				
Steel damper unit, $37\frac{1}{2} - 30" \times 9\frac{7}{8}"$ overall, built-in, including fill and smoke shelf.	5.		1	No.
Sectional flue unit, $8" \times 12"$ area and $9'-6"$ overall, as spec., in 3 sections (total wt: 600 #)	6.		1	No.
Cast iron Ashpit door + frame, $7" \times 5"$, built in.	7.		1	No.
Cast iron Clean out + frame, $8" \times 6"$, built in.	"		1	No.
MISCEL. ITEMS				
Steel bar lintel, $\frac{1}{4}" \times 2"$ and $40"$ long overall.	8.		1	No.
Clean exposed brickwork.	9.	28 SF	28	SF
END				

Figure 5.9 (cont.)

Chap. 5 / Case Study

ESTIMATE RECAPITULATION

PROJECT: Post and Beam House
LOCATION: Blaine, WA.
ARCHITECT/ENGINEER: Byles/Weston
SUMMARY BY: GMH
PRICES BY: GMH
CHECKED BY: LFH.
ESTIMATE NO. IMP-3
SHEET NO. 1 of 1
DATE: April.

DESCRIPTION	QUANTITY	UNIT	UNIT PRICE	TOTAL ESTIMATED MATERIAL COST	UNIT PRICE	TOTAL ESTIMATED LABOR COST	UNIT PRICE	TOTAL
MASONRY (04200)								
1. Firebricks in mortar:								
3" hearth	20	ea	50¢	10	60¢	12		22
4" lining	100	"	"	50	"	60		110
2. Facings in ditto:								
4" to stack	275	ea	25¢	69	50¢	138		207
4" to facing	185	"	"	47	"	93		140
3. 3" P.C. Cap	1	ea	15.00	15	25.00	25		40
4. Cement mortar	1	cy	45.00	45	—	—		45
5. Damper Unit	1	ea	40.00	40	25.00	25		65
6. Flue unit	1	ea	75.00	75	30.00	30		105
7. Cast iron units:								
Ashpit	1	ea	8.00	8	15.00	15		23
Cleanout	1	"	10.00	10	15.00	15		25
8. Bar lintel	1	"	4.00	4	5.00	5		9
9. Clean brickwork	25	SF	say	5	—	30		35
SUBTOTAL:	—			378		448		826
10. Overhead	10	%	·	38		45		83
SUBTOTAL:	—			416		493		909
11. Profit	5	%	·	21		25		46
12. Grand Total	—		·	437		518		955 ✓

Figure 5.10 Pricing (IMP–3).

GENERAL ESTIMATE

BUILDING: Post and Beam House
LOCATION: Blaine, WA
ARCHITECTS: Byles / Weston
SUBJECT: MASONRY - FIREPLACE
ESTIMATE NO. IMP-3
SHEET NO. 1 of 2
ESTIMATOR: GWN
CHECKER: LFH
DATE: April

Description of Work	No. Pieces	Dimensions l. w. h.	Extensions	Extensions	Total Estimated Quantity	Unit Price M'T'L	Total Estimated Material Cost	Unit Price Labor	Total Estimate Labor Cost
BRICKWORK.									
1. Standard Firebricks in cmt. mrtr. (nom. size: 9"x 4½"x 2½") incl. cutting. Allow 10% waste; prime cost = $400/m									
3" hearth		3·4 1·10	6 sf (×3.5)		20 /ea	50¢	10	60¢	12
4" lining		6·0 2·6	15 sf (×6.5)		100 /ea	"	50	"	60
2. Standard Red facings in cmt. mrtr. (nom. size: ditto). incl. cutting and ties. Allow 10% waste; prime cost $200/m.									
4" to stack, w/ tooled joints:		7·2 4·9	35 sf						
		3·4 2·1	7						
			42 sf (×6.5)		275 /ea	25¢	69	50¢	138
4" interior facing, w/ struck joints:		4·0 8·10	—	35 sf					
ddt. opng.		2·8 2·0	5						
hrth.		3·4 0·8	2						
			7	−7					
		(×6.5)		28 sf	185 /ea	25¢	47	50¢	93
3" P.C. Cap. 48"× 26" w/ flue-hole. Set in cmt. mrtr. on brkwrk					1 /ea	15⁰⁰	15	25⁰⁰	25
				Carried forw'd/-			191		328

Figure 5.11 Combined Estimate (IMP-3).

Chap. 5 / Case Study

GENERAL ESTIMATE

BUILDING: P+B House.
LOCATION:
ARCHITECTS:
SUBJECT: Masonry (cont.)

ESTIMATE NO. MN-3
SHEET NO. 2 of 2
ESTIMATOR: GMW
CHECKER: LFH
DATE: April

Description of Work	No. Pieces	Dimensions	Extensions	Extensions	Total Estimated Quantity	Unit Price M'T'L	Total Estimated Material Cost	Unit Price Labor	Total Estimate Labor Cost
brought ford /-							191		328
4. Cement mortar (for 530 units) allow (say)					1	45⁰⁰ /CY	45	—	—
PREFAB. ITEMS									
5. Steel damper unit 37½-30" × 9⅞" overall. Built-in including till and smoke shelf					1	40⁰⁰ /ea	40	25⁰⁰	25
6. Sectional flue unit, 8"×12" area × 9'-6" overall, as spec, in 3 sections (wt: 600#)					1	75⁰⁰ /ea	75	30⁰⁰	30
7. Cast iron ashpit door & frame 7×5," built in.					1	8⁰⁰ /ea	8	15⁰⁰	15
" Ditto clean out door & frame 8×6," ditto.					1	10⁰⁰ /ea	10	15⁰⁰	15
MISCELL. ITEMS.									
8. Steel bar lintel, ¼"×2" and 4." long.					1	4⁰⁰ /ea	4	5⁰⁰	5
9. Clean exposed brickwork					28 SF	sum	5	sum	30
SUBTOTAL							378		448
Overhead					10%		38		45
SUBTOTAL							416		493
11. Profit					5%		21		25
12. Grand Total							437		518 ✓

Figure 5.11 (cont.)

ESTIMATE RECAPITULATION

PROJECT Post & Beam House - GENERAL SUMMARY **ESTIMATE NO.** IMP-3
LOCATION Blaine, WA **SHEET NO.** 1 of 3
ARCHITECT/ENGINEER Byles/Weston **DATE** April
SUMMARY BY GMH **PRICES BY** LFH **CHECKED BY** GMH

DESCRIPTION	QUANTITY	UNIT	UNIT PRICE	TOTAL ESTIMATED MATERIAL COST	UNIT PRICE	TOTAL ESTIMATED LABOR COST	UNIT PRICE	TOTAL
1. GENERAL CONTRACTOR'S OWN WORK								
DIVISION 3 CONCRETE								
3a Concrete & Formwork				1030		1390		
3b Sundries & Insulation				180		240		2840
DIVISION 6 WOOD & PLASTICS								
6a Rough Carpentry				660		750		
6b Finish Carpentry				2740		1170		
6c Millwork (install only)				—		60		5110
TO SUMMARY							$	7950
2. SUBCONTRACTOR'S WORK								
DIVISION 2 SITEWORKS								
2a Excavation & Fill						1260		
2b Draintile & Gravel						300		
2c Asphalt Paving & Base						420		1980
DIVISION 4 MASONRY								
4a Brickwork						950		950
DIVISION 5 METALS						NOT USED		—
DIVISION 6 WOOD & PLASTICS								
6c Millwork (supply only)						1020		1020
DIVISION 7 MOISTURE PROTECTION								
7a Roofing & Flashings						1320		1320
TO SUMMARY							$	5270

Figure 5.12 Estimate Summary (IMP-3).

Chap. 5 / Case Study

ESTIMATE RECAPITULATION

PROJECT: Post & Beam House (cont.)
ESTIMATE NO.: IMP-3
LOCATION:
SHEET NO.: 2 of 3
ARCHITECT/ENGINEER:
DATE: April
SUMMARY BY: GMH
PRICES BY: LFH
CHECKED BY: GMH

DESCRIPTION	QUANTITY	UNIT	UNIT PRICE	TOTAL ESTIMATED MATERIAL COST	UNIT PRICE	TOTAL ESTIMATED LABOR COST	UNIT PRICE	TOTAL
2. SUBCONTRACTOR'S WORK (CONTINUED)								
DIV. 8 DOORS AND WINDOWS								
8a Main Sliding Door & Frame						360		
8b Wood Doors & Frames						1140		
8c Aluminum Windows & Frames						720		
8d Glass & Glazing						1140		3360
DIV. 9 FINISHES								
9a Exterior Stucco						1120		
9b Interior Plaster						2270		
9c Terrazzo Work						250		
9d Ceramic Tile Work						530		
9e Resilient Flooring						270		
9f Painting & Decorating						nil		4440
DIV. 10 TO 14 NOT USED						—		—
DIV. 15 MECHANICAL								
15a Heating & Venting						1200		
15b Plumbing & Fixtures						1980		3180
DIV. 16 ELECTRICAL								
16a Distribution						840		
16b Fixtures (install only)						60		900
					TO SUMMARY		$	11880

Figure 5.12 (cont.)

ESTIMATE RECAPITULATION

PROJECT: Post & Beam House (cont.)
ESTIMATE NO. IMP-3
SHEET NO. 3 of 3
DATE: April
SUMMARY BY: GMH
PRICES BY: LFH
CHECKED BY: GMH

DESCRIPTION	QUANTITY	UNIT	UNIT PRICE	TOTAL ESTIMATED MATERIAL COST	UNIT PRICE	TOTAL ESTIMATED LABOR COST	UNIT PRICE	TOTAL
3. GENERAL CONDITION ITEMS								
DIV. 1 GENERAL REQUIREMENTS								
1a General Conditions:								
1. Permits						300		
2. Surveys						60		
3. Bonds						110		
4. Insurance						200		670
1b Temporary Services:								
1. Electricity						70		
2. Water Supply						30		
3. Supervision						3500		3600
1c Cash Allowances:								
1. Landscaping						1440		
2. Light Fixtures (supply)						360		
3. Finish Hardware (supply)						600		
4. Carpets & Underlay						1200		3600
				TO SUMMARY			$	7870
4. GENERAL ESTIMATE SUMMARY								
1. CONTRATOR's OWN WORK				from page 1		—		7950
2. SUBCONTRACTOR'S WORK				from page 1		5270		
				from page 2		11880		17150
3. GENERAL CONDITION ITEMS				from page 3		—		7870
				TOTAL DIRECT COST			$	32970
				ADD OVERHEAD (say) 10%				3300
				ADD PROFIT (say) 5%				1650
				GRAND TOTAL BID PRICE			$	34950

Figure 5.12 (cont.)

```
                    THE POST AND BEAM HOUSE
                           BID FORM

    To the Owners:
        Having examined the Contract Documents for the above
    project, we hereby offer to furnish all labor, materials,
    equipment, and other services necessary for the proper
    completion of all construction work for the sum of:
    - THIRTY FOUR THOUSAND NINE HUNDRED FIFTY -  ($ 34,950.00 )

        We enclose herewith a bid bond in the amount of ten
    percent of our bid price. The following unit prices shall
    be used to price additions to or deletions from the work.

         ITEM (as specified)      UNIT    ADDITIONS    DELETIONS
    1. Concrete in foundations:    CY       $50.00       $30.00
    2. Formwork to foundations:    SF       $2.00        $1.00

        Upon acceptance of this offer, we will execute the Form
    of Contract, furnish specified securities, and substantially
    complete all work within 100 calendar days from the date
    of the award of the contract.
                       Institute Construction Inc.
    FIRM NAME:..................................................
    SIGNATURE:.... Jim Macdonald, President (authorized officer)
                   Federal Way, Seattle, WA          123-4567
    ADDRESS:.........................PHONE:....................
    ─────────────────────────────────────────────────────────────
    Job No: IMP-3    Waterfront Designers Inc.    Page 1 of 1
```

Figure 5.13 Completed bid.

5.2.14 Review and Conclusion

All of the main elements of measurement and pricing mentioned in this case study will be dealt with in considerable detail elsewhere in this book. The secondary elements relating to bidding, contracts, and specification are treated superficially only in this book; more exhaustive studies of these secondary elements can be made by reference to other books. The purpose of the study is simply to give an overall picture of the bidding and estimating process, in concentrated and sequential form, to give the reader a framework on which the detailed studies of the stated subject matter can be built.

QUESTIONS

5.1. Distinguish between "Notices to Contractors" and "Instructions to Bidders" by explaining what these two phrases mean to the estimator.

5.2. Identify two objectives of the estimator when preparing a checklist before starting the measurement and pricing of work.

5.3. Why should an oral bid be confirmed in writing as soon as possible?

5.4. State the meaning of the word "recapitulation." Why and how is this word used in estimating procedures?

5.5. What is the advantage to expressing the amount of a bid in writing in both words and figures? What should the owner do if the words and figures in a bid are not consistent with each other?

PART I
RESEARCH ASSIGNMENTS

The research assignments that follow are intended to give readers a channel through which to acquire knowledge of current local issues and aspects relative to the contents of the chapters contained in this part of the book. The basic objective in all of these assignments is to allow readers to compare theory with practice, by studying this and other similar books, and by encountering opinions and activities, other than those of themselves and this author, on the stated topics. A guide to content, procedure, and evaluation for these assignments has been included in Section 1.4. A review of these guidelines is recommended before starting work on the assignment projects.

CHAPTER 1: INTRODUCTION TO ESTIMATING

Prepare an annotated bibliography of published books that deal exclusively with construction estimating. Identify each book by title, author, publisher, and date. List the main features, strengths, and weaknesses of each book; identify at least three common principles or precepts advocated in all of the books; describe one unique feature of significance for estimators contained in each book. Titles listed in Appendix C of this book may be included; at least half of the titles selected should be less than five years old.

CHAPTER 2: ESTIMATING

Investigate the terms and conditions under which construction estimators work in the region where you live. Examine those who work for large construction firms, those who work for small firms, and those who are either self-employed or who work for consultant construction economists. Report on education, training, experience, and related aspects of the background of local construction estimators. Tabulate approximate salary levels and related fringe benefits.

CHAPTER 3: ESTIMATORS

Investigate the extent to which women have found positions in the construction industry in the region where you live. Specifically identify actual positions by title and job description; examine opportunities in the professional, managerial, and artisan areas of endeavor. Report on differentials (if any) in salaries and wages paid to women compared to those for men in similar positions. Comment on any special problems that women do or may encounter in the industry. Report on proportions of male to female memberships in local construction associations operating in your region.

CHAPTER 4: ESTIMATES

Make a study of the types, techniques, and quality of detailed and approximate estimates used by construction and design firms in the region where you live. Comment on the qualifications and experience of the personnel who prepare such estimates, and the time, costs, and care taken in their preparation. Identify at least one local and recent court case where an incorrect estimate of probable construction costs has led to legal problems for the parties concerned.

CHAPTER 5: CASE STUDY

Compare the detailed recommendations on procedure contained in this chapter with the actual practices of estimators who work for construction companies in the region where you live. Identify all significant differences and give valid reasons for such divergence. Include samples of checklists if obtainable from local sources. Include details of any intermediate steps not specifically identified in the case study.

PART II MEASUREMENT

CHAPTER 6
INTRODUCTION TO MEASUREMENT

6.1 GENERAL OBJECTIVES

6.1.1 Preamble

Estimating consists of two processes: **analysis,** the separation of a thing into its constituent elements, and **synthesis,** the combination of such elements into a composite whole. In the analytical process, the building project requirements are divided into **units of work,** which are the various trade sections or distinct disciplines of work to be done; these units are further subdivided into **items of work,** which are the smallest individual portions of the various types of work that can be readily classified, measured, priced, and accounted for. In the synthetic process, predictions in the form of prices are made about productivity and costs of materials, labor, and machines; these prices are then applied to the measured quantities of work to produce the anticipated total cost of the project. In Part II the analytical or measurement process is examined in detail. Part III examines the synthetic or pricing process. The general objectives are to develop understanding, reliability, accuracy, and speed in the application of both processes.

6.1.2 Analysis of Measurement

The measurement process can be conveniently studied under four headings: general, methods, techniques, and systems. The primary objective of this chapter is to present some general aspects of measurement for consideration and adoption; in succeeding chapters, specific aspects of methods, techniques, and systems are detailed for study. Before proceeding, some thought should be given to the precise meanings of the foregoing terms. The word **method** means a mode or procedure or a way of doing something in accordance with a definite plan. The word **technique** means the application of technical skill, especially in the artistic sense; this is appropriate in estimating. The word **system** means, among other things, an assemblage or combination of things forming a complex or unitary whole, or a coordinated body of methods.

6.2 MEASUREMENT IN GENERAL

6.2.1 Preamble

The cost of construction work consists of five discrete but related elements: labor, materials, equipment, overhead, and profit. The general purpose or objective of measurement is to present these elements separately or jointly in such a way that their individual or combined costs can be predicted and later confirmed. To achieve such an end, the following general principles of measurement are articulated, for the benefit of estimators in the offices of both designers and contractors. These principles will be augmented in succeeding chapters by additional precepts more suited to particular methods or techniques of measurement, specifically recommended for use by designers or contractors.

6.2.2 Accuracy

All work should be measured as accurately as the data will permit. Estimators are always faced with the necessity of having to make decisions about the probable conditions under which work will be done. If *measurements* can be put onto the firmest base possible, then the approximations and judgments that inevitably have to be incorporated into the pricing process will be minimized. Furthermore, in the event of a subsequent dispute about the quantity or cost of any item or unit of work, it should be relatively easy to remeasure the quantity to confirm the original estimate.

6.2.3 Economy

The estimator should strive to have the least number of **words** in descriptions of measured items, the least number of **dimensions** for calculation, and the least number of **extensions** and totals. A deliberate policy should be adopted to make the estimate as lean and tight as is practical consistent with the maintenance of verbal and mathematical logic. There are two good reasons for the foregoing recommendation: first, every word is capable of correct or incorrect intepretation, and every calculation may be correctly or incorrectly done; therefore, the less said and calculated, the better; and second, as time is always a pressing problem in estimating, methodology that will hasten the process while improving comprehension and accuracy is to be encouraged.

6.2.4 Standards

Work should always be measured in accordance with some **standard method** of measurement, either internally established by the individual company or organization, or externally recommended for use throughout the industry or trade. There are several good reasons to adopt such a practice: first, the work of the estimator will be made more simple, more rapid, and probably more dependable; second, communications between affected personnel will be clarified, strengthened, and hastened; and third, in the event of a dispute, there will be some precedents on which the estimator can rely.

In general, the methods of measurement used throughout this book are based on recommendations contained in the *Measurement of Construction Work* (Metric

Edition), published by and obtainable from the Canadian Institute of Quantity Surveyors (CIQS), Toronto, Ontario. These recommendations are based on similar methods used in most of the countries of the noncommunist world, and they are used in this book with permission of the CIQS.

6.2.5 Confidence

One of the prime concerns of any estimator is to develop confidence in his or her own work by adopting practices that will assist in avoiding estimating error wherever possible, and to reduce the effects of errors of measurement (or pricing) that may accidentally be incorporated into the estimate. One way of minimizing the chances of error in measurement is to adopt the practice of measuring *overall* and deducting *voids* and *wants*. By definition, **overall** means from end to end, or to the fullest extent possible. A **void** indicates an adjustment made where one item displaces another item, such as at a window space in a wall. A **want** indicates an adjustment made to compensate for deliberate overmeasurement, such as at a recess at the edge of a concrete slab. These terms are illustrated diagrammatically in Figure 6.1.

Overall measurement has a number of advantages: first, overall dimensions and the location of voids and wants are or should always be overtly stated or clearly implied by the drawings and specifications, and there is therefore seldom any need for the estimator to calculate or otherwise establish such data, by scaling or other means; and second, by measuring overall first and making deductions second, the effects of any minor errors or oversights in voids or wants will be minimized. The

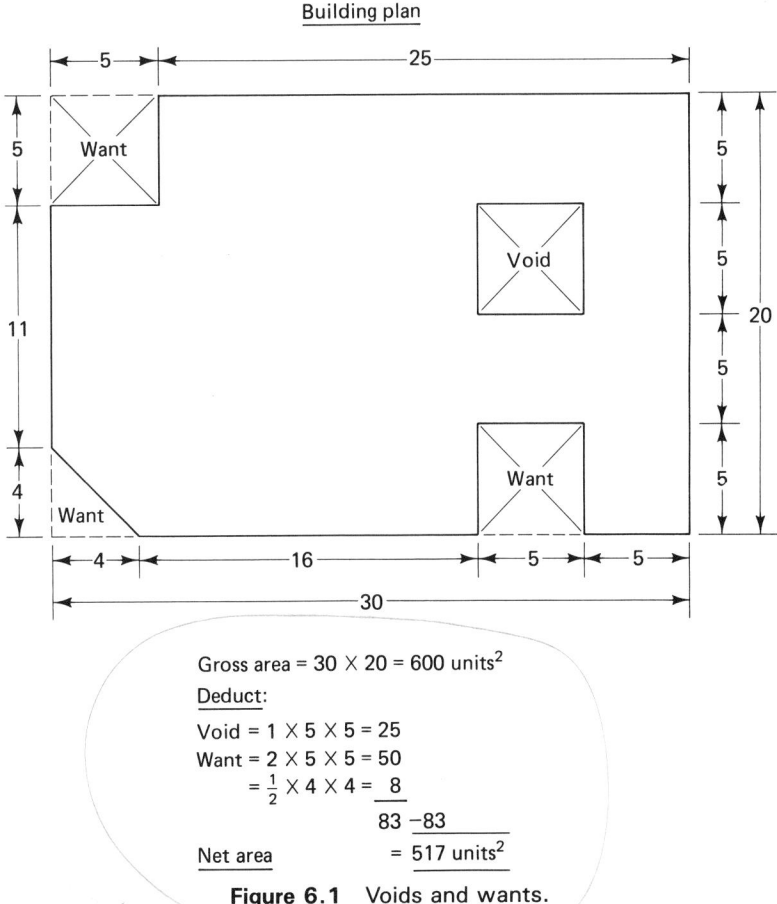

Figure 6.1 Voids and wants.

result may be a minor overmeasurement or overpricing, but this is seldom fatal to the budget.

6.2.6 Flexibility

As a general rule, it is better to measure a few more items rather than a few less. The time taken to do the actual measurement will be the same, but there are a number of benefits to be considered. First, wherever a choice can be made between measuring items of work of similar type together or separately, they are best initially kept separate. Although it is relatively easy subsequently to combine subtotals of quantities of similar items or identical prices, it is more difficult to separate and price composite items later. Second, care should be taken to organize the measurement of work in such a way that the labor, material, and equipment components of items can be priced separately or together, depending on what is most expedient.

For example, in a concrete-framed building, one might measure a certain quantity of work in foundations, more in beams and columns, and yet more in horizontal slabs. Although the *quality* (and even the *quantity*) of concrete might be the same for all these items, the *nature* of the work to be done is different. Even though the work in all four items may be measured in the same measurement unit (cubic meters or cubic yards in this case), the unit price to be subsequently applied to each of the quantities may be different. The quality of concrete in each item may be the same or different, and the waste factors may vary from item to item. The labor crew size, type, and productivity may be the same or different for each item, as may be the equipment also.

6.2.7 Simplicity

It is good practice to design and construct buildings using standard and typical components and details wherever possible. This places little limitation on creativity in design or methodology. However, it is not always possible to stick with such standards, and occasionally some nontypical work will be encountered. To keep the idea of simplicity in measurement simple, one might say that nonstandard or nontypical work should be measured and kept separate from standard or typical work, so that the *additional* costs associated with the nonstandard aspects of the work can be separately calculated and *added on* to the costs of the standard work. This technique is called the **extra-over technique,** and two examples will illustrate the principle.

1. When measuring drain tile around a building, the work is measured as if the line were straight, and the quantity is stated in lineal meters or lineal feet, measured along the centerline of the drain. The additional work involved in installing bends, branches, cleanouts, elbows, and the like, is then stated separately as numbered items, which are later priced on the basis of the *extra* cost of supplying and installing such irregularities *over* the cost of supplying and installing the regular straight pipe.
2. When measuring a concrete wall built to a batter, that is, with one face vertical and the opposite face inclined slightly from the vertical, it is simple to measure the wall as if both faces were vertical and parallel to each other, which is the typical, standard, or normal configuration for a concrete wall. The area of the sloped portion of the wall is then measured separately and the additional cost of modifying all the standard accessories and labor to produce that particular feature are calculated and applied to the nontypical quantity as measured.

Drain tile with covers. [From R. C. Smith, *Principles and Practices of Light Construction*, 3rd ed. (Englewood Cliffs, N.J.: Prentice-Hall, Inc., 1980).]

The primary benefit of this technique is that any discrepancy in the pricing of such nontypical or irregular items is contained within the *difference* in cost between the standard work and the nonstandard work, and almost always has to be *added* to the cost of the standard work. A secondary benefit is that the estimator does not usually have to make deductions in his measurements of the standard work, but can simply make a note to add an item for the nonstandard portion. A third benefit is that the extent of the additional cost can be identified for study, to assist in determining whether the nonstandard feature should be retained, modified, or deleted during the design or building process.

6.2.8 Repetition

Work should be measured in such a way as to minimize repetition of words or figures. If, for example, a number of floor areas are found to have a common length but varying widths, it is probable that the widths can be totaled or summed and the sum then multiplied once by the common length to produce the required gross area.

Wherever possible, work should be measured using dimensions that can be used more than once. Such dimensions are often called **reflected dimensions** and they have two primary advantages. First, if they were correct for the first use, they will be correct for all subsequent uses with appropriate modification where necessary; and second, they eliminate subsequent calculation, thus saving time and reducing possibility of error. Some examples of reflected dimensions are to be found in the area of formwork necessary to form an area of concrete wall, the area of gypsum drywall work requiring to be subsequently painted, or the area of plywood subfloors to be covered with carpet or resilient flooring.

6.2.9 Style

There are a number of useful exhortations to be made under the general title of style. First, estimating work should be done in a clean, quiet, comfortable, and familiar environment, with good lighting, heating, seating, and equipment arrangements. There should be adequate space to lay out sheets of drawings on desks or tables, or to pin them up on wall-mounted tackboards, as well as space to lay out estimating paper, pencils, scales, calculators, books, and other useful pieces of office equipment or documentation.

Second, it is a poor practice to scribble or doodle on the drawings or specifications from which the estimate is being made. For one thing, such scribbles may subsequently be misread as being part of the design or contractual requirements. Furthermore, such idle activity wastes time and reflects adversely on the reputation of the estimator and attitudes toward the work. In the case of the estimator employed by a construction company, a deposit often has to be paid to borrow drawings and specifications for estimating purposes, and such deposits are usually not refundable if these documents are not returned to the designer in good condition.

Third, most of the measurement process should be completed before the calculation process is started. The estimator should resist the temptation to manually do extension and summary calculations immediately following the establishment of dimensions for each item of work, except where absolutely necessary or logical with respect to other parts of the measurement, such as in the use of reflected dimensions. The reason for this comment is to try to avoid interruptions to the train of measurement thought, to avoid omissions, duplications, or other errors that such interruption can cause. With the use of electronic equipment, such as mini- or microcomputers, the calculation process may be essentially automatic, in which case this comment has less effect or importance.

Finally, it may be said that a deliberate attempt should be made to record words, figures, and diagrams in a clear, bold, and readable style, either by hand or by machine, with suitable use of headlines and underlines. A soft pencil is usually adequate for recording notes on paper, as minor errors can easily be corrected with an eraser. It is seldom necessary or advantageous to write in ink or use a typewriter for estimating. With the recent widespread use of desktop computers and printers, such advice may not now apply, as the many advantages of the use of such equipment easily offset their limitations in this regard. The various worked examples shown in this book demonstrate acceptable styles for presentation of estimating data.

QUESTIONS

6.1. State two reasons why the measurement of construction work should be done as accurately as possible.

6.2. Explain why it is a good practice to adopt a standard method of measurement wherever possible.

6.3. Distinguish between a want and a void in estimating by verbally defining these two terms. Illustrate your definition with a small sketch to show each feature.

6.4. Explain the concept of the extra-over technique as it applies to construction estimating, using a simple example to show your understanding. State one benefit that accrues to the estimator by the use of this technique.

6.5. What is a reflected dimension? How are such dimensions used to advantage in estimating?

CHAPTER 7
METHODS OF MEASUREMENT

7.1 INTRODUCTION

There are some general principles of estimating that apply to the work of all estimators; these were stated in Chapter 6. However, not all estimators need to know how to estimate all types of construction work. The object of this chapter is to present the additional information necessary to form complete methods for the use of estimators having separate and different needs.

Most estimators tend to specialize in the measurement and pricing of specific types of construction work, and at the risk of oversimplification, all estimators may be classified as belonging to one of two categories: contractor's or designer's estimators. In general contractor's offices, the estimators may be interested in the work of several trades, such as concrete work and carpentry framing, whereas in subcontractor's offices, the estimator may be interested primarily in the work of one trade, such as painting or roofing. In designer's offices, the estimator may be interested in determining the probable cost of several major elements of the building, such as the structural work, the architectural finishes, and the sitework, whereas in specialist consultant's offices, the estimator may only be interested in the cost of one element, such as electrical systems, or mechanical services, or speciality work often found in commercial or institutional buildings such as theaters and hospitals.

In the following sections of this chapter, the methods available to and used by estimators in both categories are examined in detail. The objective is to develop understanding, reliability, accuracy, and speed in measurement for the student and professional practitioner alike.

7.2 METHODS FOR CONTRACTORS

7.2.1 Preamble

Regardless of whether a contractor is bidding on public work or negotiating for private work, an estimate of some kind is essential to determine the approximate anticipated costs of the proposed construction. The estimated costs usually form an important part of any contract between the contractor and his client, the owner of the property being developed. Estimators working for general or subcontracting companies have at least two basic methods of measurement from which to choose: net measurement and materials take-off. In **net measurement,** the amount of construction work to be priced is expressed in composite units of number, length, area, volume, capacity, or mass. In **materials take-off,** each element of cost is measured separately and expressed in discrete units of time or commodity, as appropriate. Both methods are explained.

7.2.2 Net Measurement

In this method, the first idea to grasp is that it is construction **work** that is going to be measured. Work involves the application of effort in the form of labor and equipment to materials or products. To measure work, a decision is made to select the best unit to represent the nature of the work to be performed. For example, in built-up flat roofing systems, the best unit would be one in which the *area* of the work to be done is indicated, and the choice would be square meters, square feet or yards, or so-called squares, which are units of 100 square feet (sf). In excavation work, a volume unit such as cubic meters or cubic yards might be appropriate for bulk digging and trenching, and an area unit selected for trimming the earth surfaces exposed by the digging. Some work is traditionally measured in the units in which the principal component is priced; for example, concrete work is usually measured in cubic meters or cubic yards, and copper or plastic piping is measured in lineal meters or lineal feet.

Work should be measured *net in place.* The estimator should consider the required installed end result of the work to be done. The detailed process as to how that end result will be achieved will be considered more fully during the pricing process. For example, using *metric* dimensions, if a wood-framed wall measuring 5.00 m in length and 2.00 m in height has to be covered with plywood, the work to be done is 10.00 m². As each sheet of plywood measures 1.2 m × 2.4 m and thus covers 2.88 m², four sheets of plywood covering (4 × 2.88) 11.52 m² will be required. The cost of the waste factor of (11.52 − 10.00) 1.52 m² will be taken into account when the *material* component of the item is priced. The cost of cutting the standard sheet of plywood down to size will be taken into account when the *labor* component is priced. The item would appear in the estimate as 10 m² of work.

For example, using *imperial* dimensions, if a similar wall was 15 ft long and 7 ft high, the area of work to be done would be 105 sf. As each sheet of plywood measures 4 ft × 8 ft and covers 32 sf, four sheets of plywood covering (4 × 32) 128 sf will be required. The cost of the waste factor of (128 − 105) 23 sf will be taken into account with materials, and the cost to cut the plywood will be included with labor as before. The item would appear in the estimate as 105 sf of work.

7.2.3 Materials Take-off

In this method, the idea to grasp is that quantities of materials, labor, and equipment are each going to be measured *separately.* Material components are measured on the basis of their manufactured **sizes,** while labor and equipment components are estimated on the basis of **time.** The unit of measurement that will be selected for each specific will essentially reflect the obvious physical properties of that material and the way in which it is marketed. Roofing felt is measured by the roll, asphalt by weight, drywall board and plywood by the sheet, bricks and blocks by the unit or package, all in their uncut or manufactured state. With respect to the marketing aspects, a large number of building products are packaged in a certain manner, and the whole package usually has to be bought, even if only a portion of the package is needed. For example, acoustical tiles are sold by the box, and softwood lumber, in imperial measure, is usually sold in even 2 ft increments. It is therefore unlikely that one could buy two-thirds of a box of tiles or a 13 ft fir joist.

Materials should be measured **gross in place,** thus including the waste factor in the measurement of the material. The estimator must still remember to allow for the labor cost to create the waste, as standard material sizes do not always fit the actual building project conditions. It is appropriate to comment here that some estimators subscribe to the fallacy that the cost of the material waste offsets the cost of the labor creating the waste. A moment's reflection will disclose that these two costs *do not cancel* each other out; they have to be *added together.*

Materials should be measured *overall,* as described in Chapter 6, but with special care taken with the deduction of voids and wants at openings, to ensure that gross quantities are still achieved in the measurement process.

Once the material quantities have been measured and tabulated, the labor and equipment content of the estimate can be approximately predicted by making reference to productivity tables or factors, developed from data produced by publishing houses, construction associations, labor unions, and equipment manufacturers. Further discussion of the sources of such data is presented in Part III, dealing with pricing. Data can also be developed by direct observation and experience of the company staff. For example, a drywall applicator firm should know the type of crew and the length of time that crew will take to install drywall boards of a specified type and size to a particular type of supporting frame, such as a steel stud wall, finished with tape and joint compound. Similarly, a masonry company must know just how long it will take a given number of journeymen masons and helpers to build a cinder-block wall of any specified configuration, quality, and quantity. Any excavation company that plans to stay in business must know the probable output of the various pieces of equipment that it proposes to use to do the digging on any specific excavation job. Labor is usually expressed in **man-hours,** and equipment is usually expressed in **utilization-hours.**

One obvious difficulty of the materials take-off method is that the estimator must know, at the time of doing the measurement, the actual physical dimensions of all the different types and classes of materials that will be measured. Some less obvious disadvantages include slightly longer estimating time, reduced opportunity for using repetitive output rates and prices, and greater difficulties in relating the amounts of work to be done to the costs of doing that work. At the same time, it should be realized that the method does force more detailed consideration of the actual use of materials on the estimator, and properly handled, it can also produce accurate and detailed material lists, ready for ordering in the event that the bid is successful. However, as only one bid in about 10 is successful, this is a questionable advantage.

The materials take-off method is especially well suited to the work of the mechanical and electrical trades; it has a less general application to the work of structural and architectural trades. Its principal application is to be found in the simpler finishing trade work, such as flooring, drywall, and painting work in small commercial and residential construction.

7.3 METHODS FOR DESIGNERS

7.3.1 Preamble

There are a number of approaches to prediction or estimation of probable construction costs from which designers can choose. In most client/designer commissions for building design, such costs have to be produced to reassure the owner/client, who is going to be paying the bills, that the subsequent value of the project upon completion will exceed the costs to acquire the land, prepare the design, and finance the construction to bring a workable economic enterprise into being. Although the designer (unlike the contractor) is usually not expected to guarantee that his estimates are absolutely correct, there is an implied legal obligation that such estimates are realistic and based on reasonable data. Furthermore, all such estimates should state what is included and what has been excluded. For example, in addition to the costs of physical building items, owners sometimes require the professional design fees to be included in such approximate estimates; inflation rates and other contingencies that have been included should also be clearly stated. As with estimators working with contractors, the methods of measurement available to estimators working in design offices fall into two broad categories: **parameter methods** and **elemental methods.**

Before examining these two categories in detail, the meanings of these two terms should be understood. The dictionary definition[1] of the word **parameter** gives the following meaning: a variable entering into the mathematical form of any distribution such that the possible values of the variable correspond to different distributions. A number of construction economists, including Ferry[2] and others, have articulated more useful definitions for the estimator: a parameter is a quantity that is constant in the particular case considered but which varies in different cases.

The dictionary definition of the word **element** is: a component or constituent part of a whole, and of the word **elemental** is: of the nature of an ultimate constituent; simple; uncompounded. Again, construction economists, including Helyar[3] and others, have devised a more useful definition for the estimator: an element is a major component to most buildings, usually fulfilling the same function irrespective of its design, specification, or construction. Both parameter and elemental methods are explained below.

7.3.2 Parameter Methods

There are a number of parameter methods worthy of note, some because of their advantages, and others because of their drawbacks. All such methods suffer from one fundamental difficulty—the problem of establishing truly reliable predictions for the basic variable or parameter, based on historical data.

Base Unit Estimates. Such estimates use predicted **outputs** from factories, mills, or plants, using **production units** such as metric tons, liters, or other measures, or they use predicted **inputs** to garages, hospitals, schools, or theaters, using **demand units** such as parking stalls, patients, students, or patrons, as the case may be. Although the establishment of the unit quantity is relatively easy in this method, and indeed may actually be given, the initial determination and subsequent modification of an accurate and reliable unit price to be applied to the quantity is extremely difficult. The common method is to examine the costs of previous similar construction projects and to divide these costs by the appropriate number of units relative to each project. It can be seen that this method makes no allowance for variations in building size, shape, style, quality, or any other significant factor. Base unit estimates can be used satisfactorily to help develop general budget forecasts, such as a school board might wish to present to a senior level of government for the con-

---

[1]*The American College Dictionary* (New York: Random House, Inc., 1960).

[2]D. J. Ferry, *Cost Planning of Buildings,* 4th ed. (St. Albans, Herts., England: Granada Publishing Ltd., 1980), p. 256.

[3]F. W. Helyar, *Construction Estimating and Costing* (Scarborough, Ontario: McGraw-Hill Ryerson Ltd., 1978), p. 161.

struction component in an annual funding program, but prudent designers would be advised to avoid their use in the development of a construction budget for an actual building project.

Factor Analysis Estimates. In this method, the primary distribution of costs in a number of previous building projects essentially similar to the proposed project are analyzed and examined to discover which factors contributed to greater or lesser proportions of cost. The factor with the greatest cost is given a value of 1 (or 100); all other factors are expressed as percentages of the greatest factor, depending on the proportion of cost of each of these secondary factors relative to the primary factor. For example, if one were to examine the cost of bowling alleys, it may be that the primary factor of cost might be the supply of automatic-pin-setting equipment. Further examination may disclose that installation of such equipment is around 25% of the cost of supply, and to support and house the equipment costs 75% of the supply cost. The cost data developed by this approach can also be used in the base unit method, in this case by developing a price per actual lane or alley and then multiplying that price by the number of alleys in the proposed new building. Some modification of the factors is also possible in this approach, although care must be taken to rationalize any such modification on the basis of reliable data.

Specific Parameter Estimates. Some years after World War II a form of approximate estimating for building costs was developed by consultants to the magazine *Engineering News-Record*. In the ENR format, there are two main divisions: the **collection** of data and the **application** of data. The division dealing with the collection of data is divided into four subdivisions. First, the building and its developers are described in general terms. Second, standard parameters that can be measured for most buildings are listed. Third, specific characteristics of design, such as story heights and floor areas, are established. Fourth, selected design ratios are determined. The division dealing with the application of data lists most of the trade sections likely to be found in any typical building project, assigns to each trade a code number related to one of the standard parameters, to connect parameter quantity with unit cost, and thus enables total cost to be predicted. The method in general is not unlike the approach adopted by contractors to prepare a price to bid or negotiate for a construction contract, in which individual costs, allocated to specific trades or sections of the work, are totaled to form an aggregate price.

Area and Volume Estimates. One well-known form of parameter estimating involves the application of previously determined unit prices to the **gross area** or **gross volume** of the proposed new building. Provided that the proposed new building is virtually *identical* to previous similar buildings in size, shape, quality, and all other significant features, some reasonable degree of reliability can be expected with this method. However, although the measurement of area or volume is relatively simple to do, the method precludes any opportunity to give effect to the cost of particular *variations* of shape, story heights, cladding, or any other specific aspect of the new building which differs from the previous buildings used to produce the unit price. Furthermore, specific elements of buildings that cost more than others cannot be separately studied for economies under this method.

To illustrate some of the limitations of area and volume measurement used in this context, consider the manner in which perimeter length varies with area enclosed (Figure 7.1). If a square has sides each 10 m long, the perimeter is 40 m and the area contained is 100 m². If the sides are doubled to 20 m, the perimeter is 80 m and the area contained is 400 m². Note that although the perimeter has increased by 100%, the area has increased by 400%. Consider the manner in which perimeter length varies with shape. If a rectangle has all sides 10 m long, the area enclosed is

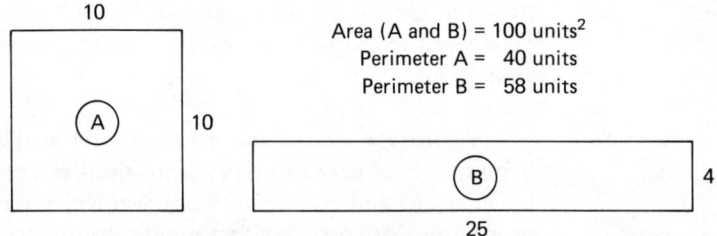

Figure 7.1 Area: perimeter.

100 m² and the perimeter is 40 m. If the shape is changed, so that the rectangle has two long sides 50 m long and the other two short sides only 2 m long, the area enclosed is still 100 m² as before, but the perimeter is now 104 m long. One can see the danger in applying a unit price derived from a square building to a proposed new rectangular building, and vice versa.

Similarly, consider how area relates to volume (Figure 7.2). A building that stands 30 m in height may contain 10 stories each 3 m high or 12 stories each 2.5 m high. If the costs predicted from the building with 10 stories were applied to the proposed new building with 12 stories, assuming that floor areas and all other significant factors of shape, cladding, quality, and so on, in each building were identical, a discrepancy in construction costs would soon become apparent, in spite of the fact that the volume of each building is the same. A further defect in this method stems from the fact that construction costs are not significantly generated by the volume contained by a building, but by the nature and quality of the structure and planes enclosing the volume and the methods selected for heating and lighting that volume. If one combines area and volume and tries to apply unit prices derived from one building equal in area to the new building but having a different height, a false and misleading result may also be produced.

One other limitation to the application of most parameter estimating methods is that they can be applied only after the primary design concepts of size, shape, quality, and style have been essentially crystallized. They therefore do not lend themselves well to aspects of cost planning or cost control.

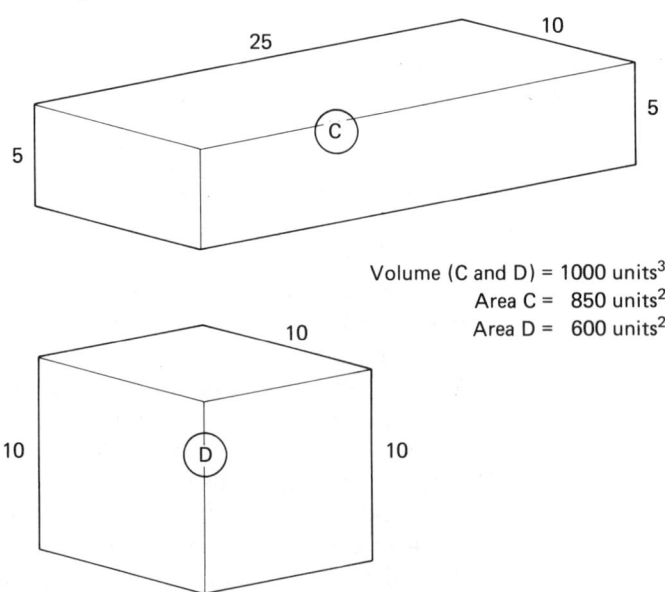

Figure 7.2 Area: volume.

7.3.3 Elemental Methods

There are two major methods of measurement involving elements of buildings, the first of which is called appropriately the **elemental analysis method,** and the other the **approximate quantity method.** Both are more time consuming and require more skill in measurement and pricing than any of the foregoing methods, but both permit a far greater degree of flexibility and modification by the designer to suit proposed project circumstances, and both produce far more accurate and reliable results on a consistent basis. Each method consists of four parts, which are described below.

Elemental Analysis. This method is most commonly used at the very earliest stages in the development of a building design concept. It can be applied to some extent even before any drawings or specifications for the building have been prepared, and reapplied at all later stages of document production.

In this method, the first part is called the **program development,** which consists of the determination, by the designer, of the proposed function and use of the property and the consequent type and style of building that will meet the owner's needs. Next, the general scope and size of the project as a whole has to be established, as well as each of its major parts, either on the basis of available funds or the minimal needs of the owner, within regulatory guidelines. From this information, the most suitable type of construction system and level of quality are considered and selected. At each stage in the development of the program, the designer/estimator should make assumptions to include things that might reasonably be expected to be found in the proposed building, such as elevators in a highrise building, refrigeration in a supermarket, or lobby and meeting space in a hotel project. An outline specification or brief description of the data on which the budget is going to be based should be prepared by the estimator during this first part of the estimating process, to serve as an initial reference point at the outset and to minimize later debate and dispute over what was agreed. Wherever possible, a site visit should be made by the estimator; failing that, the estimator should have the designer or owner produce extensive, detailed, and accurate information about the site conditions.

The second part is called **element measurement** and consists of the recording of quantities of major elements of the proposed building. As defined previously, an element in this context is a portion of a building that performs some major discrete function in any building, regardless of specific quality or style. Elements therefore have titles such as foundations, exterior cladding, floor systems, internal partitions, heating, vertical movement, and so on. More detail on element titles and how to measure them is given in the part of this book dealing with measurement technique. To give one example here, a perimeter foundation wall could be an element, and its description would include: the composition of the wall, whether of concrete, brick, wood, or steel; any work related to such composition, such as formwork, preservatives, mortar, and the like; all excavation necessary to permit the wall to be positioned correctly, waterproofing if specified, and backfill of excavated or imported material after the wall is built. The wall could be measured in square meters or square feet in its center vertical plane. Alternatively, such a wall could be described as being of a certain stated height, and measured in lineal meters or feet. If during the early stages of design, actual measurement of any particular element cannot be made, there are techniques (which will also be explained later) which permit reasonable estimates to be made of probable quantities for the missing elements. Element measurement is always simple and rapid, and as with all measurement, is always done as accurately as the given data will permit. Care should be taken to relate the quantities to the approved program.

The third part of the elemental method involves **price analysis** of each element. The physical composition of each element is analysed to discover its probable makeup, and then each component in that makeup is priced on a unit basis, with all the units being added together to produce a composite unit price for the element quantity. For example, an internal partition element may be proposed to consist of steel stud framing, covered with gypsum board drywall, taped and filled, and coated with two layers of latex paint. Each of these components can easily be priced, with the prices being added together to produce a composite price to be applied to the element quantity on a lineal or superficial basis. Again, more will be said about the pricing aspects in Part III. Decisions can now be made about the relative effects of including that particular element in this particular building project, and the element can be included as is, modified structurally or economically as necessary, replaced with another system if desired, or scrapped entirely. The data can be further manipulated to produce costs per element, costs per unit floor area or building volume, costs on a percentage-of-the-whole basis, or costs in any one of a number of other useful modes. Care should be taken to relate the prices to the probable quantities.

The final part of this method involves preparation of a **cost summary,** usually utilizing a standard form prepared for this purpose and modified by the estimator to suit the special characteristics of the particular project under study. The quantities and costs of the various elements are entered on the form in an orderly and useful manner, exemplified later. The extensions of quantity and cost are then summarized to produce the probable total cost and to show the distribution of costs over the various project parts, both in dollars and as percentages or in other ratios as preferred. The estimate is now ready for discussion by all concerned, such as the designer, the estimator, the owner, and others involved in the production of documentation.

In the elemental analysis method, some of the measurement and much of the pricing often can be done directly from the approved program before any design or working drawings or specifications have been produced. This can result in a saving of time at the budget establishment stage, an opportunity for design changes before ideas become too crystallized, and also helps to present clearer guidelines for the initial preparation of those drawings and specifications and for subsequent checking that such documentation is still in line with the original budget thinking. Further information on elemental analysis is given in Part IV.

Approximate Quantities Method. This method is more commonly used after fairly detailed design drawings and specification outlines become available to the estimator. The reader is reminded of the special use and meaning of the word "approximate" in this context. The method is not unlike the detailed method used by estimators in construction companies, in which quantities of work are measured and priced more or less on a trade-by-trade basis, although precise division of the type and cost of work allocated to the various trades is of less importance to the designer.

In this method, the first part is called the **specification** and it is used to establish quality, style, size, and price data of each of the various components proposed for use in the project. The development of standard checklists is of value in this process, to avoid accidental omission of important items.

The second part is called the **take-off** and consists of the accurate, though simplified measurement of specific significant composite items of work proposed for the various parts of the building. For example, the composition of a perimeter foundation wall would be described—whether of treated wood, waterproofed concrete block, or cast-in-place concrete with formwork—and then measured in its vertical plane in square meters or square feet for pricing as one composite item. The individual components of the wall are not separated, as would occur in a contractor's estimate. The excavation necessary to permit the placement of the foregoing foundation wall would also be separately described, measured, and priced, perhaps in two items, one to describe general reduced-level digging for a basement area, and a second to describe some additional trench digging for the foot of the foundation wall below the basement level. Such excavation items would be described and priced to include all necessary backfill, trimming, temporary shoring, and other preparation or protection liable to cost money. In the contractor's estimate, these operations would each be separately measured and priced.

The third part, **pricing**, and the fourth part, **summary**, are identical to the corresponding parts in the elemental analysis method described above.

A variation of the approximate quantity method can be useful for rapid cost estimating of simple buildings with repetitive features, such as warehouses, highrise apartment blocks, and supermarkets. The method involves either detailed or approximate measurement of a typical bay, floor, or wing, depending on time available, and then the extension of the cost of that typical portion of the building to predict the cost of the whole project. Care has to be taken with nontypical features, such as are to be found in sitework, foundations, and lobby areas of such buildings, to make all necessary allowances for specific conditions or features. Such features have to be measured and priced separately, and their costs added to the composite cost of the repetitive portion of the building.

QUESTIONS

7.1. Distinguish between net measurement and materials take-off as these two terms are used in construction estimating.

7.2. How is waste allowed for in the materials take-off method? Consider both materials and labor in your answer.

7.3. Distinguish between parameter and elemental methods of estimating by briefly defining these two terms.

7.4. In factor analysis, which factor is given a value of 1? Explain your choice.

7.5. Identify the primary drawback to the use of gross floor area as a basis for determining probable costs of a proposed building.

7.6. Briefly explain when or why one might choose to use the approximate quantity method in preference to the elemental analysis method of determining probable costs of a project.

CHAPTER 8
TECHNIQUES OF MEASUREMENT

8.1 INTRODUCTION

As stated earlier, a distinction can and should be made between the words "method" and "technique" in estimating, **method** being a mode of procedure, and **technique** being the application of technical skill. In this chapter a number of recommendations are presented as guidelines to measurement techniques, to assist estimators to properly utilize the methods which they select for estimating costs under various conditions of service. These guidelines are based on general industry practices that have been found to be both workable and reliable. Each recommendation is keyed to an illustrative example, where appropriate, to show how it works in practice. To be consistent with Chapter 7, the guidelines have been separated for study by estimators employed by contractors, who generally prepare detailed estimates, and by designers, who generally prepare approximate estimates.

8.2 TECHNIQUES FOR DETAILED ESTIMATES

8.2.1 Preamble

The recommendations contained in the following paragraphs apply with only minimal modification to both *net measurement* and *materials take-off* methods. Some also apply to the techniques recommended for the preparation of approximate estimates, and these are so noted in Section 8.3.

8.2.2 Checklist

First, prepare a checklist of work to be measured, by reviewing the contract drawings and specifications and by comparing the contract requirements with standard checklists of work commonly associated with specific trades. Such standard lists are produced by such organizations as the Construction Specifications Institute in "Masterformat" and the Canadian Institute of Quantity Surveyors in "Method of Measurement of Construction Works."[1]

The contract documents can be divided into units of work, and the units can be divided into items of work, thus determining the scope of the work to be measured and priced. The checklist will permit decisions to be made as to which parts of the work will be estimated directly by the estimator, as distinct from related parts that can be delegated to others, such as subcontractors. The checklist permits analysis to determine the most logical sequence in which to measure work; such a sequence does not always follow the sequence of building. For example, although formwork must be *built* before concrete can be poured, concrete work is usually *measured* before formwork, because of the arithmetical relationships between these two items of work. The work of *placing* concrete in floor slabs would, however, be measured before the work of *finishing* the surfaces of such slabs.

8.2.3 Format

Second, select the correct type of estimating form (Figure 8.1) appropriate to the type of work that is going to be measured. Some forms permit only measured quantities of work or material to be listed, whereas others permit measurement and pricing to occur on the same page. Some forms require that measured items be listed in a vertical format, with dimensions and extensions alongside, whereas others list items horizontally, with dimensions and extensions placed vertically below. Some trades, such as drywall and painting, prefer the estimate page to be arranged to permit the dimensions of the four walls of most typical rooms to be set down side by side in horizontal format, whereas other trades, such as plumbing and electrical, need to list lengths of pipe or conduit runs in a vertical format, with the appropriate fittings such as bends and tees listed at convenient points alongside the measured runs of pipe. Each construction company can have its own particular estimating forms designed and printed, although there are many inexpensive standard forms published for general industry use in every region of the continent. If electronic equipment such as microcomputers are used for estimating, the program can be selected or designed to present a spreadsheet of the correct configuration on the screen for the use of the estimator. A wide variety of such programs are now available for construction applications.

8.2.4 Identification

The blank spaces at the top of the first page of the estimate form or at the start of an electronic spreadsheet or file, used to identify the estimate number, the title of the project, the estimator's name, the classification of work, and so on, should be carefully completed before the start of measurement (Figure 8.2). On each subsequent page or file, sufficient detail should be entered to identify the page and its relation to and position in the complete estimate. The initials of the person responsible for overseeing or checking the measurement process should also appear on every page. The major objective of this procedure is to minimize error and to place responsibility for error at the right source. A minor objective is to identify work of acceptable quality for recognition by bonus or promotion.

8.2.5 Recording Data

On paper, it is a good practice to write legibly in *pencil*, in preference to using ballpoint pen or nib pen in ink. Minor errors will be made and can easily be corrected

---

[1]See Appendix D for addresses.

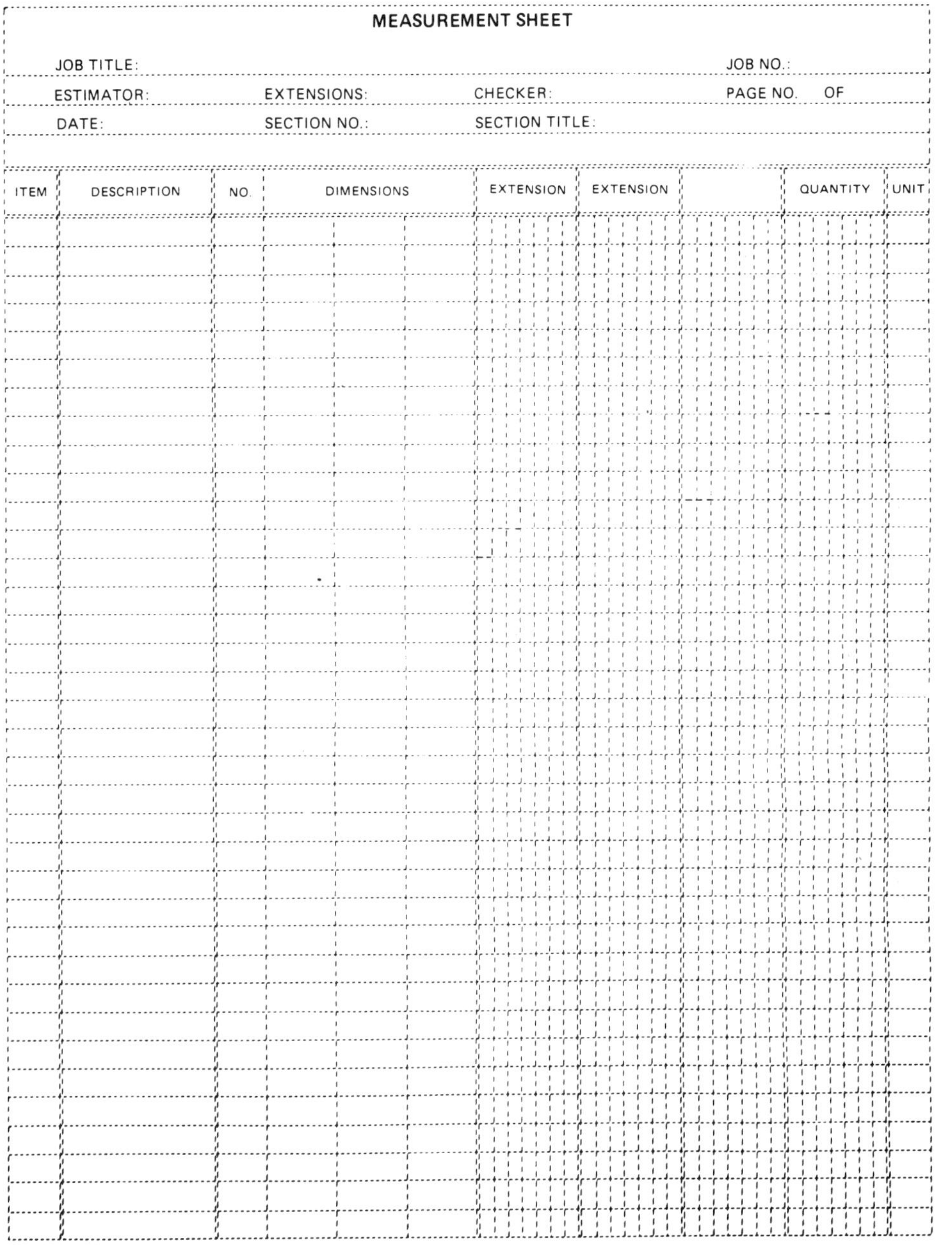

MEASUREMENT SHEET

JOB TITLE: JOB NO.:

ESTIMATOR: EXTENSIONS: CHECKER: PAGE NO. OF

DATE: SECTION NO.: SECTION TITLE:

ITEM	DESCRIPTION	NO.	DIMENSIONS	EXTENSION	EXTENSION		QUANTITY	UNIT

Figure 8.1 Selection of forms.

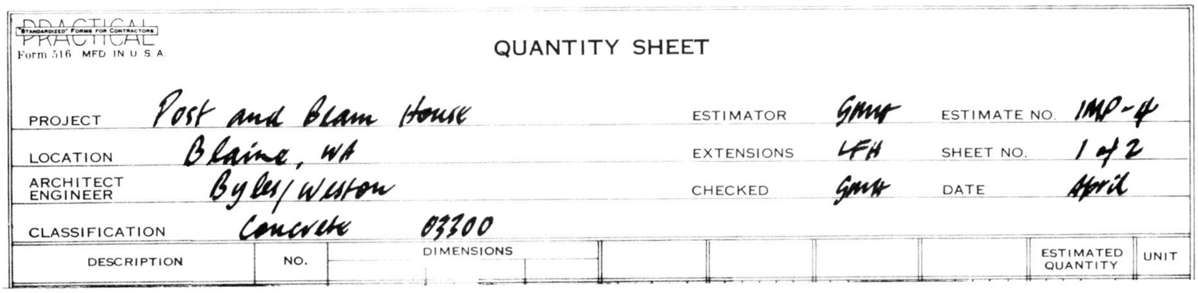

PRACTICAL "STANDARDIZED" FORMS FOR CONTRACTORS
Form 516 MFD IN U.S.A.

QUANTITY SHEET

PROJECT Post and Beam House ESTIMATOR GMH ESTIMATE NO. IMP-4

LOCATION Blaine, WA EXTENSIONS LFH SHEET NO. 1 of 2

ARCHITECT ENGINEER Byles/Weston CHECKED GMH DATE April

CLASSIFICATION Concrete 03300

DESCRIPTION	NO.	DIMENSIONS					ESTIMATED QUANTITY	UNIT

Figure 8.2 Selection and identification.

if a soft-lead pencil is used, although this should not be construed as a license to be careless or casual of habit. Similarly, on microcomputers or other electronic data-processing equipment, the program should permit simple and easy amendment of incorrectly entered words or figures before final printout. Care should be taken to use the printed or programmed format correctly, by placing all words, figures, and symbols in precisely the right column or space, to avoid descriptive figures being misread as dimensions, tens being misread as hundreds, and so on.

8.2.6 Preparation

The estimator should do any necessary or convenient *preliminary calculations* that will aid in the subsequent development of the estimate and enter these on separate sheets at the beginning of the take-off. On a small project, such calculations can conveniently be placed at the top of the first page of the measurement. Such preliminary calculations are usually purely arithmetical in nature; they are intended to give the estimator a general idea about the significant dimensions of the project, the magnitudes of the main parts of the work, and to indicate possible economies or shortcuts in the measurement stage. The temptation to start calculation of actual quantities of specific items of work at this point should be resisted. Such preliminary calculations might include the determination of significant perimeter lengths around the building, areas of main floors, walls, and roofs, or spacing of steel reinforcing bars, to name some examples. In general, the objective is to improve accuracy, rapidity, and utility of the subsequent measurement process. To determine and adjust perimeters, refer to Chapter 10; for an example of preparation, refer to Figure 8.3.

8.2.7 Classification

The **heading** for the first group or classification of items should now be selected from the checklist and written on the form or program in bold block letters, together with any notes that might apply to the group as a whole. The heading should be on a line by itself and be underlined for emphasis. The heading might be a title such as FORMWORK or ROUGH CARPENTRY or other appropriate descriptor, usually embracing a specific body of work that will be done by one crew or at least one company.

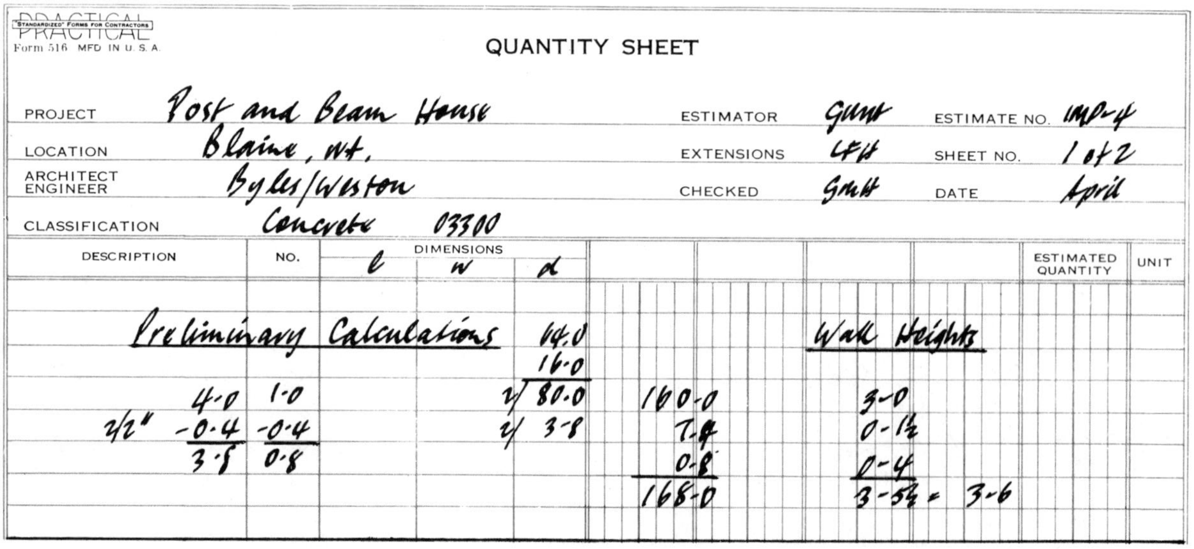
PRACTICAL "STANDARDIZED" FORMS FOR CONTRACTORS
Form 516 MFD IN U.S.A.
QUANTITY SHEET
PROJECT Post and Beam House
ESTIMATOR Gunt
ESTIMATE NO. IMP-4
LOCATION Blaine, Wt.
EXTENSIONS LFH
SHEET NO. 1 of 2
ARCHITECT ENGINEER Byles/Weston
CHECKED GMH
DATE April
CLASSIFICATION Concrete 03300
DESCRIPTION | NO. | DIMENSIONS l w d | ESTIMATED QUANTITY | UNIT
Preliminary Calculations 14.0 16.0 2/ 80.0 2/ 3-8
4-0 1-0
2/2" -0-4 -0-4
3-8 0-8
160-0 7-4 0-8 168-0
Wall Heights
3-0 0-1½ 0-4 3-5½ = 3-6

Figure 8.3 Preliminary calculations.

8.2.8 Description

The first of the items of work or materials can now be considered, by writing an abbreviated **description** of the item into the description column on the estimate form. The description, extracted from the checklist, must contain enough information between itself and the heading to permit the item to be subsequently priced. Any aspect of an item that does not have a cost component can usually be omitted. In many countries and companies, standard descriptions of common construction items have been developed and coded with cost-accounting numbers; the use of such numbers can save time and confusion. The descriptions should be entered close to the left-hand side of the description column, although some lengthy descriptions may spill over into adjacent columns, as shown in the examples. Descriptions are not usually underlined.

8.2.9 Abbreviation

Abbreviation is often used to reduce the amount of space and time taken to write descriptions. A common method of word abbreviation is to drop the vowels. For example, "window" becomes "wndw," "foundation" becomes "fndn," and "footing" becomes "ftg." Some words, such as "concrete," require care in their abbreviation, as "cncrt" is not as clear as "conc," and the word "include" cannot be rendered as "ncld" with any benefit. Another common method is to use the initial letters of groups of words to form a new word, such as "ASTM" for the American Society for Testing and Materials, and "CGSB" for the Canadian Government Standards Board. Again, care must be taken not to create ambiguous acronyms, as such words are called. The letters "NRC" could mean *either* National Research Council or Noise Reduction Coefficient. Glossaries of standard and accepted abbreviations for use in construction are available from many sources; a short list is included in Appendix A.

8.2.10 Dimensions

Every item of work consists of three dimensions and a number (Figure 8.4); these four elements can be arranged in any configuration convenient to the estimate. For example, the dimension for the *thickness* of a concrete wall can be placed in the description column, and the remaining two dimensions of *length* and *height* placed in the dimension columns and used to calculate the area for that thickness of wall. The *number* in this case would be one, for one wall.

On the next line below the description, the *dimensions* can now be entered in some logical manner and into the proper columns of the estimate form. The common practice is to put numbers first, followed by length, then width, then depth or height, one in each column. Some estimators follow a clockwise sequence for the extraction of dimensions from drawings, other may adopt some other convention; there is no hard-and-fast rule. Some drawings are set out using a grid of some type, and often the sequence of dimension taking can be related to the reference points on the axes of the grid. All entries in the dimension columns should be **side-noted** in the description column, to show where they came from, and for ease of subsequent reference in the event of a later change.

Dimensions are of two types: **direct** and **derived**. Where possible, dimensions should be read directly off the drawings and inserted into the estimate without conversion of any kind, except for minor simplification explained below. If it is not possible to read the dimension of any part of the work directly from the drawings, a small calculation should be made and shown in the description column alongside

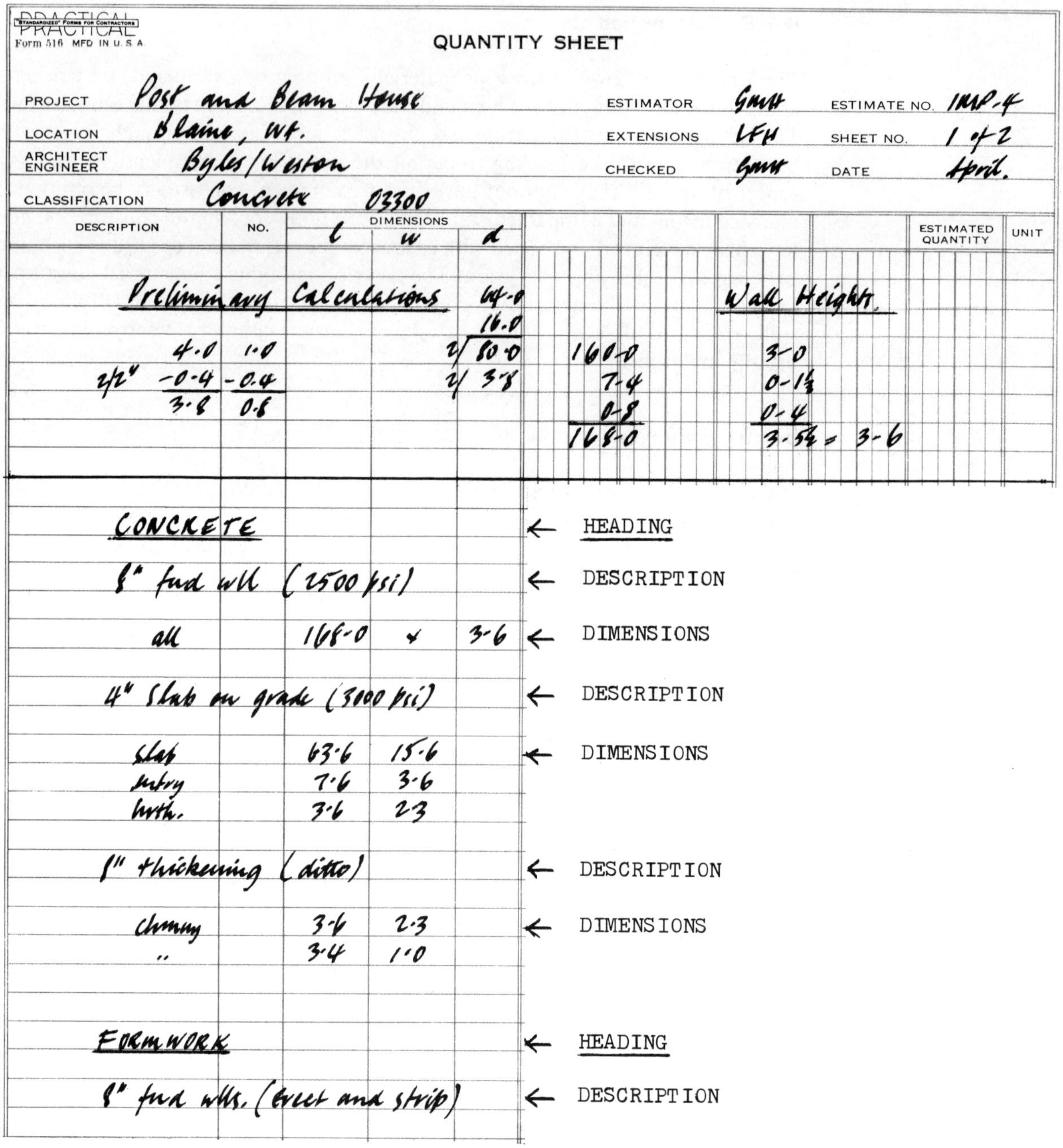

Figure 8.4 Classification and dimensions.

the dimension entry in the estimate, to show how the dimension was derived. It is not advisable to trust one's memory in this regard.

When using *metric* dimensions, enter the figures in millimeters, thus: 27500 for 27.50 meters or 270 for 27 centimeters. When using *imperial* dimensions, enter the figures in feet and inches, thus: 23-4 for 23'-4" or 9-11 for 9'-11". Do not insert the *symbol* for millimeters or the *signs* for feet and inches in the dimension columns. In imperial usage, there is no particular benefit to be derived from converting inches to decimal fractions of feet if they are not shown as decimals on the drawings; this conversion can be more readily and reliably done at the extension stage, using an electronic calculator with a constant factor.

8.2.11 Extension

Once all the dimensions have been entered in their proper places, the measurement is now ready for **extension**. In this context, the word "extension" means that the individual numbers in each dimension column can be multiplied together with dimensions in adjacent columns in such a way as to produce interim results which are placed in the extension columns of the estimate page. Lengths are multiplied by breadths to produce areas, areas multiplied by heights to produce volume, and so on. These results are then totaled to produce the final quantities of work of the various types measured. It is a good practice to underline final totals. In some companies, the extension and totaling process is done by machine or by personnel other than the estimators, for reasons of efficiency and economy. Some electronic programs permit automatic extension and totaling as dimensions are entered into the estimate.

The results and totals of the extension process are best shown as follows: when using the *metric* system, show numbers to two places after the decimal point; when using the *imperial* system, use whole numbers, rounding simple or decimal fractions up to the nearest whole number. On occasion, one will encounter a situation where the total quantity is less than 1 or the item of work being measured is extremely expensive, so that it might make sense to apply the unit price to a fraction of the whole. In such cases, care should be taken to ensure that the decimal or simple fraction cannot be misread, so that 0.50 is not mistaken for 0.05, and that ½ is not read as ⅓.

8.2.12 Units Symbols

The final step in the measurement process requires that the **unit symbol** for the work being measured be placed alongside the first extension of any series of interim results, and always alongside and to the right of the final totals for each separate item of work (Figure 8.5). Some examples of such unit symbols in the *metric* system include m for meters, mm for millimeters, kg for kilograms, and L for liters, and in the *imperial* system include ft or lf for lineal feet, sf for square feet, lb for pounds, and gal for gallons. Strictly speaking, in the imperial system, the symbols are really just conventional contractions, whereas in the metric system, the symbols are true symbols. A more comprehensive list of useful symbols for identifying items of construction work is given in Chapter 9.

8.3 TECHNIQUES FOR APPROXIMATE ESTIMATES

8.3.1 Preamble

The following recommendations apply primarily to measurement in the *elemental* approach to estimating. Some of these recommendations are essentially identical to corresponding techniques used to prepare detailed estimates; to save time and space, such techniques are simply noted as being identical and are not repeated in this section. The techniques used to establish quantities for *parameter* estimates are explained briefly in Section 8.3.7.

8.3.2 Checklist

First, acquire a copy of a standard list of elements, such as the one prepared by the Canadian Institute of Quantity Surveyors and shown in Section 18.1.3. Second, review the project program or design drawings to identify all elements that conform to the list and those that do not. Modify the list to suit the project.

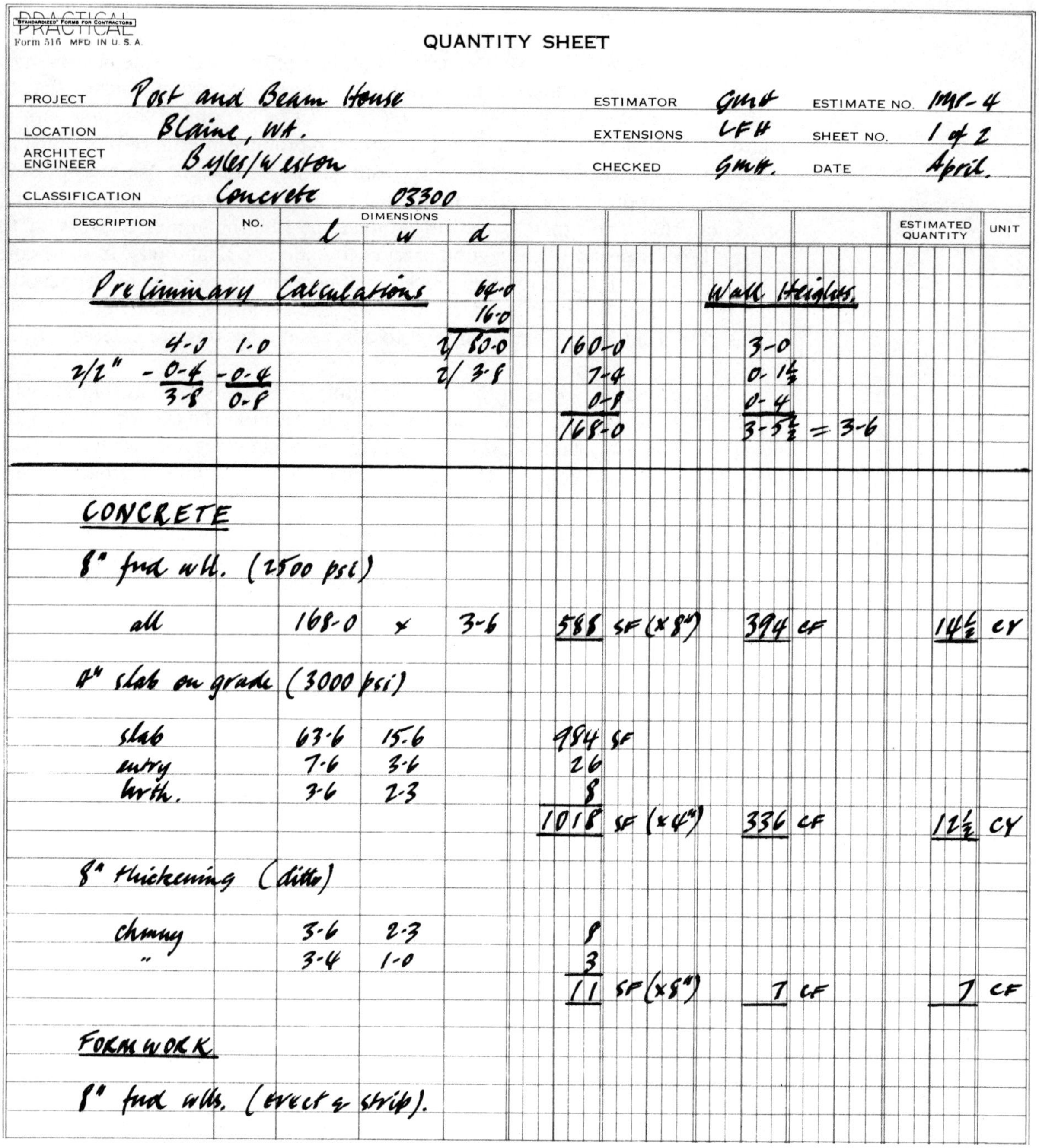

PRACTICAL "Standardized" Forms for Contractors
Form 516 MFD IN U. S. A.

QUANTITY SHEET

PROJECT Post and Beam House — ESTIMATOR GMH — ESTIMATE NO. MP-4
LOCATION Blaine, WA. — EXTENSIONS LFH — SHEET NO. 1 of 2
ARCHITECT ENGINEER Byles/Weston — CHECKED GMH. — DATE April.
CLASSIFICATION Concrete 03300

DESCRIPTION	NO.	DIMENSIONS l	w	d			ESTIMATED QUANTITY	UNIT
Preliminary Calculations				64-0		Wall Heights		
				16-0				
4-0	1-0			2/ 80-0	160-0	3-0		
2/2" - 0-4	-0-4			2/ 3-8	7-4	0-1½		
3-8	0-8				0-8	0-4		
					168-0	3-5½ = 3-6		
CONCRETE								
8" fnd wll. (2500 psi)								
all		168-0	×	3-6	588 SF (×8")	394 CF	14½	CY
4" slab on grade (3000 psi)								
slab		63-6	15-6		984 SF			
entry		7-6	3-6		26			
hrth.		3-6	2-3		8			
					1018 SF (×4")	336 CF	12½	CY
8" thickening (ditto)								
chmny		3-6	2-3		8			
"		3-4	1-0		3			
					11 SF (×8")	7 CF	7	CF
FORMWORK								
8" fnd wlls. (erect & strip).								

Figure 8.5 Extensions and unit symbols.

8.3.3 Format

Next, select or prepare a form on which the listed elements can be measured, bearing in mind that much less detail is required or available at the measurement stage than is encountered in the detailed estimate. Most standard measurement of work estimate sheets or programs are adequate for this purpose.

8.3.4 Classification

The **heading** for the classification of the first group of elements should now be selected from the element list and written on the form or program in block bold letters. The heading should be on a line by itself, and will be a title such as SUBSTRUCTURE or EXTERIOR CLADDING.

8.3.5 Description

The title of the first element can now be entered under the heading, usually in not more than two or three words, as these words are more fully defined and described in the standard element checklist used to organize the whole estimate. Sample items might be "Normal Foundations" or "Roof Finish." The specific measurement of selected elements is described in Chapter 18.

8.3.6 Remainder

The remaining aspects of approximate technique—identification, recording data, preparation, abbreviation, dimensions, extension, and unit symbols—are identical to the corresponding aspects of detailed technique.

8.3.7 Parameter Estimates

In **base unit** estimates, the technique of measurement is simply to establish the number of selected units, whether such units be students in a college, patients in a hospital, cars in a parking garage, or output of some commodity from a mill or factory. Such information is usually secured directly from the owner/client and his or her needs. In **factor analysis** estimates, measurement consists of identifying the quantity of the significant factor, usually a primary function of the purpose for which the building is being built. In **specific parameter** estimates, such as area or volume, measurement is usually taken to the outside planes or faces of significant floor, wall, or roof systems, ignoring minor projections beyond such planes as might be encountered in apartment balconies, columns or beams protruding outside the face of floor or wall planes, or chimneys or small parapet walls protruding above the general level of flat roofed buildings. The specific measurement of selected parameters is described in Chapter 18.

QUESTIONS

8.1. Distinguish between method and technique in estimating.

8.2. Give two reasons in support of the practice of making checklists before starting measurement or pricing in estimating.

8.3. What is the primary purpose for doing preliminary calculations before proceeding with measurement of work or material? Also identify one secondary purpose.

8.4. Distinguish between classification of work and description of work.

8.5. Abbreviate the following words, using the principles recommended in this book: roofing, sheathing, flashing, screeds.

8.6. Explain the meaning of the term "derived dimension." Give a brief example to illustrate your explanation.

CHAPTER 9
SYSTEMS OF MEASUREMENT

9.1 INTRODUCTION

In this chapter, both the metric and imperial systems of measurement are explained, insofar as they apply to the work of the construction estimator.

9.2 THE METRIC SYSTEM

9.2.1 Preamble

The metric system of measurement has been officially adopted by virtually every country in the world, the principal exception being the United States. In the United States, many sectors of the economy have informally adopted the system for reasons of practical and economic necessity. For example, much of the automotive, military, pharmaceutical, and container sectors of industry either use the system exclusively, or use parts of the system in conjunction with other systems. Much of the data prepared in connection with weather reports is now given in metric terms in many parts of the country; the wine industry in California now uses metric-sized bottles for its product; many other examples could be cited.

The primary resistance to the adoption of the system in the United States appears to be at the level of consumer transactions in the marketplace; the ordinary man and woman in the street resists change, even if changes are for the better, because of the uncertainty and expense with which change is often accompanied. It is not the intention of this author or this book to advocate change or the adoption of the metric system by the United States. The intention is only to present an explanation of the system for estimators who may encounter metric measurement in their work and therefore have to apply it in practice, as is the case for those who live in Canada and the rest of the world.

9.2.2 Development

The concept of metric measurement originated in Europe around the end of the sixteenth century. It was refined many times and over many years before its official adoption by France around the middle of the nineteenth century. Since then, its concepts and applications have spread worldwide and are now generally governed by conclusions reached at the international General Conference on Weights and Measures held in 1960. The refined system now used in most countries as known as the Système International d'Unités, or SI. There are some minor regional variations here and there, few of which affect the explanations given below.

The SI metric system is consistent, rational, and coherent. For these reasons, it is easy to learn and simple to use, and it was designed with these objectives in mind. Furthermore, there is no real need for estimators to learn the entire system, although it is not difficult to do that; it is only necessary to learn the parts of the system that will be encountered in daily use. At the same time, it should be stated that the reader must make a *deliberate* effort to understand the system and to *visualize* the magnitudes of the basic units. One should also resist the natural temptation to make conversions of metric units to units of other systems; to effectively *apply* metric, one must *think* metric. At first, this takes some effort, but with practice, the advantages of the system soon become obvious to any open and rational mind.

9.2.3 System Components

The SI metric system consists of three primary elements: numerical, nominal, and prefix. Each is explained separately below, and then the relationship of each to the others is shown.

The Numerical Element. In the metric system, the numerical base is 10. All arithmetic functions are either multiples or decimal fractions of 10. In estimating, the most common functions involve applications of formulas for length, area, and volume. If the SI base of 10 is combined with these formulas, the following table is produced:

$$
\begin{aligned}
\text{length} &= \text{length } or \text{ breadth} = 10 \times 1 &= 10 \\
\text{area} &= \text{length } by \text{ breadth} = 10 \times 10 &= 100 \\
\text{volume} &= \text{height } by \text{ area} = 10 \times 100 &= 1000
\end{aligned}
$$

In other words, in the SI system, each unit is smaller or larger than the preceding unit by a factor of 10 for length, 100 for area, and 1000 for volume. Simple fractions, such as 1/2, 1/3, and 1/4, should not be used in metric measurement; their decimal equivalents, such as 0.5, 0.33, and 0.25 should be used instead.

The Nominal Element. Units in any measurement system have to be identified by name. In the metric system, there are seven base units:

1. Meter (m) for length
2. Kilogram (kg) for mass
3. Second (s) for time
4. Ampere (A) for current
5. Kelvin (K) for heat
6. Candela (cd) for light
7. Mole (mol) for amount of substance

In the foregoing list, the first three units are commonly encountered in estimating, the next three units are common in design, and the last unit is used only in scientific work. There is also a series of derived units, that is, units which are derived from the base units listed above. Five of these derived units are common in estimating:

1. Square meter for area
2. Cubic meter for volume
3. Liter for capacity
4. Meters/second for speed
5. Degrees Celsius for temperature

Superscript numbers are used to indicate exponents, such as 2 for squares and 3 for cubes. Ten square meters is therefore written as 10 m² and 10 cubic meters is written as 10 m³. In other measurement systems, it is possible to confuse 10 square units with 10 units squared, the answer being 10 in the first case and 100 in the second case. In the metric system, there is no such confusion.

The Prefix Element. In the metric system, a simple and uniform prefix system is used in conjunction with the names of the base units to distinguish different quantities of such units by name. All quantities less than 1 or unity have Latin prefixes, and all quantities greater than 1 have Greek prefixes. The names of the prefixes of most interest to estimators are shown in Figure 9.1. Of these six prefixes, only c, m, and k are used daily by estimators; the others are seldom used, but are shown here to illustrate the system more fully. Because of the prefix system, however, each quantity or capacity has its own unique and descriptive name. For example, a kilometer equals 1000 meters, a hectare contains (100 × 100) 10,000 square meters, and a centimeter is one-hundredth part of a meter.

Relationships. The connections between length, area, and volume are shown in Figure 9.2. The relationship between length, area, and volume is to be found in the length of the meter, which for practical purposes can be considered to be just less than 10% longer than the imperial "yard." The reader is advised to acquire possession of a metric measuring tape and to use it! The relationship between vol-

Latin Prefixes			Greek Prefixes		
deci	(d)	means 1/10th	deka	(da)	means 10 times
centi	(c)	means 1/100th	hecta	(h)	means 100 times
milli	(m)	means 1/1000th	kilo	(k)	means 1000 times

Figure 9.1 Metric prefixes.

Linear units × 10	Area units × 100	Volume units × 1000
basic = 1 mm	basic = 1 mm²	basic = 1 mm³
10 mm = 1 cm	100 mm² = 1 cm²	1000 mm³ = 1 cm³
10 cm = 1 dm	100 cm² = 1 dm²	1000 cm³ = 1 dm³
10 dm = 1 m	100 dm² = 1 m²	1000 dm³ = 1 m³
10 m = 1 dam	100 m² = 1 dam²	1000 m³ = 1 dam³
10 dam = 1 hm	100 dam² = 1 hm²	1000 dam³ = 1 hm³
10 hm = 1 km	100 hm² = 1 km²	1000 hm³ = 1 km³

Figure 9.2 Metric relationships.

ume, mass, and temperature is to be found in the properties of water. A cube having sides of 10 cm² in area contains 1000 cm³; this volume is called 1 liter. One liter of water has a mass (or weight) of 1000 grams or 1 kilogram. One gram is about the weight of a paper clip. One thousand kilograms of any commodity weighs 1 metric ton.

The Celsius (or centigrade) temperature scale considers water just at the point of freezing to be at zero degrees, and just at the point of boiling to be at 100 degrees. On this scale, a cool spring day would be around 10 degrees, a warm summer day would be around 25 degrees, and hot weather would be above 30 degrees. The temperature of human blood is 37 degrees. A really hot bath would be just under 50 degrees. On construction sites, care should be taken if the outside temperature drops below 5 degrees or rises above 30 degrees Celsius.

Metric Symbols. Some conventions have been adopted for the proper use of **symbols** to represent units in the metric system. A few of the more useful rules, relative to the estimator's work, are appended below.

1. Symbols are not abbreviations, so they do not need periods after them, unless they occur naturally at the end of a sentence.
2. Symbols are always printed in roman type, never in italic or other form; for example: m, g, kg.
3. Symbols are always printed in lowercase type, unless the symbol is derived from a proper name; for example: g for gram, but A for ampere. The only exception is that l is sometimes written as L, to avoid the confusion between the letter "el" and the numeral "one."
4. Never leave a space between the prefix and the unit symbol; for example: km, not k m. Always leave a space between the quantity and the symbol; for example: 15 km, not 15km.
5. Symbols should be used instead of names and letters; for example, 20 cm² and not 20 square centimeters, and never 20 scm. They should always be in the singular form; for example: 20 kg, not 20 kgs.

Some conventions have also been developed with respect to the proper use of **numbers** in the SI system. A few of these recommendations, relative to the estimator's work, are as follows:

1. Use spaces, not commas, to separate long lines of figures; for example: 98 765.432 10 and not 98,765.432,10.
2. For numbers less than 1, use decimals, not fractions, and always place a zero before the decimal point; for example: 0.25 kg, not 1/4 kg, and never .25 kg.
3. Show all dimensions in millimeters and extensions to two places of decimals, unless otherwise stated; for example, a wall that is 14.5 meters long would be shown as 14 500, with the symbol *implied*.

9.2.4 Other Aspects

Measurement of time and angular measurement are unchanged in the metric system. Although there has been some discussion of the possibility of developing the 10-hour day, the 10-day week, the 10-month year, and the 100-degree circle, no such changes have been introduced anywhere in the industrialized world. In construction, time is still measured in the traditional units of seconds, minutes, hours, days, weeks, months, and years, and there are still 360 degrees in a circle and 90 degrees in a right angle, with degrees being subdivided into 60 minutes and 60 seconds as in time. The possible benefits of making changes to such measurement are not clearly outweighed by the enormous problems of making such changes, particularly in the fields of electricity and electronics, which make such extensive use of 60-cycle power.

The original length of the meter was established as one ten-millionth part of the quarter meridian of the Earth, that is, the great circle distance from the equator to either pole. Subsequent calculation of refined data produced by more accurate instrumentation showed that there was a minor discrepancy in the length of this standard, as well as the major difficulty of reproducing the standard in any practical way. As a result, the length of the modern meter is now related to specific radiation properties of a particular crystal, called krypton, so metric lengths can be measured very easily and accurately in laboratory situations. The official length of the "yard" used in the United States is expressed in metric terms relative to this standard.

9.3 THE IMPERIAL SYSTEM

9.3.1 Preamble

The imperial system of measurement, also known as the English system, the American system, the U.S. system, the standard system, or the avoirdupois system, among other names, is not so much a system as a collection of a number of essentially separate and discrete systems, some of which have no relationship to the others. Furthermore, within some of these systems there are many additional discrepancies, where, for example, the volume of the American gallon is different from the English gallon, and the former British ton is heavier than the American ton. The names of the units and the relationships among them have largely come about through custom, expediency, arbitrary rule, or some other force. Little explanation is necessary or even possible.

9.3.2 Development

The study of the historical development of measurement systems, although both interesting and instructive, is unfortunately beyond the scope of this book. However, it can briefly be said that the principal origins of the units in the imperial system are to be found in many common objects which were in everyday use and which were capable of some degree of standardization. For example, the main units of linear measure were based on the proportions of the human body, such as the width of the fingers and hands, the lengths of the arms and feet, the distance from the center of the head to the tips of the fingers, and the length of the average person's stride or pace. The word "mile" for example, comes from an earlier word, based on the Roman or Italian word for 1000 stadia, steps, or paces. Our modern word "bath," as in bathtub, come from a former unit of liquid measure. The present American gallon is based on the old English wine gallon, also known as the Queen Anne gallon, the volume of which was standardized in England in 1707, long before the War of Independence. That measure was later abolished in England, but its use continues in the United States today. Also, in America, some refinement of the measurement of weight occurred for everyday use, where, for example, the old hundredweight (which actually weighed 112 pounds!) was abandoned, and the short ton of 2000 pounds was introduced. For further information on this historical development, the reader is referred to any of the many books that have been written on systems of weights and measures.

Linear measure	
12 inches	= 1 foot
3 feet	= 1 yard
1760 yards	= 1 mile
5280 feet	= 1 mile

Area measure	
144 square inches	= 1 square foot
9 square feet	= 1 square yard
4840 square yards	= 1 acre
640 acres	= 1 square mile

Volume measure	
1728 cubic inches	= 1 cubic foot
27 cubic feet	= 1 cubic yard
(144 cubic inches	= 1 board foot)

Weight measure	
16 drams	= 1 ounce
16 ounces	= 1 pound
2000 pounds	= 1 ton

Liquid measure	
4 gills	= 1 pint
2 pints	= 1 quart
4 quarts	= 1 gallon

English measure	
14 pounds	= 1 stone
8 stones	= 1 hundredweight
20 hundredweight	= 1 ton
or 2240 pounds	= 1 ton

Mariner's measure	
6 feet	= 1 fathom
1013 fathoms	= 1 nautical mile
3 nautical miles	= 1 league
1 nautical mile per hour	= 1 knot

Dry measure	
2 pints	= 1 quart
8 quarts	= 1 peck
4 pecks	= 1 bushel

Surveyor's measure			
5½ yards	= 1 rod	1 square chain	= 484 square yards
4 rods	= 1 chain	484 square yards	= 16 perches
1 chain	= 22 yards	40 perches	= 1 rood
10 chains	= 1 furlong	4 roods	= 1 acre
8 furlongs	= 1 mile	1 acre	= 43 560 square feet
1 square mile	= 1 section	6 miles square	= 1 township

Miscellaneous measures			
1 gill	= 4 fluid ounces	1 pound	= 454 grams
1 pound	= 7000 grains	1 yard	= 0.914 metres
1 gallon	= 8.330 pounds	1 gallon	= 3.785 litres

Time and angular measurement			
60 seconds	= 1 minute	60 seconds	= 1 minute
60 minutes	= 1 hour	60 minutes	= 1 degree
24 hours	= 1 day	90 degrees	= 1 right angle
7 days	= 1 week	4 right angles	= 360 degrees
52 weeks	= 1 year	360 degrees	= 1 circle

Figure 9.3 Tables of imperial measurement.

9.3.3 Tables of Imperial Units

The various tables of units and values are simply presented for reference in Figure 9.3. Parts of them may already have been committed to memory, and the remainder may be utilized by the estimator from time to time during some measurement process.

Notes on the Tables.

1. The American measures for liquid and dry units are not the same. One U.S. dry pint = 1.164 U.S. liquid pints. The American gallon equals 0.833 British gallon or 3.785 liters. The British gallon equals 1.202 U.S. gallons or 4.546 liters. Conversely, 1 liter equals 0.264 U.S. gallon or 0.220 British gallon; therefore, 1 liter is slightly larger than an American quart and slightly smaller than a British quart.
2. The U.S. gill equals 4 fluid ounces; the British gill equals 5 fluid ounces. The British gallon has a mass of 10 pounds and a volume of 277.42 cubic inches; the U.S. gallon therefore has a mass of (10 × 0.833) 8.33 pounds and a volume of (277.42 × 0.833) 231.09 cubic inches. One pound weighs 454 grams; 1 kilogram weighs approximately 2.2 pounds (2.204 lb, to be precise).
3. One chain equals 4 rods, which equal (4 × 5.5) 22 yards or 66 feet. The area contained by 1 chain wide and 10 chains long is (66 × 660) 43,560 square feet, which is 1 acre.
4. Other parts of the Imperial system, such as troy weight and apothecaries weight, are not included here, as they have little bearing or effect on the work of the construction estimator.
5. Extensive conversion tables for metric and imperial equivalents have been excluded for two reasons: first, such tables are readily available elsewhere, on paper or in electronic form, and second, it is thought better not to make such conversions, but rather to become "bilingual," by understanding the elements and concepts of both systems.

QUESTIONS

9.1. In the metric system, is each unit of area 10 times larger or 100 times larger than the preceding unit? Explain your choice of 10 or 100.

9.2. What is the basic relationship between length, area, and volume in the metric system? What is the basic relationship between volume, mass, and temperature in the imperial system?

9.3. What is the weight of 5 liters of cold water? What is the weight of 5 gallons of cold water?

9.4. Suggest a technical reason why the measurement of time and angles have not been converted to a base of 10 or 100 in the metric system.

9.5. How many square yards are there in 1 acre? How many rods are there in 1 chain? How many gills are there in 1 U.S. gallon?

CHAPTER 10
CALCULATION

10.1 INTRODUCTION

In this chapter a brief review is made of the basic principles of calculation, to refresh the memory of the reader and to suggest some practices to be adopted and others to be avoided. For those who are so inclined, a review of the historical development of calculation throughout the ages makes for interesting reading, although such a broad study is beyond the scope of this book.

It is not necessary to have a profound knowledge of higher mathematics to become a competent estimator. It is only necessary to understand and be able to reliably apply the normal arithmetic processes of addition, subtraction, multiplication, and division, together with some simple algebraic, geometric, and trigonometric functions, all of which are described in this section. The estimator should be able to calculate on paper, by machine, and mentally on occasion, and should learn to look for relationships among numbers. He or she should spot-check critical calculations for accuracy, and become familiar with some common "rules of thumb" used in estimating, some of which are reliable and others of which produce problems for the unwary.

The word **calculation** means to ascertain by mathematical methods, to count, to figure, to compute. This definition suggests precision in the process and accuracy in the results. Calculations are never approximate; they are either correct or they are incorrect. Furthermore, the process leading to the correct result is usually as simple as that leading to an incorrect result; steps should therefore be taken to get a correct result in the first place, and to check it in the second place. The estimator has to contend with quite enough problems arising from other sources without having problems with calculation processes.

In Section 10.2 a review is made of some general attributes and functions of numbers, followed in Sections 10.3 and 10.4 by explanations of the basic arithmetic, algebraic, trigonometric, and geometric functions required to be known by estimators. Although a knowledge of calculus may be useful on occasion, it is not necessary for most estimating purposes.

10.2 REVIEW OF NUMBERS

10.2.1 Types of Numbers

Cardinal numbers are one, two, three, and so on; **ordinal numbers** are first, second, third, and so on. **Whole numbers** consist of integers, such as 1, 2, 3, and so on; **fractional numbers** consist of parts of integers, such as ½, ⅓, ¼, and so on. Fractional numbers may also be written as **decimal fractions**, such as 0.5, 0.33, 0.25, and so on. In general, a number is a collection of units, denoting a **total**.

10.2.2 Bases of Numbers

Numbers can be organized using any radix or **base**. A system of numbers using the base 10 is called a **decimal system**; this is the *basic* system in use throughout society. A system using the base 12 is called a **duodecimal system**; this system is often encountered in estimating, because of the frequent occurrence of the number 12 in the imperial measurement system.

10.2.3 Conventions

Whole numbers are conventionally written from right to left, with figures in columns representing units, tens, hundreds, thousands, and so on. Fractional numbers are written from left to right, with figures to the right of the decimal point, in columns representing tenths, hundredths, thousandths, and so on, in the decimal system, and twelfths in the duodecimal system. Examples are shown later. The comma is often used to separate groups of figures into threes, although it is also recommended to leave a space between groups of three, as shown in the following two examples:

$$98, 765. 432, 10 \quad \text{or} \quad 98\ 765.432\ 10$$

Notice that the decimal point remains in both examples. In some parts of the world, the comma is used *instead* of the point to indicate the separation between whole numbers and fractions. Care must be taken to correctly interpret all such symbolism. In estimating, it is of critical importance that all figures be placed in their correct columns.

10.3 ARITHMETICAL FUNCTIONS

10.3.1 Addition

Addition is the process of adding any series of two or more numbers together to arrive at a total. The numbers are arranged in vertical columns, so that units, tens, hundreds, and so on, in each number are located exactly above or below units, tens, hundreds and so on, in adjacent numbers. The following two examples show the arrangement, one using the decimal system and the other using the duodecimal system.

9,876.54	(meters)	7–10½	(feet and inches)
123.34		15– 5¼	
0.12		6– 8¼	
10,000.00	(total)	30– 0	(total)

Figure 10.1 Addition.

Chap. 10 / Calculation

To *do* addition, start at the right-hand side and add from bottom to top. To check addition, add from top to bottom.

10.3.2 Subtraction

Subtraction is the process of taking one number away from another number by deduction. Arrange the numbers as for addition; put the **minuend** first, the **subtrahend** second, and the **remainder** third, as shown in the following examples.

$$
\begin{array}{ll}
9{,}876.54 & \text{(minuend)} \\
\phantom{9{,}}123.34 & \text{(subtrahend)} \\
\hline
9{,}753.20 & \text{(remainder)}
\end{array}
\qquad
\begin{array}{l}
15\text{-}5¼ \\
6\text{-}8¼ \\
\hline
8\text{-}9
\end{array}
$$

Figure 10.2 Subtraction.

To *do* subtraction, start at the right-hand side and deduct the subtrahend from the minuend. To *check* subtraction, add the remainder to the subtrahend to produce the minuend.

10.3.3 Multiplication

Multiplication is the process in which a given number is taken by addition a given number of times; that is, one number is multiplied by a factor. A **factor** is one or more numbers which, when multiplied together, result in a product. For example, 5 and 6 are factors of 30. The numbers are arranged in columns, as for addition. In the two examples shown below, both have fractions to two places; therefore, the products will have fractions shown to four places.

	Decimal Example	*Duodecimal Example*	
(*n*)	1,234.56 (meters)	5'.4½" = 5- 4- 6 (feet)	
	78.91	8'.9½ = 8- 9- 3	
(*n* × 1)	12.3456	1- 4- 1- 6	(*n* × 3)
(*n* × 9)	1,111.104	4- 0- 4- 6	(*n* × 9)
(*n* × 8)	9,876.48	43- 0- 0	(*n* × 8)
(*n* × 7)	86,419.2	47- 1- 8- 7- 6	
	97,419.1296 m²	47 and 2/12 sf	

Figure 10.3 Multiplication.

To *do* multiplication, start at the right-hand side and multiply each digit in the number by each digit in the factor. Place each result on a succeeding line, then add the lines to arrive at the product. To *check* multiplication, divide the product by either factor, to arrive back at the other factor (see also, Section 10.3.5).

10.3.4 Division

Division is the process by which one number, known as the **dividend**, is divided by another number, known as the **divisor**. If the dividend is an exact multiple of the divisor, there will be no **remainder**. The following two examples show the arrangement, with and without remainders. In both examples, the four-digit number is the dividend, the three-digit number is the result, the two-digit number is the divisor, and the one-digit number is the remainder, which in one case is zero.

```
           27)4780(177        27)4779(177
              27                 27
              ‾‾‾                ‾‾‾
              208                207
              189                189
              ‾‾‾                ‾‾‾
              190                189
              189                189
              ‾‾‾                ‾‾‾
                1 (remainder)      0
```

Figure 10.4 Division.

To *do* division, place the dividend on the top line and the divisor on the bottom line. To *check* division, multiply the divisor by the result and add any remainder.

10.3.5 Explanatory Notes

The following explanation of the multiplication calculations in the *duodecimal* example given above may be helpful to those not familiar with the system.

1. Starting in the right-hand columns, 6 × 3 = 18; 18 divided by the base 12 = 1 and 6 left over; enter the 6 in the first column and the 1 in the second column.
2. Now multiply 4 × 3 = 12; 12 divided by 12 = 1 and 0 left over; add the 0 to the 1 in the second column and carry the 1 over to the third column, holding it in mental reserve for the moment.
3. Now multiply 5 × 3 = 15; 15 divided by 12 = 1 and 3 left over; add the 1 from mental reserve to the 3 left over and enter 4 in the third column and 1 in the fourth column.
4. Repeat steps 1 to 3 with the other factors 9 and 8 in succession. Note that 5 and 8 are whole numbers, not fractions, and therefore they are not divided by the base 12.
5. When all factors have been multiplied, add the columns, starting at the right-hand side. Divide column totals by 12 as necessary and carry excess results over to the column to the left. In the example, shown, there is no total greater than 12, so there is no excess to carry over from one column to the next.
6. Express the answer as whole numbers plus the first fractional number to the right of the whole number. Remember that each column to the right represents one-twelfth of the column to the left, just as in the decimal system each column represents one-tenth of its neighbor to the left. If a fraction is 6 or greater, it is usual to round the next column up by one.

It is also possible to check multiplication products by a process called **casting out the nines**. First, find the sum of the digits in each of the factors and in the product. Remove from each sum a whole number of nines. Multiply the remainders of the two factors together and again remove nines, if possible. The final remainder should equal the product. Using the decimal example given above as a model, the process is as shown below.

$$
\begin{aligned}
1{,}234.56 &= {}^{21}\!/_9 = 2 \text{ and } 3 \text{ over*} \\
78.91 &= {}^{25}\!/_9 = 2 \text{ and } 7 \text{ over*} \\
97{,}419.1296 &= {}^{48}\!/_9 = 5 \text{ and } 3 \text{ over}\checkmark \\
\text{and* } 3 \times 7 &= {}^{21}\!/_9 = 2 \text{ and } 3 \text{ over}\checkmark
\end{aligned}
$$

The remainders are 3 in each case.

Figure 10.5 Casting out the nines.

With decimal numbers, and in the metric system, there is no difficulty in dividing one measurement by another. With duodecimal numbers, in the imperial system, fractions of feet, pounds, or other units are best reduced to decimal fractions and then divided by decimal methods, as shown below.

$$\frac{6'.8''}{1'.4''} = \frac{5}{1} = 5 \quad \text{and} \quad \frac{8'.7\frac{1}{2}''}{5'.4\frac{1}{4}''} = \frac{8.625}{5.271} = 1.636$$

Figure 10.6 Reducing fractions.

10.3.6 Squaring

Squaring is the process by which a number is multiplied by itself (Table 10.1), thus:

$$3^2 = 3 \times 3 = 9 \quad \text{and} \quad 9 \text{ is the square of } 3$$
$$0.3 \times 0.3 = 0.09 \quad \text{and} \quad 0.09 \text{ is the square of } 0.3$$

Figure 10.7 Squares.

In any number having two equal factors, each factor is the *square root* of that number. In the first example above, 3 is the square root of 9; the radical symbol $\sqrt{\ }$ is used to signify the square-root process. $\sqrt{9}$ is equal to 3. The square-root process is the opposite of the squaring process. There are three ways to find the square root of any number: it can be calculated from first principles, which will not be explained here; it can be determined using any calculating machine that has a square-root function; and it can be read from a table of square-root values, of which a short, edited version is given as Table 10.2. Readers are reminded that they can expand Tables 10.1 and 10.2 to suit their own purposes.

TABLE 10.1
Numbers from 1 to 10 Squared

Number	Square	Root	Number	Square	Root
1	1	1.000	6	36	2.449
2	4	1.414	7	49	2.646
3	9	1.732	8	64	2.828
4	16	2.000	9	81	3.000
5	25	2.236	10	100	3.162

TABLE 10.2
Selected Roots from 1 to 100

Root	Number	Square	Root	Number	Square
1	1	1	6	36	1296
2	4	16	7	49	2401
3	9	81	8	64	4096
4	16	256	9	81	6561
5	25	625	10	100	10000

10.3.7 Percentages

Percentages are rates or proportions per 100 units. Therefore, to find a percentage for any given number, put the required percentage over 100 and multiply it by the number. For example:

What is 16% of 62? $^{16}/_{100} \times 62 = 9.92$
What is 62% of 16? $^{62}/_{100} \times 16 = 9.92$

Many percentages can be quickly found by converting them to simple fractions. For example:

50% = ½   33.3% = ⅓   10% = ¹/₁₀
25% = ¼   12.5% = ⅛   05% = ¹/₂₀

To quickly find 5% of any number, take ½ of ¹/₁₀, so that 5% of 76 would be (½ of 7.6) 3.8 and 12.5% of 96 is (⅛) 12. It is useful for the estimator to notice such relationships. To express any number as a percentage of any other number, such as 8 as a percentage of 32, multiply the first number by 100 and divide the result by the second number, thus: 8 × 100, divided by 32 = 25%, although in this case, one can see that 8 is ¼ of 32 and ¼ of any number is 25% of that number. If the first number is greater than the second number, the answer will be greater than 100%; for example, 15 as a percentage of 6 is 250%.

10.3.8 Averaging

Averaging is the process of finding the arithmetic mean of a series of numbers; in this context, the word **mean** indicates a value intermediate between the values of other quantities. Some examples will illustrate how the process works. In Figure 10.8, the average height is half the sum of the two end dimensions: ½ (12 + 15) = 13.5 m. In Figure 10.9, the average height is one-fifth the sum of all vertical dimensions: ⅕ (5 + 10 + 15 + 20 + 25) = 15 m.

The average can be found in this manner if each bay of the figure has the

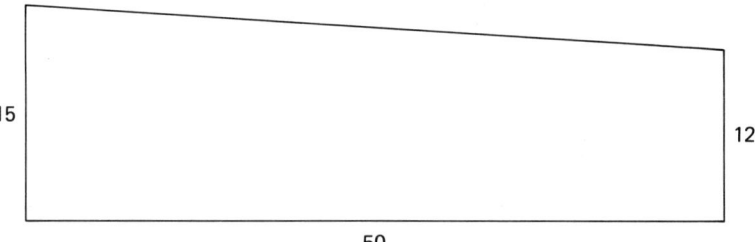

Figure 10.8 Averages.

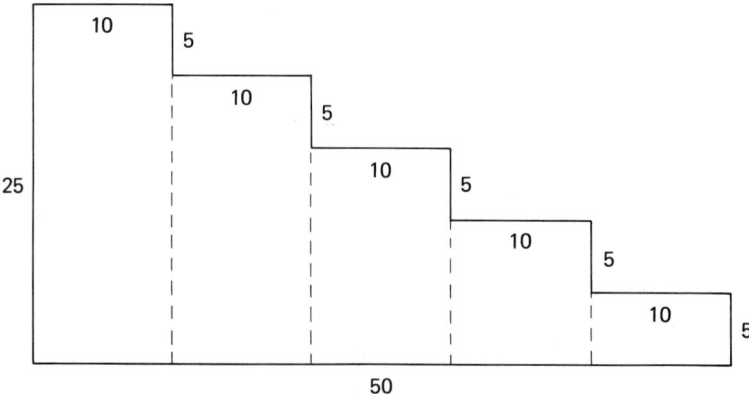

Figure 10.9 Averages.

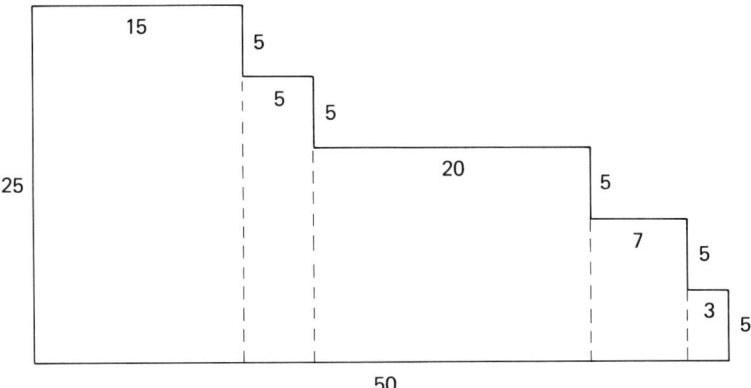

Figure 10.10 Averages.

identical width. If the bays have *different* widths (Figure 10.10), the average height of the area can be calculated as shown below.

$$\begin{aligned}
3 \times 5 &= 15 \\
7 \times 10 &= 70 \\
20 \times 15 &= 300 \\
5 \times 20 &= 100 \\
15 \times 25 &= \underline{375}
\end{aligned}$$

Total area = 860 m² and total wall length = 50 m
Average wall height = $860/50$ = 17.2 m

Figure 10.11 Averages.

When averaging areas, there have to be *one* common dimension and one variable dimension in each area; when averaging volumes, there have to be *two* common dimensions and one variable dimension in each volume. For example, if one wished to measure a series of columns, each having the same cross-sectional area but all having different heights, the average of the heights, multiplied by the typical cross-sectional area, would produce the average volume of all the columns, and that average volume, multiplied by the number of columns, would produce the total volume. If one or more of the columns has a different cross-sectional area, this process will produce an incorrect result.

It is appropriate to note the meanings of two other words in this general connection: median and mode. The **median** is the middle number in any series of numbers; the **mode** is the most frequently occurring value in any series of numbers. In the series 4, 12, 6, 4, and 24, the median is 6, the mode is 4, and the mean is 10.

10.3.9 Perimeters

The majority of buildings are more or less rectangular in plan and usually enclosed in circumference. In simplest terms, if a building were square or simply rectangular in plan, a person could start at one corner and walk right around it, ending up back at the starting point, having walked past four sides and four corners of the building. The person would also have passed through 360 degrees, or four right angles.

It is of course true that many buildings (or parts of buildings) are not simply square or rectangular; they often have six, eight, or more sides, all of which are usually at right angles and connected one to the other, thus enclosing the whole. Regardless of how many such sides or planes a building may have, a complete circular tour of the exterior will always involve passage through 360 degrees, or four right angles.

Furthermore, many buildings (or parts of buildings) are triangular or even circular in plan; however, these occasional unusual features need not divert attention from the basic facts about most buildings, and in any event, it will be explained how allowance can be made for such features if they are encountered in practice.

Regardless of how many sides there are to any building, the basic construction features composing the external walls are usually fairly consistent in their general design throughout the length of the perimeter. In addition, there are usually a significant number of items of work, such as foundations walls, perimeter drainage, wall plates, external wall studs, sill courses, sheathing, and all kinds of features around the eaves, whose length is in direct proportion to the exterior perimeter of the building. It therefore follows that the distance around the perimeter of the majority of buildings is a figure the estimator should know.

The technique to calculate a perimeter distance at any point in the exterior plane of the building is relatively simple to learn and easy to apply. The advantage of calculating and then using such perimeters is twofold: first, there is a saving of time in being able to use a useful reflected dimension of this type, and second, if the figure is correct in the first place, it will be correct for every future use throughout the estimate, thus enhancing the accuracy and reliability of the estimate as a whole.

To calculate the outside perimeter of the foundation wall for a simple square or rectangular building, add the overall length of the building to the overall width and multiply by 2. If the building has re-entrant corners, creating six, eight, or more sides, they may be ignored, as they have no effect on the overall length. If the building has recesses, the depth of each recess must be added to the overall length and width before doubling the result. The result will be the outside perimeter.

To adjust any perimeter length from one vertical plane of the building to another vertical plane closer to the center or interior of the building, *deduct* four times twice times the horizontal distance in the section between the two planes. Conversely, to adjust any perimeter length from one plane to another farther out from the interior, *add* four times twice times the distance between the two planes or perimeters. The reason for the "four times" is because of the four right angles (or four corners) included in the 360-degree "tour"; the reason for the "two times" is because at each right angle (or corner) an adjustment has to be made to the overall distance twice, once in each direction. The formula may be written as follows:

$$\text{perimeter adjustment} = 4 \times 2 \times HD$$

where HD is the horizontal distance in the section.

Refer to Figure 10.12. Plan A shows the overall dimensions, measured on the outside face of the concrete foundation wall. Section B shows a typical vertical cut through the foundation wall, with a drain tile line alongside the footing and 36 in. (or 1 m) away from it. Detail C shows an enlarged plan of one external corner, indicating the centerline, and the double adjustment which has to be made to the horizontal distance (HD) in each of two directions at each corner.

The overall length is 55 units (feet or meters). The overall width is 35 units. There are two recesses each 5 units in depth which must be taken into account. There are three re-entrant corners which can be ignored in the *perimeter* calculations (they cannot be ignored in the calculations to determine the area of Plan A).

To calculate the outside perimeter, add the overall length, width, and recesses, all on one side only, and then multiply by 2 to complete the other side, thus:

Length	55.00
Width	35.00
Recesses (2 × 5)	10.00
Half total	100.00
Full total (× 2)	200.00

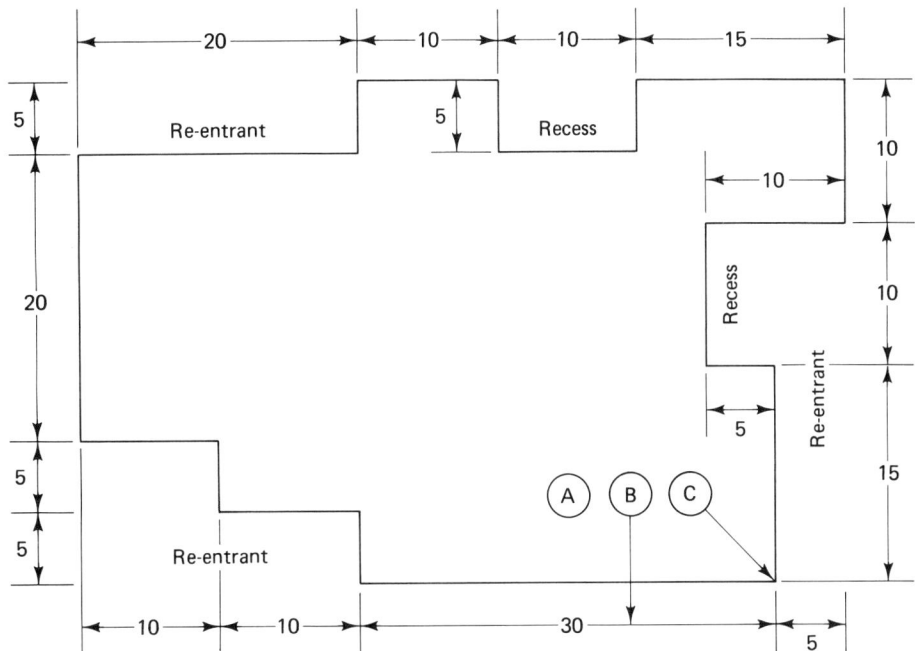

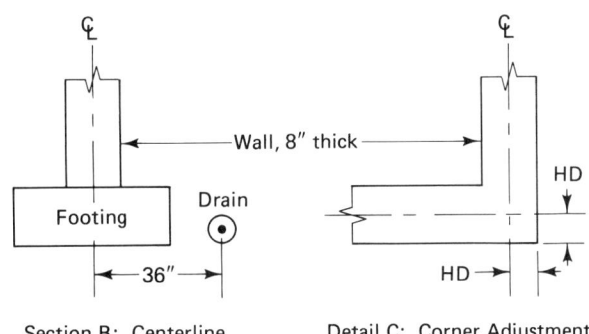

Legend: ℄ = centerline
HD = horizontal distance
OP = outside perimeter

Figure 10.12 Perimeters.

To calculate the inside perimeter of the foundation wall of this rectangular building, deduct four times twice times the thickness of the wall from the exterior perimeter dimension. If the outside perimeter was established to be 200 lineal feet and the wall is 8 in. thick, the inside perimeter would be 200 ft *less* (4 × 2 × 8 in. = 64 in.) 5'-4", or 194'-8" long. If the wall was 100 m long on the inside and 200 mm thick, the outside perimeter would be 100 m *plus* (4 × 2 × 200 mm) 1.60 m, or 101.60 m long.

To calculate the mean perimeter (along the centerline of the wall) for each of the foregoing conditions, the length in the imperial example would be 200 ft *less* (4 × 2 × 4 in.) 2'-8", or 197'-4", and in the metric example it would be 101.60 m *less* (4 × 2 × 100) 0.80 m, or 100.80 m. If there were a drain tile located 36 in. outside the centerline of the imperial wall, the length of that drain line would be 197'-4" *plus* (4 × 2 × 36 in. = 288") 24'-0", or 221'-4". In the metric example, the drain would be 108.80 m in length [100.80 + (4 × 2 × 1.00)].

If a building has perimeter exterior foundations of irregular thickness, calculate the mean perimeter at a constant distance inside the exterior face of the wall. With regard to buildings having *minor* plan or perimeter irregularities, such as triangular or circular features as are sometimes found in bay windows, at corners, and the like, the length of the irregularity can be separately calculated (or scaled, on a small job) and an appropriate adjustment made to the perimeter length. If the irregularities are *major*, to the extent that the building is essentially circular or triangular or other unusual shape, this perimeter adjustment rule is invalid and should *not* be used.

It should be possible to see the value of understanding and applying these perimeter adjustment calculations in construction estimating. Once the basic perimeter has been established, it is a simple matter to adjust that figure to produce any other relative perimeter, and that adjusted figure will automatically be correct, provided that an arithmetical mistake has not been made. As a result, many items can be inserted into the average estimate, using the perimeter adjustment technique, thus avoiding unnecessary measurement, calculation, and extension.

10.4 OTHER FUNCTIONS

10.4.1 Preamble

As has been said, one does not have to be a mathematician to be a competent estimator. A good knowledge of the basic arithmetic functions and their proper application will suffice for the majority of situations in calculation which the estimator will encounter. However, a limited knowledge of some other basic functions of algebra, geometry, and trigonometry will often prove to be helpful, and indeed, in some types of calculations, necessary. In the following paragraphs, rudimentary explanations of these basic functions are presented, primarily to show the functions of greatest use to the estimator, but also to refresh the memories of those who have studied such subject matter at some time in the past. For more detailed and advanced study, reference should be made to any good-quality math book intended for use at high school or junior college level.

10.4.2 Algebraic Functions

Algebraic functions are symbolic representations of verbal statements of mathematical problems. In other words, they are a shorthand method of expressing arithmetical concepts in the most concise and precise terms. In estimating, one may use the basic algebraic laws of **commutation** or substitution, **association** or combination, and **distribution** or dispersal. For example, if a, b, and c represent real numbers, then

$$a + b = b + a \quad \text{(commutation)}$$
$$ab \times c = a \times bc \quad \text{(combination)}$$
$$a(b + c) = ab + ac \quad \text{(distribution)}$$

The most common algebraic problem facing the estimator is the solution of linear equations to determine values for one or two unknowns. The formulas are given below.

A linear equation is written in the form $ax + b = 0$, where x is the unknown value and where a does not equal zero. Any other equation that can be reduced to this form is also a linear equation. For example, $4x + 6 = 2x + 12$ can be rewritten as $2x - 6 = 0$. In this case, $2x = 6$, and x therefore equals 3. Linear equations can also be written in the form $ax = b$, or $x = b/a$. Where there are two unknowns, they can often be expressed as two linear equations and then jointly solved to discover the unknown values. For example, if $a + b = 12$ and $a - b = 6$, then by adding these two together, $2a = 18$, and a equals 9, with b being equal to 3.

The estimator will seldom encounter any need to apply quadratic or polynominal equations or to deal with exponential or logarithmic functions. However, some study of these and other algebraic functions will enhance the intelligent use by the estimator of electronic calculating machines, which can be programmed to carry out such functions.

10.4.3 Geometric Functions

Geometric functions will be dealt with in some detail in Chapter 11. There is one other major function with which the estimator must be familiar: the **theorem of Pythagoras**, a Greek mathematician. This theorem states that in any triangle containing a right angle of 90 degrees (known as a right triangle), the area of the square on the side opposite the right angle will equal the areas of the squares on the other two sides adjacent to the right angle. In Figure 10.13, $(AB)^2 = (AC)^2 + (CB)^2$. Because of this relationship, it is possible to calculate the length of any one side if the lengths of the other two sides are known. If the length of side AB in the triangle is 5 and BC is 3 m, the length of AC can be found by rearranging the formula, thus:

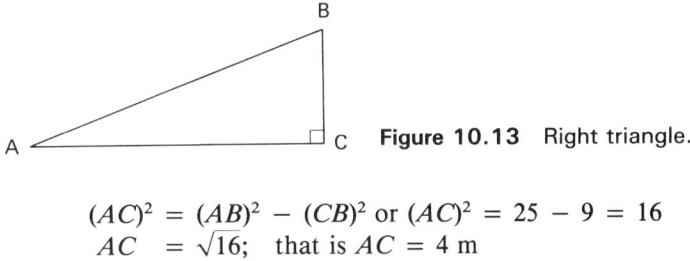

Figure 10.13 Right triangle.

$$(AC)^2 = (AB)^2 - (CB)^2 \text{ or } (AC)^2 = 25 - 9 = 16$$
$$AC = \sqrt{16}; \text{ that is } AC = 4 \text{ m}$$

It will also be helpful for the novice to study the construction of common geometric figures and to become familiar with aspects of congruency, similarity, and symmetry.

10.4.4 Trigonometric Functions

Trigonometric functions are occasionally employed by estimators to determine lengths, areas, volumes, and angles. The functions most commonly used are sines, cosines, and tangents in the right-angled triangles. These are explained below, using the triange shown in Figure 10.14, in which side AB is called the **hypoteneuse**, side BC is called the **opposite side**, and side CA is called the **adjacent side**, relative to the angle θ.

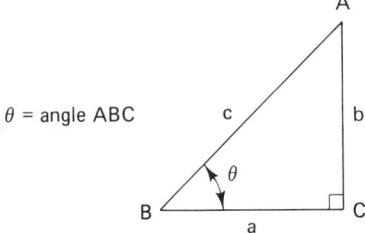

Figure 10.14 Functions.

The value of the **sine** of an angle equals the opposite side over the hypoteneuse. The **cosine** equals the adjacent side over the hypoteneuse. The **tangent** equals the opposite side over the adjacent side. These three are symbolically represented as follows:

$$\text{sine } \theta = b/c \qquad \text{cosine } \theta = a/c \qquad \text{tangent } \theta = b/a$$

The reciprocals of these are as follows:

$$\text{cosecant } \theta = c/b \qquad \text{secant } \theta = c/a \qquad \text{cotangent } \theta = a/b$$

To round out these relationships, it can also be shown that the result of a function value multiplied by its reciprocal function value is equal to 1, thus:

$$\text{sine } \theta \times \text{cosecant } \theta = 1$$
$$\text{cosine } \theta \times \text{secant } \theta = 1$$
$$\text{tangent } \theta \times \text{cotangent } \theta = 1$$

Applying the foregoing functions to estimating, it can be seen that if any one side and one angle of a right triangle are known, the remaining two sides and angle can be calculated. Furthermore, if the values for functions or reciprocals of angles are known, the lengths of unknown sides can be determined. If a triangle is not a right triangle, it can be reduced to two or more right triangles by drawing intermediate altitudes, as shown in Figure 10.15. Thus any triangle is capable of solution, provided that any two properties of its subsidiary right triangles are known.

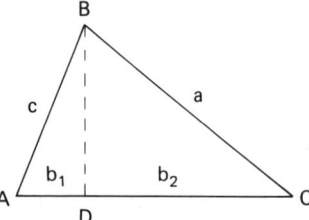

Figure 10.15 Functions.

The abbreviations for sine, cosine, and tangent are sin, cos, and tan, respectively. Figure 10.16 shows the relationships among the functions. Knowing either two angles and one side or two sides and one angle in any of the relationships, the remaining sides or angles can be determined. Values for trigonometric functions can be established by direct calculation, by use of an electronic calculator programmed with such functions, or by reference to tables of precalculated values given in most standard math books. For solutions to more complicated problems, reference should be made to a trigonometry textbook or a math tutor.

$BD = a \sin c$ and $c \sin A$; therefore, $a \sin C = c \sin A$ and $\sin C/c = \sin A/a$, and $\sin A/a = \sin B/b$ and $\sin B/b = \sin C/c$. So

$$\frac{\sin A}{\sin B} = \frac{a}{b} \qquad \text{and} \qquad \frac{\sin B}{\sin C} = \frac{b}{c} \qquad \text{and} \qquad \frac{\sin C}{\sin A} = \frac{c}{a}$$

Figure 10.16 Relationships among the trigonometric functions.

10.5 *MENTAL CALCULATIONS*

It is useful for the estimator to be able to do simple and rapid calculations mentally. Such a facility will help to reduce dependence on the machine and to enhance the confidence that the estimator needs to have in the results of calculations made and used in estimates. A few techniques for small numbers are given below; many others are available. Practice will develop this skill.

To multiply a number that ends in a zero by a number that does not end in zero, such as 40 × 26, set the zero down on paper, mentally multiply 26 × 4, and set the result alongside the zero, thus: 1040. To multiply two numbers both of which end in zero, such as 40 × 260, set down two zeros and then proceed as before, thus: 10,400.

To multiply a number with another number of equal value, such as 27 × 27, raise one number to end in a zero, thus: 30 and lower the other number by the same amount, thus: 24. Multiply 30 × 24 as before, which equals 720, and add the square of the figure used to change the value, in this case 3; 3 × 3 = 9 + 720 = 729.

To multiply two numbers of unequal value, such as 24 × 28, make one number the same as the other and proceed as above. So 24 × 24 = 576; to this result, add the difference multiplied by the lower number, which in this case is 4 × 24 = 96 + 576 = 672.

To add columns of whole numbers, look for combinations that make up to round numbers, such as 20, 50, 80, and so on. When adding fractional numbers, convert them all to decimals, or reduce them all to their common denominator, and then add the numerators, thus:

Add these: $\frac{1}{2} + \frac{3}{4} + \frac{5}{8} + \frac{7}{12}$
Reduce to: $\frac{12}{24} + \frac{18}{24} + \frac{15}{24} + \frac{14}{24}$
Equals: $\frac{59}{24}$ or 2 and $\frac{11}{24}$ or 2.458

When calculating percentages, look for simple fractions, such as $\frac{1}{2}$, $\frac{1}{3}$, $\frac{1}{4}$, and so on, as described earlier in this chapter.

It is not only instructive to study and practice mental calculation techniques, but it can also be enjoyable. There are many books that contain advice on this topic, dealing not only with the mathematical issues, but also with aspects of mental discipline and training.

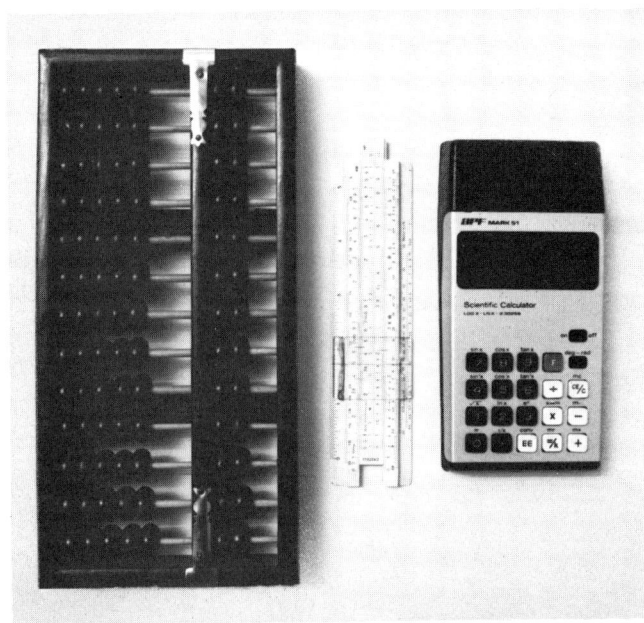

Calculating machines.

10.6 CALCULATORS AND THEIR USE

Manual, mechanical, electrical, and electronic calculators are invaluable aids to the estimator in the performance of his primary duties, provided that their place and function are clearly understood. To illustrate, some studies have shown that most people tend to accept machine-produced results as being correct, even in test situations where the participants knew that there might be some problems with the machines or their operation. The estimator must develop habits that will reduce his dependence on machine results and improve reliance on his own judgment.

In general, calculations should be arranged with respect to their utility in the estimate, not on their ease of operation on the machine. Many valuable intermediate results are not made available to the estimator who uses a machine that will only produce end results. At the same time, it should be said that many machines can produce results that might not otherwise be obtainable, because of time, complexity, or other constraint on the estimator.

Calculating machines used for estimating purposes are best equipped with a printout feature, so that entries and results can be checked by eye. All significant numbers should be on paper, although minor intermediate calculations can often conveniently be done mentally or on a small pocket calculator. Good machines, properly maintained and utilized, can unquestionably speed up the calculation processes which are such a large part of the estimator's life. But it should be remembered that the final responsibility for the accuracy and worth of the estimate lies with the estimator, not with the machine.

10.7 SIGNS AND SYMBOLS

Estimators have to know and use a number of conventional signs and symbols to indicate units, dimensions, characteristics, values, and so on. Some of the more common ones are listed in Table 10.3. It will be observed that some of these have more than one meaning; therefore, care must be taken with their use. Other symbols, used in connection with the metric and imperial systems of measurement, are shown in Chapter 9 and in Appendix A.

TABLE 10.3
Symbols

Operation or concept	Symbol	
Addition	Small vertical cross	+
	Period, called a dot	.
Subtraction	Small horizontal dash	−
Multiplication	Small diagonal cross	×
	Oblique solidus or stroke	/
Division	Long horizontal line	—
	Oblique solidus or stroke	/
	Reverse bracket	)
A number	Small double-cross	#
A ratio	Colon	:
Equal to	Small pair of dashes	=
As above	"Ditto" sign	"
Each item	"At" symbol	@
Take note	Asterisk	*
Approximate	Two wavy lines	≈
Greater than	Right arrowhead	>
Smaller than	Left arrowhead	<

Chap. 10 / Calculation

QUESTIONS

10.1. Distinguish between cardinal numbers and ordinal numbers. Give one example of each type.

10.2. Explain the term "duodecimal." Give examples of two duodecimal numbers, representing the lengths of the sides of a small square in imperial measure. Multiply these two numbers together to produce the area contained by the sides.

10.3. Mentally calculate 37.5% of 80. Mentally square the result. (Do not use a machine or pencil and paper in this exercise; it is not necessary!)

10.4. Distinguish between the median and the mode in any series of numbers. Illustrate your verbal answer with an appropriate series.

10.5. Define the value of the trigonometric sine of an angle, in words. Identify the reciprocal of a sine. If you multiply a sine and its reciprocal together, what is the numerical value?

10.6. Identify the problem in using an oblique solidus (or stroke) to represent multiplication in estimating. Suggest a solution.

CHAPTER 11
MENSURATION

11.1 INTRODUCTION

11.1.1 Preamble

The estimator is concerned with the lengths and thicknesses of things, with perimeters and areas, with volumes and capacities, with the whole and with the part, with the regular and the irregular. The study of this aspect of life is called **mensuration**, which embraces that branch of mathematics that deals with the determination of length, area, and volume. As with calculation in Chapter 10, a review of the historical development of mensuration beyond the bare outlines contained in this chapter should prove to be of interest, use, and enjoyment to the reader.

In relation to the topic of construction estimating, the topic of mensuration can be conveniently considered under two broad categories: first, lengths and areas, and second, volumes and capacities. In each category, the subject matter can be classified as follows.

11.1.2 Lengths and Areas

Quadrilateral shapes: parallelogram, quadrilateral, rectangle, square, trapezium; all having four straight sides.
Triangular shapes: acute, equilateral, isosceles, right, scalene; all having three straight sides.
Circular shapes: arc, chord, circle, circumference, diameter, ellipse, radius, sector, segment; having some straight and some circular sides or shapes.
Irregular shapes: hexagon, octagon, polygon, and other more irregular shapes; having either straight-line or curved-line sides or a mixture of both.

11.1.3 Volumes and Capacities

Cubical shapes: having surfaces essentially normal or at right angles to each other, such as in foundations, walls, and slabs.

Spherical shapes: having surfaces essentially rounded or elliptical, such as in domes and barrel-shaped solids.

Conical shapes: having surfaces like cones, such as in column bases and capitals.

Pyramidal shapes: having surfaces like triangles, arranged on a square or rectangular base, such as in ornamental work.

Cylindrical shapes: having rounded surfaces parallel to a central axis, such as in columns and tubes.

Prismoidal shapes: having surfaces essentially flat, with triangular cross-sectional configuration, such as in earth embankments.

Irregular shapes: having no distinctive typical shape, but usually capable of reduction to one or more of the foregoing shapes.

The foregoing categories and classifications are by no means exhaustive of all possible configurations, but they do represent most of the common shapes that the estimator may encounter in the majority of construction projects. The special words used to describe specific shapes are defined and arranged alphabetically in the following list. Formulae for calculating length, area, or volume of these common figures are given in Figures 11.1, 11.2, and 11.3.

11.2 DEFINITION OF TERMS

Acute: a triangle in which all internal angles are less than 90 degrees.

Arc: any part of a uniformly curved line.

Chord: that part of a straight line between two points of its intersection with a curved line.

Circle: a closed plane curve having all points equally distant from a center point.

Circular: of or pertaining to a circle; having the form of a circle.

Circumference: the length of the outer boundary of a circle.

Cone: a solid generated by straight lines joining a fixed point to points on a plane circle, whose plane does not contain the point.

Conical: resembling a cone.

Cube: a solid bounded by six squares of equal area, the angle between any two adjacent faces being 90 degrees.

Diameter: a straight line passing through the center of a circle and terminated at each end by the circumference.

Name	Length	Formula
Perimeter	Sum of sides	Σn
Circumference	Pi × diameter	$\pi \times D$
Arc	Center angle over 360 × circumference	$\theta/360 \times \pi D$

Figure 11.1 Common lengths and perimeters.

Chap. 11 / Mensuration

Name	Shape	Area	Formula
Parallelogram		Base × vertical height	$b \times h$
Rectangle, square		Base × height	$b \times h$
Quadrilateral		Sum of two triangles	$2(\frac{1}{2} \times b \times h)$
Trapezium		$\frac{1}{2}$ sum of parallel sides × vertical height	$\frac{1}{2}(a + b) \times h$
Triangle		$\frac{1}{2}$ × base × altitude	$\frac{1}{2}(b \times h)$
Circle		Pi × radius squared	$\pi \times r^2$
Ellipse		Pi × $\frac{1}{2}$ major axis × $\frac{1}{2}$ minor axis	$\frac{\pi}{4} \times (b \times l)$
Sector		Center angle over 360 × circle area	$\frac{\theta}{360} \times r^2$
Segment		Area of sector less area of triangle	$\frac{\theta}{360} \times r^2$ minus $\frac{1}{2}(b \times h)$
Hexagon (6)	l	Side squared × 2.60	$l^2 \times 2.60$
Octagon (8)	l	Side squared × 4.83	$l^2 \times 4.83$
Polygon (n)	n	Divide into n equal triangles and total	$n \times \frac{1}{2}(b \times h)$
Irregular polygon	h, A, B	Draw base AB; erect (h) ordinates at ends and at equal spaces; draw midordinates. Then area equals sum of lengths of midordinates within the figure times the distance between any two ordinates.	

Figure 11.2 Common plane shapes.

Ellipse: a plane curve such that the sums of the distances of each point in its periphery from two fixed points are equal.

Equilateral: having all sides equal in length.

Hexagon: a polygon having six sides and six angles, all not necessarily equal.

Isosceles: a triangle with any two sides equal.

Normal: a plane or line at right angles to another.

Octagon: a polygon having eight sides and eight angles, all not necessarily equal.

Name	Shape	Formula
Cube		$V = l \times b \times h$
		S = sum of side areas
Sphere		$V = \frac{4}{3}\pi r^3$
		$S = 4\pi r^2$
Cone[b]		$V = \frac{1}{3}Bh$ (or $\frac{1}{3}r^2 \times h$)
		$S = \pi r(1 + r)$ (l = slope length)
Pyramid[b]		$V = \frac{1}{3}Bh$
		$S = 4 \times \frac{1}{2}bl + B$ (l = slope)
Cylinder		$V = \pi r^2 h$
		$S = 2\pi r^2 (h + r)$ (h = vertical height)
Prism[a]		$V = \frac{1}{6} \times L \times (A_1 + A_2 + 4A_3)$
		S = sum of side and end areas
Polygonal	—	V/S, combinations of above

[a] A_1, A_2, end areas; A_3, area at midpoint of length.   [b] B, base

Figure 11.3 Common volumes.

Parallelogram: a quadrilateral, the opposite sides of which are parallel.
Polygon: a plane figure having more than four sides and angles.
Prism: a solid whose ends or bases are triangles, quadrilaterals, or polygons, and whose sides are parallelograms.
Prismoidal: shaped essentially like a prism.
Pyramid: a solid having a triangular, square, or polygonal base and triangular sides that meet at a point.
Quadrilateral: a plane figure having four sides and four angles.
Radius: a straight line extending from the center of a circle to the circumference of that circle.
Radii: the plural form of *radius*.
Rectangle: a parallelogram with all its angles 90 degrees.
Right: having the axis perpendicular to the base.
Right angle: an angle of 90 degrees.
Right triangle: a triangle with one right angle.
Scalene: a triangle having three unequal sides.
Sector: a plane figure bounded by two radii and the arc of an ellipse or circle.
Segment: a part of a circle cut off by a straight line.
Sphere: a solid generated by rotation of a semicircle about its diameter.
Spherical: shaped like a sphere.
Square: a four-sided plane figure having all sides equal and all angles right.
Trapezium: a quadrilateral plane figure in which no two sides are parallel.

Chap. 11 / Mensuration

Triangle: a plane figure bounded by three straight lines meeting at points to form three angles.

11.3 FORMULAS

A clear understanding of each of the foregoing definitions will assist the estimator to produce rapid and reliable measurements, using these common shapes, when appropriate. Figures 11.1, 11.2, and 11.3 give formulae for calculating lengths, areas, surfaces, and volumes of these common forms using the terms described in Table 11.1.

TABLE 11.1
Legend for Figures 11.1, 11.2, and 11.3

b = breadth	l = length	c = circumference
h = height	r = radius	π = pi (3.14)
S = surface area	V = volume	θ = angle
B = base area	D = diameter	n = number

QUESTIONS

11.1. Verbally distinguish between a trapezium and an octagon. Illustrate each figure by a small but accurate sketch.

11.2. In what essential respects does a prismoidal shape differ from a pyramidal shape? Which of these two shapes most closely resembles a conical shape?

11.3. Define any two of the following three geometrical terms: ellipse, isosceles, scalene.

11.4. State the formulas for the volume and surface area of a sphere. Calculate the volume and surface area of a sphere having a diameter of 2.00 m.

11.5. Verbally explain how you would calculate the approximate area of an irregular polygon. Draw a small sketch of such a figure to illustrate your explanation.

PART II
Research Assignments

The research assignments that follow are intended to give readers a channel through which to acquire knowledge of current local issues and aspects relative to the contents of the chapters contained in this part of the book. The basic objective in all of these assignments is to allow readers to compare theory with practice, by studying this and other similar books, and by encountering opinions and activities, other than those of themselves and this author, on the stated topics. A guide to content, procedure, and evaluation for these assignments has been included in Section 1.4. A review of these guidelines is recommended before starting work on the assignment projects.

CHAPTER 6: INTRODUCTION TO MEASUREMENT

Investigate the topic of *style* in estimating. Interview local estimators and record their opinions on the attributes of estimating style mentioned in this chapter. In particular, review the working conditions of estimators, and their practices in connection with the use of (1) colored pencils to check-mark drawings and specifications, and (2) stationery (or computers) to do take-offs, extensions, and pricings. Include examples.

CHAPTER 7: METHODS OF MEASUREMENT

Investigate at least two local architectural or engineering design offices to discover the methods used by them to determine probable costs of projects. Compare or contrast these methods to the ones described in this book. Inquire into their sources of cost data for such purposes and (if possible) their record of success in terms of accuracy of prediction of bid prices. Report on the education, training, and experience of the personnel involved in this work in the offices visited.

CHAPTER 8: TECHNIQUES OF MEASUREMENT

Make an examination of estimating techniques used in various parts of the construction industry, such as in the architectural, structural, mechanical, and electrical fields, in the region where you live. Compare the North American experience and practice with that of any other continent, such as Australia, Europe, or India.

CHAPTER 9: SYSTEMS OF MEASUREMENT

Investigate support for or resistance to the adoption of the SI metric system by the construction industry in the United States or Canada, depending on where you reside. Support your views with excerpts from recent reports on this issue prepared by public and private agencies, such as national and local governments and responsible construction trade and professional associations. Report on the status of metric products and projects in your region, as well as local industry opinion on this topic. Compare your findings with the experience of the construction industry in one other country that has made a recent conversion to metric measurement (such as Australia, China, Fiji, Great Britain, Japan, or New Zealand).

CHAPTER 10: CALCULATION

Investigate manual, mechanical, electrical, and electronic calculating devices used by construction estimators in the area where you live. Discuss advantages, disadvantages, reliability, and costs of the various devices on which you report. Also examine the techniques employed by estimators for mental calculations, and comment on the extent, accuracy, and reliability of such activity.

CHAPTER 11: MENSURATION

Prepare a short review (say, 5000 words) of the historical development of mensuration, beyond the brief outline contained in this book. Identify the primary stages of significant development of mensuration concepts of particular interest to the construction estimator. Report on trends of practice in this subject area, with specific reference to the use of computers and related equipment, programmed to facilitate measurement and calculation of simple and complex geometric shapes and figures.

PART 3   PRICING

CHAPTER 12
INTRODUCTION TO PRICING

12.1 GENERAL OBJECTIVES

12.1.1 Preamble

As was stated in Chapter 1, estimating consists of two parts: *analysis*, or measurement, and *synthesis*, or pricing. This part of the book deals with the synthetical process of *pricing*. It can be argued that the study of pricing should precede the study of measurement, because the objective of measurement is to permit pricing to be done, and that one should therefore know how to price work in order to know what work to measure and how to measure it. It can also be argued that the measurements have to be done before the prices can be calculated or applied to the quantities, in spite of the fact that the majority of prices used in estimating processes can be calculated independently of probable quantities, measured for any specific project.

As a matter of practical fact, the measurement and pricing processes are interdependent to such an extent that it is immaterial which is tackled first; ideally, they should be considered simultaneously. However, it is expedient to separate them for the purposes of explanation, study, and practice, in both the academic and the industrial milieu. As most novice estimators start to learn measurement first and then progress to pricing, that chronological convention has been followed in this book. Before examining the pricing process in detail, it may be appropriate to give some consideration to the meaning of the word "cost" and to some of the basic factors of cost with which the estimator must become familiar, as well as to some issues that affect the cost of buildings in general.

12.1.2 Terminology

The historical origin of the word **cost** appears to be connected to medieval English and old French words which meant "to stand together," which is rather interesting from the estimator's point of view. The dictionary definition[1] of "cost" is the price paid to acquire, produce, accomplish, or maintain some thing; also, an outlay or expenditure of time, money, trouble, or labor; and also to estimate or to determine the cost of anything. The definition of the word **price** is the sum of money or the amount for which anything is offered for sale, bought, or sold; also that which must be done, given, or undertaken in order to obtain a thing. The definition of the word **value** is that property of a thing because of which it is esteemed, desired, or useful; its worth, merit, or importance; its force or significance. It will be obvious to the reader that the *cost* to produce an item of work is not necessarily the *price* for which that item will be sold, and that neither the cost nor the price necessarily reflect the *value* of the item to the buyer or the seller.

There are of course, several other words that are used in similar contexts, words such as "charge" and "expense." Care should be taken to fully comprehend the precise meanings of such words, and in particular, the many meanings of the word "cost" when used in conjunction with other words, such as direct cost, indirect cost, reproduction cost, replacement cost, development cost, financing cost, interim cost, and so on.

12.2 FACTORS OF COST

12.2.1 Preamble

The composite cost of every item of work on every construction project consists of a number of factors, each of which contributes a greater or lesser proportion of the composite whole. Recognition and understanding of the costs associated with each of these factors will assist the estimator to make better judgments: first, about what has to be measured; second about how to measure it; and third, about how to price the entire item. There is no set number of clearly defined or universally agreed-upon factors; various authors present their views of such factors in a number of different, though similar configurations.

12.2.2 Direct Factors

In this study, four primary factors will be considered, each of which has two secondary considerations.

1. Production
   a. Materials have to be produced from some raw origin, such as clay, oil, sand, or timber.
   b. Materials have to be cut, shaped, fastened, and assembled into building products.
2. Delivery
   a. Products have to be packaged for protection.
   b. Packages have to be transported from factory to site.
3. Acceptance
   a. At the site, products have to be received and checked for compliance, errors, or damage.
   b. Products often have to be temporarily and safely stored at the site before installation.

---

[1]*The American College Dictionary* (New York: Random House, Inc., 1960).

4. Installation
    a. Products have to be properly installed in their correct positions in the project.
    b. Products have to be properly maintained during the construction period and process, until the contract is complete.

In addition to and in conjunction with the foregoing primary and secondary aspects, there are a number of tertiary factors which also have to be taken into consideration in pricing.

12.2.3 Indirect Factors

1. Taxes, permits, fees, guarantee costs, insurance, and financing costs.
2. Promotional and instructional literature, marketing and publicity, and other business expenses.

The estimator can see why it is important to enquire just what is or is not included in any price quoted for any reason in connection with any building products required for any construction project.

12.3 SOURCES OF COST DATA

12.3.1 Preamble

Some sources of cost data are obvious, other are obscure; some are reliable, others are less so; some are immediately useful, others require conversion or other modification before use. However, all such sources, for the purpose of discussion and study, can be conveniently classified under two broad headings: actualities and probabilities. In this context, the word **actualities** means that the cost data sources are factual, current, and adequately established for all practical purposes; the word **probabilities** means that data are statistically derived or historically projected from some formerly factual data. In the following paragraphs, some sources listed under one heading may well be able to provide some data appropriate to the other heading. Furthermore, some data sources identified here in fairly general terms will be reintroduced in more specific terms where appropriate, in later parts of the book.

12.3.2 Actualities

Company Records. The first source for reliable factual cost data should be the cost records of the construction company for which the cost of the proposed new work is being estimated, provided that these records have been properly prepared and maintained. Data on labor rates and productivity, material costs and waste factors, equipment utilization and downtime, and overhead and other costs should be available from the accounting department or personnel of every well-organized construction company. The form of such data will be explained later.

Supply Houses. Subcontractors often prepare firm price quotations to do parts of the work. Suppliers often publish price lists and discount rates to provide materials, systems, or services. Rental agencies often advertise rates or fees to rent equipment for any project.

Organizations. Labor unions often publish tables of wage rates and other data relative to fringe benefits and productivity. Bonding and insurance companies quote fees and other costs which they will charge for protection of various types. Government offices and other agencies, such as public utilities, make available timely and accurate data on taxes, fees, permit costs, minimum wage rates, unemployment insurance costs, and other social security benefits having a cost factor.

Consultants. In various regions, there are companies known as construction economists, quantity surveyors, project consultants, and by other titles, who specialize in keeping in touch with construction costs and trends in building developments. Such firms are usually prepared, for a fee, to investigate and report on local, national, and even international construction costs. These consultants often have a close relationship with owners and developers, and also have access to specific types of cost data peculiar to particular construction projects either under way or in the process of development.

Difficulties with Data. In general, it can be said that it is often quite difficult to get accurate, reliable, factual cost data in the construction industry, partly because of the highly competitive nature of the industry and also because of the secrecy that surrounds much cost data as a result. It can also be said that some difficulties arise in part because of the fact that some cost data are generally poorly handled by many people who should literally know better, but have neither the skill nor the inclination to improve on the quality of production of such data. Other factors include the sheer amounts and types of data available and the complexities encountered at every stage of the design/build process. Furthermore, it is well known that costs themselves are in a state of flux in Western economies, and are constantly changing in response to economic pressures in the market place. For example, at the time of this writing (Winter 1985), the inflation rate in North America was about 5% per annum, whereas the rate in Argentina was around 300%, and in Israel an astonishing 400%! The worth of any data obtained under such circumstances is questionable, at best.

12.3.3 Probabilities

Data Books. The first source of probable or statistical cost data are the cost data books, published privately for regional or national markets. These books come in a number of formats, some of which are most useful to estimators and appraisers, whereas others are of more interest to designers and developers. Some of these books cater to special-interest groups, such as subcontractors and consultants who specialize in providing mechanical services, such as heating, ventilating, and air-conditioning. Many of these books are highly reliable and well organized for ready reference, whereas others are of lesser quality and utility. Such books are generally arranged in sections, with lists of materials, components, or systems appropriate to the section shown alongside corresponding average costs for materials, installation, equipment, and other features. In some of these books, costs are given on a national-average basis for work of a stated type, quality, and magnitude, with tables of factors to modify the average data to suit regional differences. Some also present useful tables, showing data on crew sizes, productivity factors, and related labor cost detail. A number of data books are identified in Appendix C.

Cost Reports. A second source of probable data is provided by a number of cost-reporting services to which one can subscribe on a monthly or annual basis. In some countries, such as Australia and Great Britain, such services are often provided by professional associations of construction economists or quantity surveyors to their members, and incidentally to the public at large. In other countries, such as Canada and the United States, most of these services are provided on the basis of private subscription. A few such services are identified in Appendix C. The data are usually tabulated on a national or regional basis, with modifying factors given by the publishers to permit adjustment of the data to suit local conditions and circumstances. The data and the factors are periodically updated by the publishers, many of whom also provide periodic cost analyses of model buildings as a service to subscribers. Some of these publications tabulate prices quoted to them by manufacturers of building products or systems, as of a certain date and showing any appropriate special restrictions, such as minimum quantities, delivery rates, or size-range limitations.

Government. A third source of data are government offices and agencies from which statistical data can be obtained. Many of these data are not in a form that can be immediately utilized by the contractor or designer, but they can often be used to indicate trends or confirm suspicions with respect to bankruptcies, defaulted loans, housing starts, inflation, interest rates, population trends, unemployment, and so on, as these might affect the construction industry in any state or province. Also, many government agencies publish reports about construction projects proposed, under way, or just completed, often with cost analyses of greater or lesser detail and accuracy attached, for the benefit of taxpayers at large. Many government and public agencies require that detailed estimates of cost be provided to them before they will approve or commission the design and construction of public buildings; some of these data can often be acquired by interested parties. Taken as a whole, government represents a fairly large proportion of the total construction activity in any region, and as a result, has access to a sizable amount of fairly current and accurate cost data, although not all such data are made available to the public, for one political reason or another. Private construction cost consultants are frequently involved in the preparation of data used by and issued from government offices, and thus themselves become a source of construction cost data of this type.

Technical Press. One other more general source of cost data is the national and local technical press, which consists of a large variety of newspapers, journals, magazines, and other forms of print media, many of which contain articles on construction costs as well as detailed advertising presentations that deal with economic aspects of building technology. The quantity of such material is very extensive, there being literally hundreds of publications produced each month in North America and elsewhere, all vying for a certain proportion of the total market of readers. The quality and reliability of such material is at best uneven, but taken as a whole, some useful conclusions can sometimes be drawn from a review of selected items or articles dealing with specific issues of interest to a particular owner, contractor, designer, or developer. The principal libraries in most medium-sized cities and college or university centers will in all probability subscribe to a number of such publications representing the national and local press relative to the construction sector of society. Also, many construction associations and similar organizations publish weekly or monthly newsletters for the benefit of their members, containing data on projects out to bid, contracts awarded, results of collective agreement processes, and other cost information of general interest.

Precaution. A general caveat to bear in mind is that cost data are, by definition, historical in nature, and the very fact that they appear in print means that they are already obsolete or at least in the process of becoming obsolescent. Although careful application of selected data can produce meaningful figures for inclusion in budget studies and general predictions regarding probable construction costs, much more care must be taken with the development of costs intended for use on any actual construction project.

12.3.4 Other Issues

In addition to the foregoing introduction to the specific factors of cost and the sources of cost data, some thought should be given to a number of the more general factors that influence construction costs in industrialized society. Some of the following factors are dealt with in more detail as they arise in context later in the book; they are presented at this point in outline form to give the reader a brief glance at a very broad field or background against which the more detailed studies are highlighted. For the purpose of this short review, two major influences on cost will be described: the organization of the construction industry and the nature of building construction.

The Construction Industry. The organization of the construction industry has an effect on construction costs. Buildings are designed and constructed by teams that are assembled before and often abandoned after each project, unlike many other industries. Buildings are mostly built outside and are therefore exposed to the weather, with all the uncertainty which that brings. The materials, labor force, and equipment have to be brought to each site, compared to other industries which are more centrally and permanently located in one spot. Although there is a trend toward prefabrication and standardization of components, most buildings of a commercial or institutional nature are still largely a "one-off" proposition, with resultant inefficiency because of the difficulty of developing mass production.

The construction industry is highly diversified or fragmented, with a tremendous proliferation of associations, companies, contractors, designers, inspectors, suppliers, testing agencies, unions, and other forces, each with its own special interests and power bases, pulling in many directions at once. There is little or no central or coordinating core or power emanating either from government or from the industry itself. Furthermore, the industry is used by government to regulate sectors of the national or local economy as a matter of political necessity or expediency, which presents many difficult and indeed grave problems for the industry and its members. It is relatively easy to shut the industry down, but it is not so easy to start it up again on a moment's notice.

It can be said that all of the tensions identified in the foregoing paragraphs are not necessarily bad or restrictive. Such loose organization provides for a great deal of flexibility and freedom, and the industry and its members are highly in-

General urban development.

novative as a result. The construction industry attracts people who know about risk and are prepared to take some risks, and it is ruthless about weeding out those who cannot handle the pressures associated with much of the operation of the industry. It is a large and vital force in the economy of any nation, with an almost limitless number of opportunities for diversified careers for enterprising beings having appropriate attitudes, knowledge, and skills.

Building Construction. The first factor that affects the cost of any building is its location. In general, buildings in the more developed regions, such as California or Ontario, cost more to build than do buildings in Montana or Manitoba, the main reason being the higher cost of living in the more developed regions. Within any region, buildings in isolated areas will cost more than similar buildings built in the more populated parts, because of additional transportation and accommodation costs. The quality of the site and the access to it will also affect construction costs.

Second, the type of building and its related use and occupancy will affect construction costs. Data have been developed to show that, when building types are classified, building costs both between and within each classification can be stratified. For example, commercial buildings can be classified separately from institutional buildings; commerical buildings can be categorized into apartment buildings, offices, stores, warehouses, and so on; institutional buildings can be categorized into banks, courthouses, hospitals, schools, and so on. Once such a classification has been made, comparisons of unit costs for the different types of categories of buildings can be made, as well as predictions about the probable costs of any proposed new building of any one specific type, use, or occupancy.

Third, the physical realities of the proposed building affect cost. In general, the larger the building, the higher the probable total cost, although the unit costs per floor or per square meter or foot may decline. Taller buildings of the same floor area as lower buildings generally cost more to build. The more complex the shape of the building, the higher the costs will be. The design and construction of exterior cladding and interior partitions greatly influence costs, as do the selection of systems for the structural, mechanical, and electrical aspects of the building. The general level of quality required to be achieved is also a factor of cost, and ranges from

Commercial high-rise building.

Commercial low-rise building.

simple utilitarian function such as found in a service building like a warehouse, to the highest level of luxury, such as found in first-class hotels or exclusive clubs.

A fourth factor of cost is the type of contract proposed for use and the resultant placement of financial risk on one or other party to the contract. Closely related to this contract factor is the factor of time available to do the work and the time of year when the work will be done. Other factors include financing, inflation, taxes (or exemption therefrom, as in some hospital work), the need for union or nonunion labor, and many other lesser influences which will be detailed in various other parts of the book. It may be said, however, that this is not a book on construction economics, which is a separate subject in its own right, and of which measurement and pricing form only one part.

QUESTIONS

12.1. Distinguish between cost and value. Give one example to illustrate each of your definitions.

12.2. Four factors of direct cost are presented for study in this chapter. Identify any one of the four and briefly discuss its attributes.

12.3. Identify three separate difficulties encountered in attempts to establish factual cost data.

12.4. Identify and briefly comment upon two ways in which the organization of the construction industry has an effect on construction costs.

12.5. The construction industry is highly diversified. Comment on two advantages and two disadvantages that such conditions permit.

CHAPTER 13

ELEMENTS OF PRICING

13.1 INTRODUCTION

13.1.1 Preamble

The pricing portion of the estimating process consists of three major segments: the *establishment* of appropriate data, the *computation* of unit prices based on these data, and the *application* of such prices to the measured quantities of work. The first segment will be examined in this chapter, followed by some selected issues related to pricing. The other two segments are presented in Chapter 14.

13.1.2 Terminology

Before beginning a detailed study of the elements of pricing, it is appropriate, to avoid ambiguity and confusion, to examine the meanings of some words and phrases commonly used in connection with this part of the subject matter.

Unit Rate, Unit Price. There is no universal agreement among estimators about the precise meanings of the terms "unit rate" and "unit price." The expression **unit rate** has two common meanings: average *output* for a measured unit of work, or *basic cost* to the contractor to perform a unit of work. The expression **unit price** usually means a composite figure, consisting of a basic cost plus a percentage markup for overhead and profit. In this book, "unit rate" refers to *output,* and "unit price" refers to *cost*.

Item of Work, Unit of Work. An **item of work** is a part of the total construction project which can be distinguished from all other parts, and for which costs can be determined separately; some examples might be a concrete beam as distinct from a concrete slab, or a wood-framed wall as distinct from the sheathing which covers that wall. A **unit of work** is a collection of items of work, generally comprising the work of one distinct trade, such as masonry or roofing. More generally stated, a unit of work consists of the subject matter of one section of the trade specifications in a project manual. A common synonym for an "item of work" is a *pay item;* a common synonym for a "unit of work" is a *specification section.*

Wage Rate, Labor Rate. **Wage rates** are those paid to the workers for doing work on an hourly basis; **labor rates** are those used by the contractor to calculate labor costs, and include wage rates plus all other payroll burden, such as fringe benefits and statutory requirements. Some estimators use wage rates in calculating unit prices for labor; such estimates should theoretically equal actual payroll disbursements. Most estimators prefer to use the more comprehensive labor rate, because it is more representative of the costs of the work being done. Note the slight ambiguity in the use of the word "rate" in this context, wherein it means *cost* and not *output.*

Plant, Equipment, Tools. These three words require some comment because of the special meanings ascribed to them by estimators. **Plant** means relatively large, immobile machinery, and often the buildings or other fixtures housing such machinery, such as is found in concrete ready-mix plants, gravel-pit operations, and large job-site shacks and storage buildings found on long-term building projects. **Equipment** means heavy but relatively mobile machinery, such as a small electric concrete mixer or a motorized conveyor such as a roofer might use to raise materials

Plant: a tower crane. [Form S. W. Nunnally, *Construction Methods and Management* (Englewood Cliffs, N.J.: Prentice-Hall, Inc., 1980).]

Equipment: a compressor. [From S. W. Nunnally, *Construction Methods and Management* (Englewood Cliffs, N.J.: Prentice-Hall, Inc., 1980).]

from the ground to roof level. Power trowels and concrete buggies are classified as equipment. **Tools** means lightweight manual devices, such as hammers and chisels, as well as hand-held machines, such as electric saws, sanders, and drills. Plant and equipment are usually provided by the construction company, whereas tools are usually provided by the tradesmen themselves for their own use.

Mobilization. **Mobilization** refers to the costs of transporting a piece of equipment from the contractor's storage yard or other source, such as a rental agency, to the construction site, setting it up ready to do work, and returning it to the source upon completion of work. If the equipment has to be transported using another vehicle, such as a flatbed truck or van, the costs of the second vehicle must also be calculated and charged to the project generating these costs. Specifically, the costs of mobilization should be charged to the item of work precipitating the necessity to provide the equipment under consideration.

Tools: a powder fastener. [From R. C. Smith, *Principles and Practices of Light Construction,* 3rd ed. (Englewood Cliffs, N.J.: Prentice-Hall, Inc., 1980).]

Delivery: hip trusses. [From R. C. Smith, *Principles and Practices of Light Construction,* 3rd ed. (Englewood Cliffs, N.J.: Prentice-Hall, Inc., 1980).]

13.2 ESTABLISHMENT OF DATA

13.2.2 Preamble

This first segment can be divided into two components: **factual data** which involves the determination of costs of specific types of materials, labor, equipment, and overhead expenses; and **productivity data,** which involves the determination of materials coverage and waste, crew sizes and outputs, equipment types and capacities, and the general risks involved. Each of these topics in these two components is examined in further detail below.

13.2.2 Factual Data

The four topics under this category are materials, labor, equipment, and overhead, the costs of each of which are influenced and determined by the normal pressures of supply and demand in the market in any region at any time. Such cost data are therefore not permanently fixed; they can only be established as of a given point or finite period of time. This is sufficient for the purposes of most estimators and designers, provided that care is taken with the definition of the point or period of time.

Materials. Most contractors or designers can readily determine the cost of the purchase and delivery of specific types of materials[1] in known quantities to designated sites. Material prices are usually established by direct quotations in the architectural and structural fields and by reference to price lists and discounts in the mechanical and electrical fields. In some cases, one has to phone around to get such quotations or prices; in other cases, manufacturers and suppliers circulate current prices for use by designers and contractors who are known to be working on current building projects. In most cases, a particular economic and legal deal will be struck between a supplier and a contractor for the provision of finite amounts of specific materials or products within a given time at a given location. For minor or standard items, common to most projects, such as nails, screws, ties, and the

---

[1]The word **materials** used in this context refers to unprocessed, semiprocessed, and fully processed materials, products, systems, components, and equipment, permanently installed in a building as part of a construction contract.

like, an over-the-counter price is established by normal market forces, and the contractor simply purchases what he needs when he needs it, and pays the going price, subject to possible discounts.

Labor. Contractors and designers can determine the current cost of labor in any particular region and for any particular trade, such as carpentry, masonry, painting, or general laboring work, as described in Chapter 12. Labor costs consist primarily of three major components: *basic wage rates, legislated requirements,* and *fringe benefits.* The first component represents the actual hourly rate of pay agreed between the employee and the employer. The second component covers additional amounts of money required to be paid by the employer to cover such social security items as pensions, unemployment insurance, holiday pay, and the like. The third component deals with the monetary effects of negotiated working conditions, achieved through bargaining between the employee and the employer (individually or collectively), such as seniority differentials, shift work, danger pay, and so on. As indicated, each of these three major components has more subcomponents, each of which require further investigation and determination to establish the correct and complete labor cost for any specific trade in any specific region at any specific time. Recourse to the sources previously mentioned should normally produce all the necessary data. Fringe benefits and social security requirements are referred to by some authors as the "payroll burden."

Equipment. Realistic costs for equipment[2] are more difficult to estimate than are material or labor costs. One reason is the incredible variety of complicated equipment that exists for use in the construction industry. A second reason is the variety of ways in which the use of such equipment can be secured, such as by purchase, rental, lease, or loan. A third reason concerns the large number of variables that enter into the calculations of costs, such as depreciation, investment, salvage value, tax benefits, and so on. A fourth reason involves variables in the output or productivity of a specific piece of equipment and the skill of its operator.

In general, though, costs of equipment can be considered under two broad headings: owning costs and operation costs. **Owning costs** include three major components: investment, maintenance, and depreciation. **Operating costs** include running costs, repairs, and operator's wages and expenses. Each of these components of equipment cost is examined in more detail in Chapter 14.

Overhead. Overhead can be conveniently divided into two major categories: direct and indirect cost. **Direct** (or job) **overhead costs** arise out of the fact that administrative work has to be done at each site where construction work is under way. **Indirect** (or operating) **overhead costs** arise out of the fact of being in business as a contractor, whether or not construction work is going on. Therefore, there will be one continuous set of (possibly fluctuating) indirect overhead costs, in addition to a number of intermittent sets of (possibly variable) direct costs, arising out of a number of construction contracts.

Direct costs consist of administrative items such as building permit fees, site offices, hoists and scaffolds, and so on, necessary to run a construction job site. Such costs are to be distinguished from costs of actual construction work, such as excavation, carpentry, or masonry. Indirect costs consist of administrative items such as head-office rental, salaries for head-office staff, advertising for the company as a whole, legal and insurance costs, and so on, necessary to run a business. Such costs are to be distinguished from costs arising out of any specific construction

---

[2] The word "equipment" used in this context means the machines that the contractor uses to do work; it does not refer to other equipment permanently installed in a building as part of a construction contract.

contract. Direct costs are estimated by preparing a list of such items and determining their individual costs from appropriate sources, usually external to the company. Indirect costs are usually determined internally by the company accountant on an annual basis and are then distributed on a proportional basis over each of the construction contracts under way in that company in that year.

13.2.3 Productivity Data

The four topics in this category are materials, labor, equipment, and risk. The development of reliable data for these factors is to a large extent determined by the exercise of judgment on the part of the estimator. Like factual data, productivity factors are not permanently fixed, although they can also be established within acceptable margins with respect to points or periods of time.

Materials. Most construction materials are marketed in standard sizes, containers, or unit quantities, such as sheets or boards, cans or drums, or in loose form, such as sand and gravel. When considering some prefabricated materials, such as plywood or concrete blocks, the estimator has to decide how much work can be done with whole units and how much with part units (resulting in some waste if whole units have to be cut). When considering products such as paint or asphalt, the rate of coverage is going to vary depending on the type of material supplied, the type of surface treated, the type of equipment used, the skill of the applicator, and the general environmental conditions at the place of work. Each product has its own characteristic factors of coverage and waste, and it is possible for the estimator to acquire or to prepare tables or charts to indicate such characteristics for common materials likely to be encountered in the work of any trade.

Labor. Labor productivity is probably the most difficult factor of the four to predict with any accuracy. In general, there are two primary reasons for the difficulty: first, the large number of variables that have to be taken into account, and second, the somewhat arbitrary decisions that have to be made by the estimator about many of these variables. To give some structure to this part of the study, some of the more important aspects of labor productivity have been classified under the following headings.

General Factors: These embrace such primary issues as the location and magnitude of the construction project as a whole, the type and complexity of the project, the time of year and the weather conditions encountered at the site, the availability of qualified tradesmen, and the quality and quantity of organization and congestion at the place of work, among other things. To elaborate briefly on some of the foregoing issues, for a construction company located in a downtown area, the labor on a rural job will probably be less productive than on an urban job, if only because of the travel time involved. Efficiency on a large job can be greater than on a smaller job, because of the opportunity for personnel to develop learning curves and to use the results, and the general advantages of larger production runs of every sort. Institutional work is generally less conducive than commercial work to high labor productivity, often because of higher complexity and intricacy, as well as more stringent contractual requirements. A better quality of work is to be expected in the spring and fall in most regions, compared to the height of summer or the depth of winter. If sufficient numbers of the right sort of tradesmen and helpers are locally and readily available, productivity will almost certainly be enhanced. Proper management and direction of the work force is a key factor in raising labor productivity. Secondary issues embrace such things as the general level of business activity in the region, worker morale, pay rates, absenteeism, personal compatibility, the training and education of supervisors, building codes, and union regulations, to name a few of the more significant ones.

Specific Factors: These embrace such primary issues as the *nature* of the work of each trade, the *amount* of work to be done, whether that work has to be done on a continuous or intermittent *basis,* the *time available* to do the work, the complexity or interest of each work task or item, and the logistics of organization to keep workers fully and productively occupied. For example, the work of a mason is clearly different from the work of a carpenter; the tools of the trade are different, the materials to be worked are different, the training of the workers is different, and the actual construction work to be done is different. The quantity of work has a significant effect on productivity: if there is a very small amount, there may be little or no benefit gained from learning how to perform the task efficiently; if there is a very large amount, some benefits of efficiency may be lost through boredom or fatigue on the part of the workers. Variations in time will have some effect on productivity; to some extent, the shorter the time, the more intense the activity, and this may lead to greater or lesser productivity.

Distribution of Time: Within the work of any specific trade, such as masonry or carpentry, there are *proportions* of time which are clearly more productive than others, and such proportions themselves are going to vary, for a number of reasons, such as the actual type of work being performed by the tradesmen. For example, in carpentry work, the framing of a standard stud wall is not as complex as the framing of a sloping roof structure with hips, valleys, gables, and so on; productivity on the wall will probably be higher than on the roof, everything else being equal (which it never is). Several interesting studies have been made of this topic of labor productivity in recent years, most of which show that the work of any trade can be considered under the following three categories:

1. *Fully productive time.* The work of the trade or related work is actually being performed by workers.
2. *Semiproductive time.* Planning or other arrangements to do some actual work are considered or discussed.
3. *Nonproductive time.* Activities such as rest breaks, delays, and other diversions.

To simplify consideration of the significance of the results, assume that the semiproductive time can be apportioned between the fully productive and the nonproductive time. Some studies suggest that productive time for some trades can be as low as 25% of the total time, although average figures tend to be around 40 to 50% for most trades working on commercial and institutional projects. In particular, carpenters can achieve just over 50% productivity, but masons can achieve just under 75% productivity, under typical and normal conditions of work. Several points arise out of these features of productivity which are of importance to the estimator:

1. Each trade has its own characteristic factors.
2. The proportion of productive to nonproductive time is a reflection of the nature of the work and its organization.
3. The estimation of probable time to do work of every type must include an allowance for the appropriate proportion of semiproductive and nonproductive time.
4. Improvement of productive time is largely a function of management, not of labor.

At least two conclusions can be drawn from the foregoing review of labor productivity: first, the estimator must consider allowing for what appears to be up to two hours of pay for one hour of productive work, depending on the type of work under consideration; and second, any improvement made to the proportion of productive time will have a double benefit on the cost of having the work done. It should be noted in passing that labor productivity factors contained in most construction estimating data books are composite figures which include the fully productive, semiproductive, and nonproductive components of the worker's time. Such data would have to be analyzed by investigation and site observation to discover the actual distribution of time for the labor content of any specific item of work. The responsibility of management to make such investigation and observation is clear if improvement of efficiency and reduction of cost is an objective of any company.

Equipment. Construction equipment falls into three main categories: **production equipment,** such as backhoes, bulldozers, and power shovels; **ancillary equipment,** such as hoists, buggies, and scaffolds; and **administrative equipment,** such as site offices, furniture, telephone pagers, and the like. Production equipment actually does the quantities of work predicted in the estimate; ancillary equipment assists or requires the production equipment to produce; and administrative equipment, although not directly associated with the construction process, nevertheless is required to permit that process to occur in an efficient manner.

Productivity of construction equipment of every type depends primarily on three factors: the *nature* of the work to be done, the type and capacity of the *machine* selected to do the work, and the skill and experience of the machine *operator.* As with labor productivity, productive and nonproductive proportions of time have to be taken into account by the estimator. These can be considered under the following headings:

Potential Production: For every machine, a maximum number of working days are available in any one year. Of these possible days, the machine may not be used on all of them. On the days that it is used, it may not be used for a full working day. Finally, it may not be used for the full 60 minutes out of each hour it is actually in operation on any one day. Factors can be developed and tabulated to represent potential production for any machine, for use by the estimator.

Actual Production: When any machine is in actual operation, doing the work for which it is intended, further inefficiencies may develop, resulting in a reduction from maximum potential. Some causes of such reduction are adverse working conditions, poor weather, lack of skill of the operator, difficulty of the actual work being done, and so on. Factors can also be developed and tabulated to adjust potential production figures to predict probable actual production figures for any machine, for use by the estimator.

The factors to be applied to any machine can be determined from data gathered primarily from two sources: first, the *theoretical* performance, based on data provided by the manufacturer of the machine; and second, the *actual* performance, based on direct observation of the machine in action in the field, over a reasonable period of time. Data from both sources require confirmation by a cost-accounting process.

Risk. As in other fields of life, risk in construction is a function of the quantity and quality of appropriate information. The more and better the information, the less the risk, and vice versa. However, the construction industry is generally a risky one, because complete and reliable information is seldom available. In the following paragraphs, a number of factors representing opportunities and difficulties with which the estimator must deal, relative to the topic of risk, are presented for review and study.

Opportunities: With respect to affairs within the control of contractors, they have good control over the *internal processes,* such as estimating, cost accounting, contract management, office management, purchasing, and so on. They can review their present and proposed *work schedules* for the time frame under consideration, to assess what work is under way, what other work is just about finished, and how much more work is needed to maintain any desired level of company performance. They can study the *market* in general and the competition in particular when bidding for any specific project; data can be assembled on the numbers, names, bidding practices, and past performances of the *competition,* as well as their current commitments where known. The level of *profit* desired or expected from any specific project can be established with some certainty in advance, by following sound estimating practices and by making a review of present needs and past performances; this topic is developed further in a succeeding paragraph. Furthermore, contractors have considerable control over the development and introduction of innovative *construction methods* to make their operations more efficient.

Difficulties: With respect to affairs beyond the control of contractors, they have little control over general *market trends,* which fluctuate throughout the year as a result of national and international political and fiscal policies. There is little that the individual contractor can do to prevent *labor unrest,* resulting in strikes or lockouts in the industry at large, brought about by union activity, association action, or government intervention. A common risk that contractors face is *delay* in material and equipment deliveries; some protection against this risk can be achieved by including a damage clause in supply contracts. Uncertain *weather conditions* also pose risks for contractors; it is only on comparatively large projects, extending over a lengthy period of time, that any benefits can be derived from studying weather patterns for previous years in the region where the project will be built. The best that most contractors can do with regard to the weather is to hope that it will be more or less normal for the period and to arrange insurance to cover unusual conditions. Another common and potentially damaging risk is *escalation of cost,* through inflation, price changes, scarcity of supplies, or other cause, because most construction contracts include a clause that excludes claims based on escalation. As every construction project gives rise to some event that was unpredictable either in nature or magnitude, most contractors include a *contingency allowance,* which is a sum of money that may cover most or all of these additional unforeseen costs, should they arise.

Profit Margins. The margins of profit in the construction industry can vary all the way from 0 to 100% and beyond. In general, however, profit margins usually lie somewhere between 5 and 25% of the sum of the direct and indirect costs, with the lower end of the scale applying to projects for which good information is available, and moving up the scale as the quantity and quality of information declines. Profit margins are not intended to allow for poor estimating practices or other mistakes of fact or judgment in the assembly of the figures for the final bid or price negotiated for the construction contract. Profit is the residue left to the contractor after all costs have been met and should generally equate to about 20 to 25% of risk capital, before taxes, where *risk capital* is defined as the amount of money required to operate the construction company as a business. At present, such risk capital should be in the region of 10 to 12% of total annual sales of the company. For example, if a company has an annual volume of business of $1 million, the risk capital would be around $100,000. The annual profit to be earned should therefore be around or above $25,000, to provide a proper return on the capital invested.

Thus, if the company completed, say, three trouble-free construction projects in one year, each worth approximately $300,000, the theoretical profit on each job should be at least $10,000, or 3.33%. In practice, a figure of 5% would probably be included. The most appropriate profit margin for any construction company should be established by the company management and accountants; margins that are too low or too high will inevitably lead to problems, one because of diminishing returns on capital, and the other because of diminishing volume of business.

Allowances for contingencies can take two forms: first, they can be overtly stated as a condition of the contract; second, they can be hidden within the overall stipulated or lump-sum price for the contract as a whole. Allowances for contigencies to cover unforeseen events can vary from 0 to 5%, with figures of 2 and 3% being common. There are two problems with contingencies: first, owners do not like additional amounts of money being tied up in the contract financing; and second, if such sums are to be defined in the contract, contractors may take deliberate steps to make sure that the funds represented by such contingency sums are spent to their advantage. Designers should be particularly careful with the formulation of contract conditions governing the expenditure of such funds.[3]

13.3 RELATED ISSUES

13.3.1 Preamble

A brief review of two issues related to the content of the foregoing paragraphs is included at this point: a configuration of the factors making up the *components* of construction cost, followed by an explanation of three aspects of price *discounting*.

13.3.2 Components of Cost

Earlier, all of the factors of cost, such as production, fabrication, and delivery, were presented for study. These factors can be arranged in a number of ways. One useful way for the estimator is to look at the total amount of money produced by these factors for any item of work, and to rearrange that amount into the five principal or major components of cost, each shown below with two major subcomponents.

1. Material costs
   a. List prices and quotations, which are based on market factors of supply and demand
   b. Discounts for cash, trade, or volume reasons
2. Labor costs
   a. Labor rates, which include wage rates, negotiated fringe benefits, and statutory requirements
   b. Productivity factors based on data showing amounts of work able to be performed in amounts of time
3. Equipment costs
   a. Owning costs, including purchase price, investment costs, depreciation, and maintenance
   b. Operating costs, including fuel, transportation, repairs, and operator's labor cost

---

[3]For further detail on this topic, see Glenn M. Hardie, *Construction Contracts and Specifications* (Reston, Va.: Reston Publishing Co., Inc., 1981), Chap. 7.

4. Overhead costs
    a. Direct costs, including costs of administration of one or more construction sites
    b. Indirect costs, including costs of running the company as an enterprise
5. Profit
    a. Net profit (or loss), which is the difference between all income and all expenditure, expressed in dollars
    b. Profitability, which is the return on invested capital, expressed as a percentage

Of the foregoing components, the first four are treated in more detail in Chapter 14; profit was dealt with earlier in this chapter.

13.3.3 Discounting

By definition, the word **discounting** means to deduct from, to reduce from a given amount of money. In the construction industry, there are three distinct categories of discounting practices encountered, for three separate reasons.

1. *Cash discount.* Cash discounts are rebates allowed by suppliers to contractors and other customers who pay their bills on time or within an agreed period. The purpose is to encourage customers to settle their accounts promptly, with two benefits to the supplier: he has the use of the money at an earlier date, and he does not have to spend more money to remind the customer of his obligation to pay the bill. To encourage the customer to keep up the good work, the supplier will probably reduce the customer's bill by the amount of money that the supplier is saving, often amounting to as much as 5% of the bill.
2. *Volume discount.* The volume discount is a rebate granted, as the name suggests, to contractors or customers who place large orders for the supply of materials, services, or equipment.
3. *Trade discount.* The trade discount is offered to contractors simply because they are in the business of contracting. If a person walks into a retail hardware store and orders a bag of cement over the counter, he will be quoted a price at one level. If the same person walks into the same store and orders the same bag of cement on behalf of a construction company known to the proprietor of the store, he will be quoted a price at a different and lower level. This discount is a technique used by supply houses to promote the goodwill of their company in the marketplace.

Contractual Issues. In many construction contracts, owners occasionally feel that as they are the ones who are paying the contractors, they should receive the benefit of all discounts. It is possible to arrange or negotiate for this feature to be included in the contract documentation. It is difficult, though, to convince many contractors of the validity of this line of thought.

As a general rule, cash rebates, discounts, or savings rightfully belong to the contractor as a reward for being a good business manager. If the owner attempts to remove this benefit from the contractor, there may be no incentive left for the contractor to be prompt with payments to the suppliers, and any potential savings will be lost to the owner anyway.

With volume discounts, the owner can argue that the reason the contractor can place large orders is because of the magnitude of the job. The contractor might respond by saying that it is because he is able to place large orders that he is able to undertake large projects on behalf of the owner. There are some difficulties that can arise in this regard, because many contractors order supplies in bulk as a matter of course, and then distribute them to the various projects on which they are working. This practice can make it difficult to calculate the proper amount of discount to be credited to one project which is using only a portion of the total order. Conversely, a large total order may be placed as a series of small orders, none of which qualifies for a volume discount.

With trade discounts, many owners feel that such discounts should be credited in their favor, as it is their projects that permit contractors to be in business in the first place. Contractors can argue that it is they who are in the trade, not the owners, and therefore they should be entitled to retain the benefits of the trade.

Great care should be taken with respect to the clarity of composition of clauses included in construction contracts dealing with the retention of discount benefits, to avoid difficult and expensive problems from arising at the settlement of accounts. Such clauses should also generally apply only to trade and volume discounts, not to cash discounts, for the reasons given above.

QUESTIONS

13.1. Differentiate between unit rate and unit price. Give examples of each to illustrate your definitions.

13.2. Fill in the blanks with common synonyms in the following two sentences:
(a) An "item of work" is a _____ _____,
(b) A "unit of work" is a _____ _____,

13.3. What are mobilization costs? How are they developed? Where should they be charged?

13.4. Name and briefly comment on the three basic components of labor costs.

13.5. Distinguish between direct overhead costs and indirect overhead costs. Give one example of each category.

13.6. State two reasons why factors for labor productivity are so difficult for estimators to establish accurately.

13.7. List at least three reasons why labor productivity varies so much between the work of different trades. What conclusions can be drawn from your reasons?

13.8. Define "ancillary equipment." Suggest how costs for such equipment might be (a) developed, and (b) accounted for.

13.9. Construction equipment is frequently inefficiently utilized. Suggest at least four sources of inefficiency that may apply to any piece of such equipment.

13.10. Define "profit." Show how profit may be established by calculation, for any specific project.

CHAPTER 14

TECHNIQUES OF PRICING

14.1 INTRODUCTION

14.1.1 Preamble

In Chapter 13 the general *elements* of pricing were presented. In this chapter the techniques of *computation* of unit prices and their *application* to measured quantities of work are explained. Examples are included to show the theories in practice; further applications are illustrated in Part IV.

14.1.2 Basic Technique

In this second segment, the basic technique used to calculate a unit price for an item of work or an element of a building is simple: the anticipated total *cost* of that work or element is *divided* by the number of *units* of the work or element in the project. The technique can be applied in one of two ways: divide the *total* cost of the total work by the total number of units; or divide the total cost of a selected or representative *portion* of the total amount of work by a known or predicted standard number of units of such work. The difficulty in applying the technique arises out of the difficulty of establishing a reliable total cost for the work or element, and that issue is addressed in detail in the following sections of this chapter.

Using the first technique, a roofing company, required to install 10000 m^2 or sy of built-up roofing, might find that its total labor cost will be $50,000; the unit price for labor will therefore be $5/$m^2$ or sy.

Using the second technique, an excavation company, required to excavate an indeterminate amount of clay, might find that its total equipment cost to remove 1000 m^3 or cy will be $5000, and the unit price is therefore $5/$m^3$ or cy. This price can then be applied to estimates of work which are close to 1000 m^3 or cy, although some adjustment will become necessary as the quantities vary distinctly above or below 1000 m³ or cy. A table or chart could easily be prepared, showing a range of unit prices for quantities of work of varying magnitudes.

One question which novice estimators frequently find puzzling is that if one can figure out the total cost of doing some amount of work, why bother with the purely arithmetic task of dividing the cost by the units to find the unit cost? The answer is twofold: first, by calculating costs on a standard unit basis, one can compare one unit cost with another for similar or related work, or compare one job with another, or compare costs for a standard job with a nonstandard job; and second, such units of cost represent average costs for doing specific types of work, and they can therefore be used in the cost-control and cost-accounting processes, to check on the amounts of money being spent in the field and on the accuracy of the estimate in the first place. The estimated rate of expenditure may be exceeded in the field for a predictable period of time, but it should decline to a level below the average before more than half of the work of the item or element is complete. More will be said about this effect and process later.

14.2 COMPONENTS OF COST

14.2.1 Preamble

As stated in Chapter 13, there are five major components of construction cost: materials, labor, equipment, overhead, and profit. In the following paragraphs, the first four of these components will be further elaborated by discussion and exemplified by illustration; the last component, profit, was developed adequately in Chapter 13 to suit the objectives and context of this book. The illustrative examples are given first in metric units and then in imperial units in each case. It would be a useful exercise for the reader to consider working out each example in the alternative units, for practice and guidance.

14.2.2 Materials

The costs of purchase, delivery, taxes, and all other factors to acquire sufficient quantities of the proper materials to adequately complete the required work must be determined. Other factors include such things as waste, absorption, losses, and accessories, such as adhesives, anchors, nails, screws, trim, and the like. The total material cost is then divided by the number of units to establish the estimated unit cost.

Metric Example. Find the material cost of masonry to provide a facing brick veneer, complete with mortar, in front of a wood-framed and wood-sheathed structural wall (Figure 14.1). Assume a standard area of 10 m² and assume a standard facing brick having a face dimension of 190 mm long by 50 mm high and a width or thickness of 90 mm. Also assume a standard mortar joint 10 mm thick, that the mortar extends through the entire thickness of the veneer facing, and that there is no mortar on the concealed rear face of the veneer.

If the joint dimension is added to the brick face dimensions, each brick and joint covers (200 × 60) 0.012 m² of wall surface. Therefore, 10 m² of wall surface will require (10/0.012) 833.33 bricks with joints. The area relationship of a brick to its joint is (190 × 50) 9500 mm² to [(190 + 60) × 10] 2500 mm², or 3.8 to 1.0; the proportion of 10 m² covered by bricks is therefore [(10.0/4.8) × 3.8] 7.92 m² and covered by mortar is [(10/4.8) × 1.0] 2.08 m².

The *volume of mortar* will therefore be (2.08 × 0.09) 0.19 m³, because the thickness of the veneer and therefore the width of the mortar between the bricks is

Chap. 14 / Techniques of Pricing

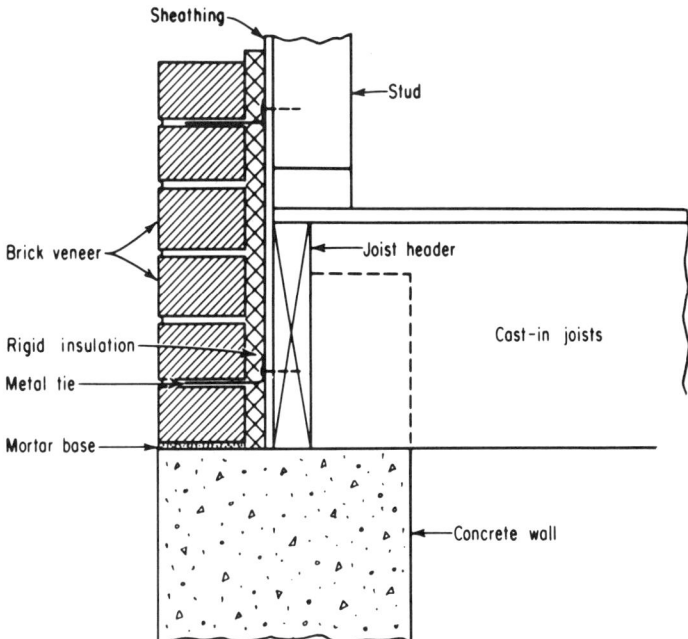

Figure 14.1 Section through veneer wall. [From R. C. Smith, *Principles and Practices of Light Construction,* 3rd ed. (Englewood Cliffs, N.J.: Prentice-Hall, Inc., 1980).]

90 mm or 0.09 m. The *number of bricks* will be (7.92/0.0095) 833.33, as already calculated above, because the face area of one brick is (190 × 50 mm, i.e., 0.19 × 0.05 m) 0.0095 m². If a waste factor of 5% is added to each of these quantities, the materials required for the 10 m² of veneer will be 0.2 m³ of mortar and 875 bricks of the specified size.

If the cost of 1 cubic meter of mortar and 1000 facing bricks is known, the material costs for the veneer can be calculated. If the veneer has to be tied to the structural backing with galvanized steel ties, spaced say one per square meter, a minimum of 10 ties will be required, plus one for waste or loss, for every 10 m² of veneer to be installed. The total cost can then be divided by 10 to find the material cost per square meter, to use on a current estimate, and to compare with previous estimates.

Imperial Example. Find the material cost to supply a specific paint on a particular concrete wall surface. If one-quarter (0.25) gallon of the paint will cover 100 sf of a known standard type of concrete wall, and the specific wall has an absorption rate of 5%, (0.25 + 0.013) 0.263 gal will be required for every 100 sf of wall. If a further 5% of paint is lost by dripping, evaporation, residue in the cans, or other cause, then (0.263 + 0.013) 0.276 gal will be required. If the paint costs $15 per gallon, the unit cost for the material will be [(0.276 × 15)/100] just over $0.04, or 4 cents per square foot.

14.2.3 Labor

First, the type of work to be done has to be clearly established; second, a decision is made about the composition of the work crew; third, productivity factors for that crew to do that work have to be determined from company or published records; fourth, the hourly labor rates to employ such a crew are added up to produce a total hourly cost; finally, the total hourly cost is divided by the productivity factor to produce the estimated unit cost.

Metric Example. Find the labor cost to build a brick veneer wall, 100 mm thick, using clay bricks 190 mm long by 50 mm high, with 10-mm struck-cement mortar joints. A journeyman mason, with a helper, should be able to lay about 700 such bricks in a 7.5-hour working day. Each brick and joint covers (200 × 60) 0.012 m^2 of wall surface; 700 bricks cover (700 × 0.012) 8.4 m^2 of wall surface. Therefore, the crew of two can build at the rate of (8.4/7.5) 1.12 m^2 per hour. If the hourly labor rate for the mason is $20 and for the helper is $15, the labor unit cost will be [20 + 15)/1.12] $31.25 per square meter.

Imperial Example. Find the labor cost to paint a specific type of surface on a particular concrete wall, using a brush to apply one coat of latex paint. From labor productivity records, it seems that one painter should cover about 300 sf per hour doing this type of work. If the painter works by himself, with no helper, and his labor rate is $20 per hour, the labor unit price for this item will be (20/300) $0.07, or 7 cents per square foot.

14.2.4 Equipment

The development of unit prices for construction equipment is complicated by two factors: there are so many types of equipment from which to choose, and there are so many ways to calculate the probable cost of owning and operating any selected piece of equipment. Nevertheless, there are some general aspects of cost that apply to most pieces of construction equipment, with minimal modification. These are presented below under the four headings of owning, operating, operator, and mobilization. Of these four, the first is the most difficult to present, comprehend, and properly calculate in every case; to assist the reader, a short preamble is given on some aspects of money as a commodity.

The difficulty in determining accurate owning costs for equipment arises from two primary sources: the time value of money, and whether money is owned or borrowed. Although it is not the intention here to digress into a detailed discussion of accounting principles and practices, some explanation and understanding of money and interest is necessary for the calculation of realistic costs for equipment.

To own a piece of equipment, a company can either use its own money or it can borrow some or all of the amount from some other source, such as a bank or finance company, excluding the unlikely possibilities of winning a lottery or a windfall gift appearing by magic. If the company buys the equipment with its own money, that money is no longer available for investment elsewhere and the interest it could have been earning should be added to the cost of owning the machine.

Interest has been defined by accountants as money paid for the use of money, like rent is paid for the use of property. If the company borrows the money, it will have to pay interest to the lender, and that money also has to be added to the cost of ownership, as before. Interest payments can be calculated in several ways: in a lump sum at the start or at the end of the loan period, on a regular or intermittent basis throughout the period of the loan, in even or uneven amounts at specified points of time, or otherwise as the parties may agree.

The effects of tax provisions, inflation, and other variables may also have to be considered, although exactly how these are introduced into the calculations is beyond the scope of this book. In most calculations involving interest, reference has to be made to interest tables, which show the effects of different rates of interest of different types on principal amounts of money borrowed for stated periods. Such data are obtainable in printed or electronic form. Two simple interest examples are shown below.

EXAMPLE 1:

What will $1000 accumulate to in 10 years if invested today at 10% compound interest? From tables, the compound interest factor for 10% for 10 years is 2.594, so the answer to the question is ($1000 × 2.954) $2594.

EXAMPLE 2:

What amount of money has to be invested today to accumulate to $2594 in 10 years? From tables, the present-worth factor for 10 years at 10% compound interest is 0.3855, so the answer to the question is ($2594 × 0.3855) $1000 (approximately).

With respect to the time value of money, there are essentially three factors to consider: the interest rate, a series of payments, and a series of receipts, all over a period of time. If any two factors are known, the third can be computed, and all three should result in an arithmetical balance. The primary object of making such calculations is to determine two things: that the original *investment will be recovered,* and that a sufficiently attractive rate of *return on the investment* will also be received. A secondary objective is to be able to make comparisons between similar propositions, to determine the best deal from a monetary point of view. Decisions regarding the selection and use of equipment for construction operations is usually made by comparing the operational suitability and the financial considerations of any one machine with its closest competitors.

Using a compactor. [From S. W. Nunnally, *Construction Methods and Management* (Englewood Cliffs, N.J.: Prentice-Hall, Inc., 1980).]

Owning Costs. The steps to calculate the hourly owning costs of a piece of construction equipment are as follows:

1. Find the total delivered price, including attachments and investment costs, such as interest and inflation.
2. Deduct the cost of all items that will be replaced during the expected life of the machine.
3. Deduct the estimated scrap value at the end of anticipated useful life of the machine.
4. Divide the residual cost by the number of years of life expectancy to find the annual cost.
5. Add all other annual costs, such as permits, storage, insurance, inflation, and the like.
6. Divide the total annual cost by the estimated number of hours of predicted annual use to find the hourly owning cost.

Note that inflation can be included in *either* item 1 or item 5.

Operating Costs. The steps to calculate the hourly operating costs of a piece of construction equipment are as follows:

1. Find the annual costs of fuel, lubricants, grease, filters, and other consumable items using data provided by the manufacturer.
2. Calculate the replacement cost of all replaceable items by dividing replacement cost by anticipated annual life of the items.
3. Include the manufacturer's recommendations for probable repairs, usually as a percentage of cost.
4. Divide the total annual operating cost by the estimated number of hours of anticipated use to find the hourly cost.

Operator Costs. The steps to calculate the hourly operator costs of a piece of construction equipment are as follows:

1. Find the equipment operator's labor costs as already explained in the section dealing with labor cost.
2. Similarly find the costs of any support personnel, such as laborers or guidesmen.
3. If the operators do no work other than to operate the equipment, the total annual labor costs are divided by the anticipated hours of use to find the hourly cost.
4. If the operators do other work, besides operating the machine, only the hours spent operating the machine should be included in the calculation.

Mobilization Costs. The steps to calculate the hourly equivalent cost of the intermittent mobilization process are as follows:

1. Find the number of hours to be used transporting the piece of equipment from storage to the site and back to storage.
2. Find the hourly cost of transportation of the equipment together with any other vehicles involved in the move.
3. Find the number of hours the piece of equipment will be located at the site, whether in use or not.

Chap. 14 / Techniques of Pricing 127

4. Divide the cost of transportation by the number of hours at the site to find the hourly mobilization cost.

The hourly costs, each calculated in the manner described above, can now be added together to produce a price per unit of *time* for a given piece of equipment. That price can now be converted to a price per unit of *measurement,* such as cubic meters or cubic yards or other appropriate unit, by dividing the *hourly cost* by the equipment *productivity rate.* An allowance for overhead and profit still has to be added, either to each individual unit price or to the estimate as a whole.

Example. Just as each piece of equipment has its own variables, each company accountant has his or her own method for dealing with these variables, with the result that it is not practical to give one definitive way of calculating equipment costs which will work in all cases. This example, based on a medium-sized backhoe, does no more than indicate the results that will be achieved by application of the methods suggested in this section; other methods will produce other results, all of which can be compared with each other. In the example, an assumption has been made that the money to buy the machine will be borrowed from a lending institution, that the rate of inflation is 8%, and that the rate of 12% will provide for both interest and income tax costs. Other assumptions are that the backhoe will have a life expectancy of five years and an annual predicted utilization of 1500 hours. Calculations are shown in Figure 14.2.

Other methods of calculating costs will produce other results. One alternative method to the one shown in Figure 14.2 is given in Figure 14.3, in a deliberately simplified form, to show its basic structure. It is strongly recommended that any method adopted in actual practice be reviewed by a certified public accountant before being utilized in a business enterprise, to avoid costly blunders of omission.

The owning and operating costs in Figure 14.3 add up to ($10.80 + $6.84) $17.64, compared to Figure 14.2, in which these two items add up to ($7.07 + $10.18) $17.25. Neither method is right or wrong; each gives different but nevertheless similar results. The implication is that various methods can be employed to test the validity (and the arithmetic) of any specific method adopted to determine such costs for equipment.

The definitive test of unit prices thus calculated for equipment is to compare these prices with current market prices being charged by equipment rental companies operating in the region where the work is to be done. Many cost accountants suggest that construction companies should charge their own equipment on their own account books at a rate close to the market rate. Any benefits will then improve the profit margin of the company operations.

14.2.5 Overhead Costs

For the contractor, overhead costs fall into two major categories: *direct* overhead costs, generated by administrative items arising out of the work of a particular construction contract, and *indirect* overhead costs, generated by administrative items arising out of the fact of being in business as a contractor.

Direct Overhead Costs. These items are generally to be found in Division 00 and 01 of any construction contract organized according to Masterformat. They include the costs of such things as temporary services, site offices, hoardings and barricades, permits, fees, salaries of superintendents, and so on. They are priced first by making a list of contract requirements and then making enquiries to determine prices for the various items. Some of these items, such as fences and services, can be measured and priced like any other construction work. Other items, such as

Find the net hourly cost to own and operate a backhoe that costs $20,000 to buy new today.

A. Owning Costs

1. *Total price*

Purchase price	$20,000	
Sales tax (5%)	1,000	
Preparation and delivery	500	
Investment costs (20%) [1]	$21,500 × 2.488 [2] = $53,500	

2. *Deduct replacements*

Set of tires	1,200	
Bucket kit	1,000 = 2,200	

3. *Deduct salvage value* (say)   1,300   3,500

4. *Net costs*

   Residual cost $50,000

   Annual cost ($50,000/5 years)   =   10,000

5. *Annual expenses*

   Permits and fees   3%
   Storage and other   3%
   Total expenses [3]   6% of $10,000   =   600

6. *Owning costs*

   Annual basis [4]   = $10,600
   Hourly costs ($10,600/1500 hr)   = $7.07/hr

Notes:
[1] 20% includes 12% for interest and taxes plus 8% inflation.
[2] 2.488 is a factor taken from compound-interest tables.
[3] Expense percentages are assumed here.
[4] Other methods will produce other results.

B. Operating Costs

1. *Consumables*

   Fuel (assume 75-hp engine × 1500 hr × 0.2
       liters/hp/hr × 32 cents/liter)   = $ 7,200
   Grease and filters (say, 25% of fuel cost)   1,800

2. *Replacement parts*
   Tires (2 sets × $1,250 per set) $ 2,500
   Bucket (2 kits × 1000 per kit)   2,000
   Total cost over 5 years   4,500
   Annual cost ($4500/5 years)   =   900

3. *Repairs*

   Major (say, 50% of depreciation: $10,000)   =   5,000
   Minor (say, 03% of consumables: 9,000)   270
   Parts (say, 10% of tires and buckets: 900)   90

4. *Operating costs*

   Annual cost   = $15,260
   Hourly cost ($15,260/1500 hr)   = $10.18

Figure 14.2 Calculation of hourly costs.

C. Operator Costs

1. Driver/operator 1 @ $20.00/hr
2. Laborer/helper 1 @ $15.00/hr

    Hourly cost ($20 plus $15) = $35.00

D. Mobilization Costs

1. Cost to transport backhoe to and from site (say, 1 hour each way @ $50.00/hr) $100
2. Time to complete work at site (say) 50 hr

    Hourly cost ($100/50 hr) = $2.00

E. Total Hourly Cost of Backhoe

   Sum of items A to D:

   ($7.07 + $10.18 + $35.00 + $2.00) = $54.25

Note:
[1] The extra $50 included in the replacement cost of the tires is to allow for a price increase after 2 to 3 years.

A. Owning Costs

1. Total price as before = $21,500

    Deduct replacements and salvage 3,500
    Initial investment cost $18,000
    Annual investment cost ($18,000 × 60%) [1] = 10,800

2. Depreciation over 5 years ($18,000 × 20%) 3,600
3. Maintenance (50% of depreciation) [2] 1,800
4. Annual owning costs = $16,200

    Hourly costs ($16,200/1500 hr) = $ 10.80

B. Operating Costs

1. Consumables as before = $ 9,000
2. Replacement parts as before 900
3. Minor repairs and parts as before 360
4. Annual operating costs = $10,260

    Hourly costs ($10,260/1500 hr) = $ 6.84

Notes:
[1] The worth of the investment can be shown to be 100% at the start of the first year and to be 0% at the end of the fifth year, giving an average annual value of 60%.
[2] The amount set aside for major repairs is lower in this method than in the preceding one.

Figure 14.3 Calculation of hourly costs: alternative method.

the costs of field offices or bonds, may require quotations from suppliers or rental agencies. There are several items, such as first-aid costs and survey or layout costs, for which an allowance or percentage must be determined by each individual contracting firm, relative to its own requirements and standards. The costs of supervision can be determined by establishing the superintendent's salary on a monthly basis and then multiplying that salary by the estimated number of months it will take to complete the project.

Indirect Overhead Costs. These costs reflect the general cost to the contractor of simply being in business as a enterprise. They include such things as rent, heat, light, phone, insurance, salaries and fringe benefits for head-office personnel, publicity, legal fees, taxes, and all the other activities and expenses that it takes to run a successful company. These costs are usually determined on an annual basis by the company accountant and are expressed as a percentage of the gross annual dollar volume of the company business. The value of each project can then be expressed in proportion to that annual figure, and the appropriate percentage of the annual overhead costs can thus be allocated to each project to cover the total expenses for the year.

14.2.6 Profit

The topic of profit was dealt with in sufficient detail in Section 13.2.3. To recap briefly, profit is the assessment of risk involved in a construction project, expressed as a percentage to be added to the estimated costs required to perform the work of the construction contract.

14.3 APPLICATION OF PRICES

14.3.1 Preamble

The third and final segment of the pricing component in estimating concerns the issues relative to the *application* of the calculated *unit prices* to the measured *quantities of work*. A number of aspects relative to application are presented below for study, consideration, and discussion.

14.3.2 Quantity

Most unit prices are affected by the general economic law which states that the larger the quantity of any given commodity, the lower will be the costs to produce each unit of that commodity. An examination of cost data in the construction industry shows, however, that once a unit price for any specific type of work has been determined, minor variations in quantity will have negligible effects on the total costs of the work. To say the same thing in another way, although the precise quantities of work will almost always vary from job to job, the unit prices that apply to such work will remain fairly uniform unless some element within the unit price itself changes between applications. One beneficial result of this phenomenon is that estimators can establish unit prices on a fairly reliable and uniform basis, independently and in advance of specific jobs being bid at any particular point in time. To give an example, if a unit price has been calculated for the labor involved in laying floor joists in a three-story 100-suite wood-framed apartment building in an urban area, the same price can be used to estimate the cost of laying floor joists in a four-story 80-suite wood-framed building in a rural area, but with additional amounts to allow for extra transportation or accommodation of the labor force to either site. The unit price for the productive part of the work need not be changed, provided that it has proven to be accurate.

14.3.3 Process

A second aspect of the application of unit prices concerns the manner in which costs for various types of equipment can be charged to various construction contracts under way. Simply stated, *production equipment,* such as power trowels and pneumatic drills, is charged by calculating a unit price and applying it directly to the measured quantities of an item of work. *Ancillary equipment,* such as table saws and concrete buggies, can be charged either by calculating an hourly rate and applying it to the estimated number of hours of use, or by adding a percentage (produced through cost-accounting techniques) to either the total material or total labor costs. As a side note, it may be useful for the estimator to consider that the costs of powered equipment are an extension of or an improvement on manual labor, and thus apply any percentage to cover such cost to the labor content of the estimate. The cost of *administrative* equipment is usually included in the general percentage or allowances for overhead costs (again produced through cost accounting) and added at the end of the estimate. Costs or unit prices for *small tools,* provided by the contractor for the use of his work force, are calculated as for production equipment and then added to the labor unit price for the items of work to which such costs apply, usually as a small percentage in the region of 2 to 3% of the labor cost. Costs of tools owned by the tradesmen and provided for their own use should be included in the basic labor rate for the trade.

14.3.4 Technique

The application of unit prices is a simple matter, provided that a number of precautions are taken, such as selecting the correct price for each item of work or element of the building, making appropriate modifications where necessary, and taking care with arithmetical functions. The *quantity* of work for each item is listed in a summary and then *multiplied* by its own particular *unit price* to produce the estimated *cost* for that item. Examples are shown in Part IV. An example of a Pricing Form is shown in Figure 14.4.

14.3.5 Conclusion

The estimated costs for each item of work are then added together to produce totals for each page of the estimate. The total for each page is then carried to a summary sheet, where the totals of all the pages are added together to produce the total estimated cost for the project under review. Appropriate allowances are then added as necessary to cover contingencies, overheads, and profit, to establish the *final price* to be bid by a contractor to an owner or the *budget figure* to be recommended by a designer to a client. The reason it is advisable to carry the total for each page of the estimate separately to a summary sheet is to avoid the problems that will arise if an error or change occurs in the extensions or additions in any one sheet and is then carried forward through all successive sheets of the estimate. Only the incorrect page and the summary page will have to be amended if the foregoing technique is adopted. An example of an Estimate Summary is shown in Figure 5.12.

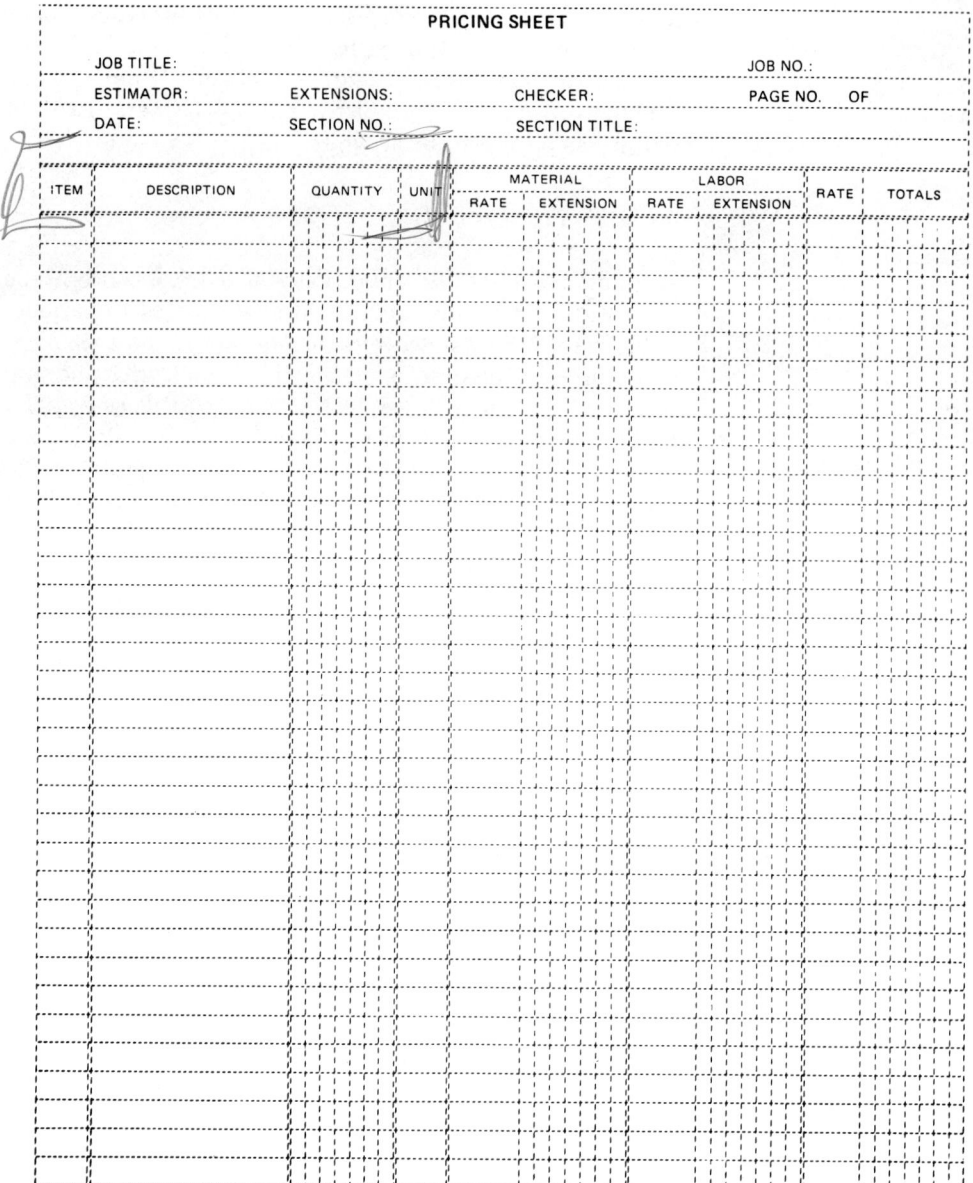

Figure 14.4 Pricing sheet.

QUESTIONS

14.1. Describe the basic technique used to calculate a unit price for an item of construction work. Illustrate your explanation with a simple worked example.

14.2. Identify two advantages that accrue to the estimator who takes the trouble to establish construction costs on a unit basis.

14.3. Suggest an approach that the estimator might adopt to determine the appropriate size of a work crew to do any particular item of work.

14.4. Define interest on money. Mention three ways in which interest can be calculated.

14.5. State the three primary factors involved in establishing the time value of money. Identify two primary objectives for such establishment.

14.6. Briefly outline the steps involved in calculating the owning costs of a piece of construction equipment on an hourly basis.

CHAPTER 15
BIDDING AND COST CONTROL

15.1 INTRODUCTION

15.1.1 Preamble

No book on construction estimating techniques could be considered complete without some discussion of *bidding procedures* and *cost control*. In this chapter some background material on these two important topics is presented as a guide to more detailed study which novice estimators may wish to undertake at a later point in their careers.

15.1.2 Objectives

The intention is not to present an exhaustive study or definitive examination of either bidding or control, but rather to show the importance of these two topics and their relationships to the estimating processes.

15.2 BIDDING PROCEDURES

15.2.1 Preamble

To have a piece of real estate developed, the owner must take some steps to bring about a contract for the construction work. This topic has been dealt with in detail in many other texts[1]; in this book, the main issues will be summarized as being either primary or secondary elements.

---

[1] For examples, see Appendix C.

15.2.2 Primary Elements

There are three primary elements or stages in most bid procedures:

1. *Inviting bids by public announcements or private invitations.* Bids may be invited either formally or informally, by publishing notices to contractors in general in the public press and by mailing, phoning, or directly inviting contractors in particular to submit a bid. Bids on public projects usually have to be accompanied by some form of bid security, such as a bond or a check. In private negotiations, bid security is seldom requested by owners or designers, the reason being that the bidding contractors are well known to the owners and their designers. Contractors considering making a bid on any project usually engage in some form of bid strategy, which involves assessing the chances of being successful in the bid, having regard to the names of the owners and the designers, specific competition for the project, the general state of the local construction and money markets, and the amount of work that the company already has in hand at the time of making the bid, among other considerations.

2. *Setting out specific and uniform instructions for the guidance of the bidders.* Instructions to bidders usually consist of a written list of specific directions, so that all bidders are bidding on the same basis, with respect to such elements as time and place fixed to receive bids, periods for which bid prices have to be held firm, and types of security or documentation that must accompany each bid. These instructions usually accompany a blank Bid Form, in which the blank spaces have to be filled in by each bidder, as well as copies of all the other documentation (such as drawings and specifications) proposed for the contract. The Bid Form should be arranged by the owner or designer to exactly dovetail or coincide with the proposed Form of Agreement.

3. *Awarding the contract to the most suitable or best-qualified bidder.* After reviewing all the bids, the owner usually signifies overt acceptance of the most appropriate one and rejects all the others. At that instant in time, the contract comes into being, provided that all of the elements of contract exist (these being mutual agreement, genuine intention, consideration, legal object, and capacity). Shortly afterward, the information in the Bid Form is extracted from it and entered into the Form of Agreement, which is then signed and sealed by both parties, and the work of the project can then get under way. For further information on the contents of contracts, see Section 15.4.

Within each of these primary elements, there are a number of secondary elements.

15.2.3 Secondary Elements

After measurement and pricing have been completed and the estimate summarized, with allowances for overhead and profit added in, the contractor is in position to make an offer to the owner to do the work described in the contract documents. The final estimate should be prepared in such a way as to permit the bidder to enter the correct data in each of the blank spaces of the Bid Form prepared by the owner or designer.

In some localities, subcontractors can use a system of bid depository to minimize the possibility of their bids or quotations being peddled around town for keener prices. Some owners will insist that the bidding contractors submit a bid bond or certified check along with their bids, to restrict the urge of the low bidder to withdraw a bid. There are also about as many possibilities for irregular bidding as there are owners and contractors, so it is not uncommon to encounter bids that are very low or very high, or late, or incomplete, or identical, or deficient in some way. Policy should be developed to handle such events.

There are also many good reasons for rejecting either some or all bids. They may all be over budget, there may be some restrictions or qualifications placed on some bids but not others, there may be too few bids to make a realistic choice, the owner may want to make some last-minute changes, and so on. However, the intention of calling for bids is usually to award a contract to one or other of the bidders, and it is on that expectation that most contractors will take the time and trouble to prepare a bid. If all bids are rejected, there should be a good and obvious reason, if damage to the reputation of the owner and designer is to be avoided.

One key issue for the owner and the designer to bear in mind is that the *number of variables* in the requested bid data should be kept to a minimum. If a lump-sum contract is proposed, one variable will naturally be *price*; if five contractors bid, one might expect five different prices. Everything else being equal, the contract would simply be awarded to the low bidder. If each contractor is asked in the Bid Form how long he will take to complete the contract, a second variable enters— *time*. The owner is now faced with 25 possible combinations of price and time. The low bidder may take longer to complete than the second low bidder, and the third bidder might offer to finish the job in half the time of the second bidder, which might not be to the owner's advantage, and so on. If the bidders are also asked to quote on a variety of alternate, separate, and unit prices in addition to the basic lump-sum price for the entire work, a situation very quickly develops where it is almost impossible for the owner or his designer to decide which is the best deal in the circumstances.

From the contractor's point of view, such variety can often be used to advantage. The total lump-sum price may be artificially depressed to create a good first impression, but some of the alternate, separate, and unit prices are fattened up a bit, the objective being to first win the bidding with the low overall price and then to make sure that these other more profitable variables come into play during the contract stage.

It might be noted that the items in the Bid Form can be listed in any sequence, although it is customary to follow the general form of a business letter, with the main part of the offer contained in the first paragraph and supporting data assembled in some logical sequence afterward. In the example shown in Figure 15-1, the first paragraph (or article) could be elaborated for a larger or more complex project, the second paragraph could have space for many more addenda (if more are anticipated), the table of unit prices could be as extensive as the designer feels necessary (in some projects such tables take up an entire page), and many more elements can be incorporated either directly or by attaching appendices, as shown in the penultimate paragraph. The last paragraph can be expanded to allow for witnesses and for the seal of the bidding company to be affixed, if this is considered to be necessary. Figure 15-2 shows the completed form ready for submission by the contractor to the owner.

15.3 COST CONTROL

15.3.1 Preamble

This section includes a short discourse on terminology appropriate to cost control and accounting, relative to construction, followed by a commentary on the control processes, and concludes with a brief explanation of planning, scheduling, and value engineering relative to construction projects.

```
                    PROPOSED OFFICE BUILDING
                        PUGET SOUND PLAZA
                         BELLINGHAM, WA.
                            BID FORM

    Bidder's Name:                                    Date:

    Address:

    To the Owners: Plaza Holdings, Inc., Bellingham.

    1. THE STIPULATED SUM
         Having carefully examined the site and the proposed
    contract documents for the above project, we hereby offer
    to furnish all labor, materials, equipment, and services
    to complete the entire work, including payment of all fees,
    taxes, and other costs for the stipulated sum of:


                                    Dollars ($              )

    2. ADDENDA
         The following addenda have been received, prior to bid:
       Number:    Date:          Number:    Date:

    3. UNIT PRICE LIST
         The following unit prices will be used to adjust the
    amount of the contract, if changes are ordered by Owners:
```

ITEM OF WORK	UNIT	DELETIONS	ADDITIONS
1. Machine Excavation	CY		
2. Hand Excavation	CY		
3. Formwork to Foundations	sf		
4. Concrete in Foundations	CY		

```
    4. SUBCONTRACTORS
         Subcontractors proposed for this Project are listed in
    Appendix "A" attached to this Bid Form.

    5. AUTHORIZATION
       Signature of Bidder:                        Title:
       Name (please print):                        Phone:

    Job # 85.04    Waterfront Designers Inc.    Page 1 of 1
```

Figure 15.1 Blank Bid Form.

15.3.2 Terminology

This aspect of building technology has a number of words and phrases that require careful consideration in order to distinguish sometimes subtle differences between them. Ten of the most common of these terms have been listed and defined below, classified into two groups, relative to their significance to designers and contractors, respectively. Some of these terms are used elsewhere in the book using the meanings presented below.

Chap. 15 / Bidding and Cost Control

```
                    PROPOSED OFFICE BUILDING
                       PUGET SOUND PLAZA
                        BELLINGHAM, WA.

                            BID FORM
```

Bidder's Name: Ravenna Contracting Incorporated   Date: 15/3/85

Address: PO Box 3870, Blaine, WA. 98542

To the Owners: Plaza Holdings, Inc., Bellingham.

1. THE STIPULATED SUM

    Having carefully examined the site and the proposed contract documents for the above project, we hereby offer to furnish all labor, materials, equipment, and services to complete the entire work, including payment of all fees, taxes, and other costs for the stipulated sum of:

    TWO HUNDRED SIXTY SEVEN THOUSAND FOUR HUNDRED AND FIFTY   Dollars ($ 267,450.00-------)

2. ADDENDA

    The following addenda have been received, prior to bid:

    Number: 01 Date: 10/3/85   Number: --- Date: ---

3. UNIT PRICE LIST

    The following unit prices will be used to adjust the amount of the contract, if changes are ordered by Owners:

ITEM OF WORK	UNIT	DELETIONS	ADDITIONS
1. Machine Excavation	CY	3.50	4.50
2. Hand Excavation	CY	5.00	6.00
3. Formwork to Foundations	sf	1.00	1.20
4. Concrete in Foundations	CY	22.50	35.50

4. SUBCONTRACTORS

    Subcontractors proposed for this Project are listed in Appendix "A" attached to this Bid Form.

5. AUTHORIZATION

    Signature of Bidder: *Lynne Macdonald*   Title: President

    Name (please print): Lynn Macdonald   Phone: 123-4567

    Job # 85.04   Waterfront Designers Inc.   Page 1 of 1

Figure 15.2 Complete Bid Form.

Designers

Cost Analysis: involves examination of data produced by various cost sources to discover significant elements of the data.

Cost benefits: indicate the worth of the project to society, and are usually measured over the economic life of the project, from design to demolition.

Cost containment: concerns measures or actions which may be taken by designers to ensure that contractors do not go over budget.

Cost planning: involves the allocation of funds by designers to cover anticipated costs of specific portions of the building.

Cost in use: those costs experienced by the owner of the building in excess of the initial land, design, and construction costs.

Contractors

Cost accounting: involves segregating and attributing costs of construction work to specific items of work or to job expenses.

Cost coding: involves the assignment of numbers, letters, or other symbols to items of cost for cost-accounting purposes.

Cost control: involves measures or actions that the contractor may take to ensure that actual costs do not exceed anticipated or estimated costs.

Cost estimating: process in which calculated unit prices are applied to measured quantities of work to predict probable total costs of a proposed construction project, in either complete or partial form.

Cost of work: sum of all direct and indirect costs incurred by the contractor during the construction of a building project under a contract.

Broadly stated, to the designer, cost *planning* is the prediction of the future, cost *containment* is the examination of the present, and cost *analysis* is the study of the past. Similarly, to the contractor, cost *estimation* is the prediction of the future, cost *control* is the examination of the present, and cost *accounting* is the study of the past.

15.3.3 Process

The most accurate estimates of the amounts of cost and work may be of little value if they are not or cannot be confirmed in practice at the job site. In most successful construction and design firms, the relationship between the estimating process and the subsequent confirmation process is very close. This process is essentially cyclical; when a construction company decides to bid on a construction project, it usually prepares an estimate, which if successful, leads to a contract in which the work is subsequently performed. This causes money to be spent to buy materials, pay wages, rent equipment, and so on, and the manner in which this money is generated and spent must be under some degree of control and subject to some accounting and analysis later. This data can then be used in the preparation of future estimates for future projects, thus completing the cycle. The data can also be used for improving operational efficiency, for charting the financial directions of the company, for rewarding particularly diligent employees, and for any number of other purposes.

The process may be simply visualized in the form of a wheel, with company management at the center hub, company policies forming the radial spokes, and company procedures, such as estimating, bidding, planning, purchasing, working, controlling, and accounting, arranged in linear fashion around the peripheral rim or circumference. As the wheel turns, the company will progress in a given direction. If any part of the hub, spokes, or rim is weak or missing, problems may be anticipated when that part of the wheel is required to sustain some load. Furthermore, if the load is not properly balanced around the rim, some imbalance will be experienced by the company in its business affairs. An approximate parallel may be drawn between procedures used by successful construction companies to *control* costs and successful design firms to *contain* costs. Many of the more enlightened architects and engineers practice cost planning of construction projects, not only on behalf of their clients, but also in their own interests and to protect and enhance their own reputations and indeed, peace of mind. As stated above, cost planning involves consideration of the allocation of amounts of money to cover the cost of the various parts of the proposed design. The process involves making comparisons of the economic effects of a variety of solutions to the various design problems that confront the designers, with a view to choosing the best, though not necessarily the cheapest, solution.

Cost containment involves the development of practical strategies to ensure that once a suitable cost plan has been determined and established, the designers, consultants, and contractors involved in the project are required to stay within the cost guidelines suggested by the adopted plan. The process consists of running cost checks during the development of working drawings, specifications, and other proposed contract documentation, of selecting the correct form of contract for the project, of putting restraints on and inducements to the contractor, of reconciling contractor estimates with designer budgets during the contract award stages, and incorporating workable and efficient change-order procedures to permit the necessary amount of variation which is inevitable on every construction project.

Cost control is a specific aspect of cost engineering which is a discipline of fairly recent origin, one that is concerned with the utilization of engineering judgment and experience in the application of scientific principles of cost. There are two general objectives for cost control: to provide management with *timely* and *accurate* cost information. There are three particular objectives: to complete the project within the predicted *time, cost, and quality* standards set by the contract.

There are three primary elements in a cost control program: the original *estimate*, a means of *adjusting* amounts of money, and a means of *checking* the results of work done and for taking remedial action, where necessary. There are three secondary elements in cost control: *planning, scheduling,* and *value engineering.* **Planning** is the determination of activities and events and their configurations; **scheduling** is the determination of timing and sequence of activities and events; **value engineering** is the process of prioritizing the costs of construction on a rational basis.

15.3.4 Engineering

In *planning*, the entire project is analyzed to identify all work activities and events associated with direct cost; the source of data for such analysis is the estimate. The activities and events are then organized into a logical configuration, showing items that must precede one another, as well as items that can occur simultaneously. The configuration is usually graphical, prepared either by hand or by computor. There are many forms in which plans can be presented; one uses arrows to represent activities and nodes to represent events; another uses horizontal and vertical axes, to indicate the position of activities and events, in the form of a bar chart. Flexibility and ease of modification are key issues at this stage. Activities and events can be identified by numbers, letters, combinations of both, or some other symbols, for reference and use in subsequent computations. The work of the general contractor can be shown separately from the work of subcontractors or suppliers, if this should be beneficial.

In *scheduling*, two aspects of time are examined: duration of activities and timing of events. Each activity on the plan should be assigned a duration, based on the estimate, with the intention of modifying the time allocated, if necessary. The points of time when events have to occur must also be established. Earliest and latest start and finish times for all activities are estimated on a provisional basis. Activities or events that must occur simultaneously or sequentially are given special consideration. Times are then worked through from beginning to end and from end to beginning, to discover any free or "float" time. Activities and events that show no free or float time are of critical importance, because any adjustment to such items will affect all items before and after them.

In *value engineering*, each item of the estimate is examined to decide whether the estimated costs can be reduced prior to the start of work in that item. The technique involves the application of four separate, though related processes: (1) information is gathered on crew sizes, costs, materials, taxes, techniques, time, timing, weather, and any other aspect considered germane; (2) all possible solutions are considered, using the technique of "brainstorming," in which all ideas related to solutions are presented and listed without interim value judgments as to their worth or practicality; (3) evaluation is undertaken to disclose the best solution, usually in terms of time and money, but also in terms of efficiency and progress; and (4) the preferred solution is implemented, followed by a second process to confirm that results are equal to expectations. Value engineering is most successful when a team approach is adopted, with the team consisting of the estimator, possibly a superintendent or construction manager, and a foreman or lead hand of the actual crew who will be doing the work. It is also often advantageous to include a lay person, who may have no special knowledge of construction affairs but who may bring fresh insights to possible solutions to solve the problems at hand.

15.4 CONTENTS OF CONTRACTS

15.4.1 Preamble

The topic of estimating cannot be separated from the topic of construction contracts. At the same time, the subject of construction contracts is worthy of study in its own right; to do it justice would require a separate book. Fortunately, many such books are already available, the titles of some of which are listed in Appendix C. Readers are urged to make reference to a number of such books to amplify and reinforce their knowledge of construction contracts. The subject matter is so large and diverse that no single book can deal comprehensively with every issue or possibility. Similar advice applies to the study of construction specifications.

The specific content of each construction contract is naturally going to vary from all other contracts, because of the variables of parties, places, time, money, quality, complexity, and many other factors. However, a number of general issues that are common to the majority of properly arranged construction contracts can be identified, and some of the more important of these have been identified below as a guide to further study and discussion. The topics are broadly based on the subject matter to be found in the standard forms of construction contracts in common use in the United States and Canada. They are included here because reference to many of them is made at various places throughout the rest of the book.

As a general principle, it may be said that the contractor only has to do the things that are indicated by the contract documents. It therefore follows that a clear understanding of such contract documents is of considerable importance, not only to discern what has to be done, but also to become aware of things that do not have to be done. As a general precept, it is good practice to use standard forms of construction contract wherever possible, and to minimize oral and written modifications, although some modification is *always* required. To obtain a clear understanding of the contents of any construction contract requires careful reading and correct interpretation of the owner's needs as presented by all verbal and graphical information that form the proposed contract documents.

The topics in the following list have been arranged in alphabetical order for ease of reference and to give each one equal weight in terms of importance.

15.4.2 Contents

Addenda. These are changes to the proposed contract documents issued by the designer during the bidding period, before the contract has been awarded. Although addenda may affect the estimated cost of a lump-sum contract, the amounts of any cash adjustments occasioned by such addenda are not usually disclosed by contractors who are bidding. The singular form of the word "addenda" is "addendum," following the Latin rule for such changes.

Bonds. There are four types of bonds commonly used in connection with construction contracts. The **Bid Bond** protects the owner in the event that a contractor withdraws a bid before it can be accepted by the owner. The **Performance Bond** protects the owner in the event that a contractor cannot complete his obligations to perform the contract. The **Labor and Materials Payment Bond** protects the owner in the event that a contractor neglects to pay his workers, his subcontractors, or his suppliers for work done or goods delivered to the project. The **Maintenance Bond** protects the owner if the contractor does not return to the site to rectify defects or deficiencies occurring during the contract guarantee period, which usually lasts for one year. In every case, the risk to the bonding company is limited by the amount of the bond, regardless of the type. Of the four bonds identified in this paragraph, the Performance Bond is the most frequently encountered; almost every contract requires a guarantee of performance. The cost of bonds is covered by premiums or fees, related to the amount and type of the bond.

Changes. These are amendments issued to alter the contract after construction work is under way. They usually deal with errors in the design or the initiation of new and additional work to suit changed requirements of the owner. Unlike addenda, the cost effect of a change order is almost always known to the owner before the change will be approved by the designer.

Disputes. Conflicts will occur on almost every construction project and in virtually every construction contract. There should therefore be some simple mechanism included in the terms of the contract to permit resolution of disputes. Some methods of resolution include negotiation of the differences between the parties by the parties, arbitration by impartial referees, and litigation or recourse to the courts. There are a number of techniques which can be incorporated into contracts that will minimize the likelihood of disputes arising and which minimize the gravity of disputes if and when they do arise.

Documents. The documents are the physical evidence of the existence of the contract and of its scope and nature. The documents are usually presented in two stages, first as bidding documents and then as contract documents. Both stages are essentially the same, but they are not identical. Construction documents usually consist of the Bid Form, the Agreement, the Conditions, the Specifications, and the Drawings.

Drawings. In general, the drawings should show the style and scope of the work to be done. They should be drawn (manually or electronically) to conform to industry standards with respect to graphical and symbolic conventions used to represent construction work. They should also be organized for ease of reference within themselves and with other parts of the contract documents.

Insurance. There are three main categories of insurable interests in construction contracts: first, the owner's interests, which normally cover physical damage to the work in progress and protection from third-party claims; second, the contractor's interests, which cover the risks that he is undertaking on behalf of the owner, as well as the additional risks that he faces by running his own business; and third, the designer's interests, wherein risks may arise from possible damage suits as the result of faulty design or other professional malpractice, as well as normal business risks.

Interpretation. Words and other symbols used in construction contracts usually have the customary meanings ascribed to them by trade practices in the region where the work is being done. It is therefore common to use and encounter technical jargon in construction contracts, and such meanings should be understood and agreed upon by the parties if disputes are to be avoided. In many regions, there is legislation affecting the interpretation of words used in contracts.

Law. The contract (and therefore the parties) must conform to all applicable federal and regional legislation. Arrangements should also be included for the orderly succession of responsibilities in the event of incapacity of one or other of the parties to the contract, through death, bankruptcy, or some other cause.

Parties. Normally, the owner and the contractor are the only parties to a prime construction contract. Others, such as designers, subcontractors, suppliers, inspectors, and many more, who will be involved in some way as work proceeds, should have their rights and responsibilities clarified in the contract.

Payments. A schedule should be arranged in the contract to permit periodic payments in agreed amounts to be made by the owner to the contractor. The holdback provisions of applicable lien legislation should be identified, as well as provisions to permit changes in part or all of the contract amount.

Progress. The rights of the parties to start and to stop work should be clarified. The rights of the designer, if any, to check on progress or quality of work should be specified.

Protection. Safety measures to protect the work force and the general public from danger or damage are required in most legal jurisdictions. Also, safety measures to prevent or minimize damage from theft or vandalism are normally included by prudent contract writers.

Specifications. The specification of the quality of the materials and workmanship of the various trade sections of the project usually form the bulk of the verbal or written portion of most construction contracts. Such specification sections are best formatted to conform to the recommendations of the Construction Specifications Institute (CSI, or its Canadian counterpart, CSC), by using Masterformat for organization of the Divisions of Work, and the Three-Part Format for the Sections of Work.

Time. There are should always be specific definition of the time elements in every construction contract, with particular attention being paid to commencement, termination, delays, extensions, and the meaning of the word "day."

Work. The work of any construction contract usually means the total construction of the project as specified in or implied by the contract documents. The documents should therefore be clear, comprehensive, and correct. Note that the obligations of most construction contracts do not go beyond the documentation.

15.4.3 Elements

Simply stated, the primary elements of a valid contract are mutual agreement (offer and acceptance), genuine intention (usually shown by the documentation), consideration (being the amount of value that is exchanged between the parties), legal object (agreements for illicit purposes are not usually enforcable in court), and capacity of the parties to contract.

15.5 *PARTIES TO CONTRACTS*

15.5.1 Preamble

A detailed examination of the titles and relationships of all parties to all construction contracts is beyond the scope of this book. However, to assist those who have little knowledge of this complex and interesting subject, the following diagrams have been included to show basic configurations that commonly exist between and among the various parties who become involved in such contracts.

On a first or superficial view of the construction industry, one might get the mistaken impression that it is all a well-organized and smoothly flowing thing, with all parts carefully arranged to interface accurately one with the other, and all cast or controlled in some sort of immutable frame. Nothing could be further from the truth. Intuition, innovation, inspiration, and risk taking are key characteristics leading to success in the industry. Flexibility of thought, attitude, and action are prime prerequisites for continued success. Constant change is an occupational hazard (or challenge) of the whole business.

The possibilities for permutations of the properties of personnel and resources are infinite. When one starts to consider the relationships between the parties to construction contracts, one would do well to remember that these relationships involve people and money and the reasons why people and money like to get together. That is the central issue in the industry, and that is the central fascination of the industry.

15.5.2 Configurations

Although infinite flexibility is the key aspect of these relationships, they can be organized into graphical patterns which represent the most common configurations usually encountered in most construction contractual relationships. For the purpose of discussion, two basic patterns are presented. These two patterns represent extremes or poles, the first of which is titled the Traditional Arrangement and is shown in Figure 15-3. The second pattern is titled the Management Arrangement and is shown in Figure 15-4. It should be noted that although titles are used in these patterns which suggest that separate persons are involved in the various transactions, these titles actually represent functions. More than one of these functions could in fact be performed by one person. For example, an owner could design and build a project for himself, without reference to architects, engineers, or contractors; all significant relationships might only be with suppliers in this case. A designer may own a piece of land, for which he may design a building that will be built by a contractor. And many contractors develop land on a purely speculative basis, by performing all functions, including real estate sales, all by themselves.

It is also important to realize that contractual relationships may exist between

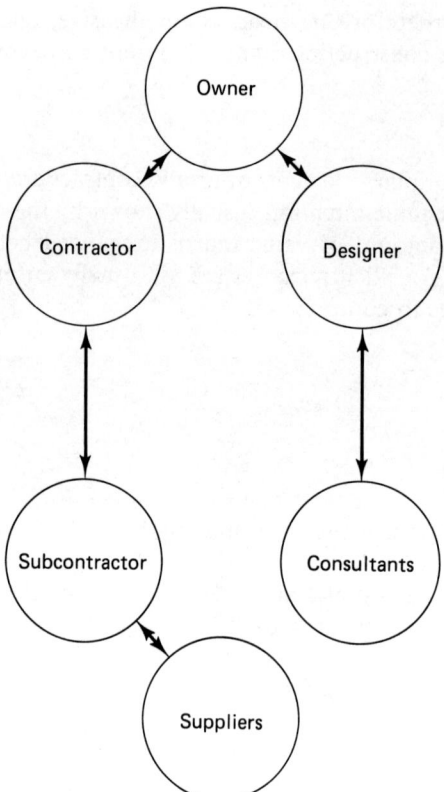

Figure 15.3 Contractual relationship: traditional arrangement.

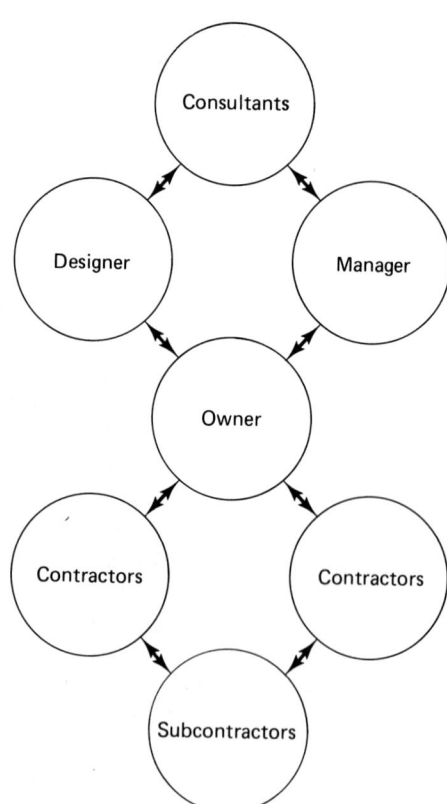

Figure 15.4 Contractual relationship: management arrangement.

some of the persons or parties shown in the patterns, but not necessarily between all of them. The exclusive relationship between parties in any construction contract is called **privity of contract**; the word "privity" has its root in the modern English word "private." It suggests a special relationship between the contracting parties, to the exclusion of all other parties and all other relationships that the contracting parties may have with other people or companies under other contracts. An example related to the homeowner might be to indicate the separation of interests between the contractual relationships which the homeowner has with, say, the telephone company for phone service, the mortgage company for home financing, the utility company for electricity or gas, the insurance company for protection, and so on.

In Figures 15-3 and 15-4, the double-headed arrows showing connections between the parties represent *contracts*, each one of which may be of a *different type*. Some contracts may be for *services* of design, whereas others may be for *implementation* of construction work or *provision* of supplies. Some contracts may be of stipulated or *lump-sum* type, whereas others may be one of several variations of the *cost-plus-fee* type. The reader should make a review of this aspect of construction contracts if the foregoing comments introduce new concepts. The reader will also realize that while only two configurations have been shown, a number of other possibilities also exist. Furthermore, in the attempt to keep the patterns simple and understandable, some peripheral relationships have been deliberately excluded; for example, most designers and contractors have contractual relationships with their employees, such as draftspersons, tradespeople, and so on, and owners have relationships with lawyers, bankers, real estate consultants, and so on; these are not shown.

QUESTIONS

15.1. Why is it unlikely that bid security would be required in negotiations between an owner and a prospective contractor to build a small private project, such as a single-story commercial store or a two story residential apartment block?

15.2. Distinguish between Instructions to Bidders and Notices to Contractors. Give examples of two items that might occur in both.

15.3. Why is it important that at least one of the bids requested for a construction project be accepted? Identify the main risk that accrues to owners or designers if they make a practice of rejecting all bids.

15.4. In the development of the bidding documentation, what is the key element for the designer to bear in mind relative to the bidding process and the subsequent award of a workable contract?

15.5. Distinguish between cost containment and cost control. Which is of more interest to the contractor?

15.6. In general terms, describe the primary obligations of the contractor in a written construction contract.

15.7. Identify and briefly describe any two of the five primary elements of contract.

15.8. Where an owner has retained a designer to act on his behalf, is it good practice to have the designer enter into a contract with the contractor? Justify your answer.

PART III
RESEARCH ASSIGNMENTS

The research assignments that follow are intended to give readers a channel through which to acquire knowledge of current local issues and aspects relative to the contents of the chapters in this part of the book. The basic objective in all of these assignments is to allow readers to compare theory with practice, by studying this and other similar books, and by encountering opinions and activities, other than those of themselves and this author, on the stated topics. A guide to content, procedure, and evaluation for these assignments has been included in Section 1.4. A review of these guidelines is recommended before starting work on the assignment projects.

CHAPTER 12: INTRODUCTION TO PRICING

Investigate the effects on construction costs brought about by building location, type, quality, and contract, respectively. Illustrate your observations with examples taken from local design offices and actual projects recently built in the region where you reside. Interview owners, designers, and contractors to gauge and report on their opinion on the significance or otherwise of the four primary elements identified for study in the first sentence.

CHAPTER 13: ELEMENTS OF PRICING

Investigate discounting practices in the region where you live. Identify and define each of the types of discounts utilized by construction suppliers and contractors and give examples of typical discounting rates for each type. Report on the frequency of occurrence of such practices and on opinions held by people who are thus involved. Include examples of actual articles or paragraphs on discounting extracted from local current construction contracts. Express a personal opinion on discounting practices.

CHAPTER 14: TECHNIQUES OF PRICING

Using the models shown in this chapter and other texts as a guide, investigate and report on the total hourly costs for an actual piece of light construction equipment, such as a concrete power buggy, a power float, or a concrete mixer, involving an operator. Compare your results with current rental rates charged by equipment companies in your region. Do not assume any values; confirm all assertions by reference and include sources of data.

CHAPTER 15: BIDDING AND COST CONTROL

Examine and report in detail on bid depository practices and procedures in Canada or the United States, depending on where you live. Describe the actual system under review, briefly outline its historical development, and give a résumé of informed local opinion both for and against the concept. Describe how such systems are funded and comment on their degree of legal status or authority. Include as an appendix to your report descriptive documentation acquired from a bid depository authority.

PART IV APPLICATIONS

CHAPTER 16
INTRODUCTION

16.1 GENERAL OBJECTIVES

16.1.1 Preamble

As stated in Part I, the scope of this book covers only the estimation of structural and architectural work required by designers and contractors. In Part IV, application of the principles and precepts of estimating is developed, first by the presentation of additional and more detailed methods and techniques of measurement and pricing of selected trades and elements; second, by the inclusion of worked examples to show the theories in practice; and third, by the opportunity to develop the skills of estimating by practice in a series of graduated exercises.

16.1.2 Materials and Methods

It was also stated in Part I that this was not a book on construction materials and methods, and that these topics would not be included, except where necessary. However, some basic explanation of the materials and methods selected for illustration of estimating theory and practice is appropriate at this juncture, to enable the reader to more fully comprehend why particular materials or methods are measured or priced in a certain manner. By engaging in such discussion, explanation, and study, a secondary objective will be achieved—increasing the reader's knowledge of specific terminology used in connection with the work of particular building trades, types, or aspects of construction work.

16.2 DISCUSSION OF SELECTED TRADES

16.2.1 Preamble

The trades selected for study are identified as follows:[1]

Masterformat	Title of trade
02200	Earthworks
03100	Concrete formwork
03300	Cast-in-place concrete
06100	Rough carpentry
09250	Gypsum wallboard

These five trades have been selected because first, their occurrence is common and fairly typical in construction projects in every region of North America; second, their general nature is relatively simple and well understood by most people in the industry; and third, they involve measurement of every type: number, length, area, volume, and mass. Understanding and mastery of these trades will assist the student toward competence in measurement of most other normal and traditional trades.

16.2.2 02200 Earthworks

The excavation process consists of a number of major components, each of which is classified and described briefly in the following list.

1. *Clearing:* removal and disposal of small trees, shrubs, and bushes, as well as minor debris lying around the site
2. *Stripping:* removal and storage or disposal of topsoil or loam
3. *Excavating:* mass or bulk digging to reduce ground elevation levels to lower levels, suitable for the building
4. *Trenching:* digging out small amounts of material to create space for foundations and service pipes, usually at a level just below that of mass or bulk excavation
5. *Trimming:* removal of minor projections or filling of minor hollows in horizontal, vertical, or sloping surfaces
6. *Protection:* provision of shoring (Figure 16.1), fencing, barricades, and other devices to prevent damage to the excavation work, the workers doing the work, and the passing public
7. *Dewatering:* removal of water from many sources, such as ice, snow, and rain, underground streams, burst pipes, and high water tables at the building site
8. *Backfill:* packing native earth or soil or other imported material into trenches, pits, and around foundation and basement walls
9. *Grading:* moving earth and soil around to produce level or contoured surfaces, to conform to the proposed grades or elevations for the new building
10. *Finishing:* final preparation of the graded surfaces ready to receive paving, landscaping, seeding or sodding, or other treatment
11. *Disposal:* transportation and dumping of excavated material in excess of the need for backfill; in some localities, burning small amounts of debris on site is permitted

---

[1]For a discussion of Masterformat, see Appendix B.

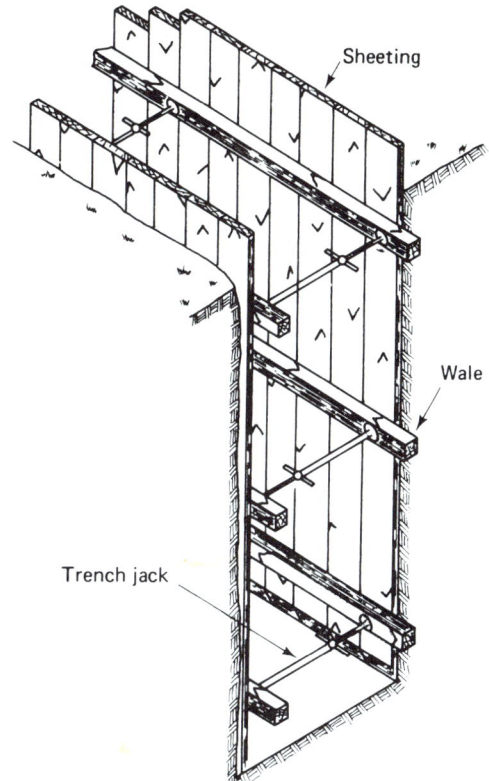

Figure 16.1 Temporary shoring. [From S. W. Nunnally, *Construction Methods and Management* (Englewood Cliffs, N.J.: Prentice-Hall, Inc., 1980).]

Preparation for excavation work which involves major demolition of existing buildings or the cutting down of large commercially valuable trees is usually handled by a contract separate from the excavation contract.

Materials to be encountered in excavation are naturally very varied and can include boulders, clay, conglomerates, glacial till, hardpan, loam, rock, sand, shale, silt, and other geologic categories. Underground services and buried debris are also frequently discovered during excavation of a site, and, on occasion, archeological artifacts.

When solid earth materials are loosened by digging, they occupy more space than before, and when loose materials are compacted, they take up less space (Figure 16.2). This phenomenon is called **swelling and shrinking,** and each category of material has its own factors. For example, if ordinary soil is known to weigh 1250 kg/m^3 in its natural solid state in the ground and is found to weigh 1000 kg/m^3 in its loose state after excavation, the swell factor is (1250/1000) 1.25, or 25%, and the shrink factor is (1000/1250) 0.80, or 20%. Note that the two factors are not the same. In contrast, solid rock may bulk up by as much as 50% but may have *no* effective shrink factor.

16.2.3   03100   Concrete Formwork

The purpose of formwork is to mold the plastic concrete into specific shapes in specific positions as required by the contract documents. Concrete formwork is, of course, not usually made of concrete, any more than plastic concrete is made of

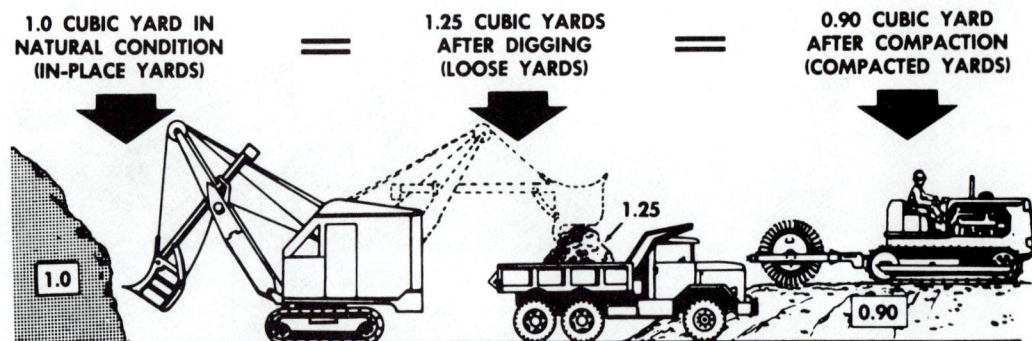

Figure 16.2 Bulkage and shrinkage. [From S. W. Nunnally, *Construction Methods and Management* (Englewood Cliffs, N.J.: Prentice-Hall, Inc., 1980).]

plastics. The processes involved in concrete formwork include design, fabrication, erection, removal, and repair, each of which is described briefly below.

Formwork *design* involves two primary considerations: hydrostatic pressure and stability. *Hydrostatic pressure* varies in proportion to the depth or thickness of the plastic concrete and to the amount of time between the placement and the initial set of the concrete. *Stability* is a function of a number of factors, such as live and dead loads, center of gravity of the formwork assembly, and type and quality of materials and workmanship.

Formwork *fabrication* consists of the selection, measurement, cutting, and assembly of components, such as plywood or fiberglass, supporting steel or lumber, and assorted pieces of hardware and equipment, into convenient units for site installation (Figures 16.3 and 16.4). Such units may be in standard panels or custom-made to suit specific features of the job.

Formwork *erection* involves placing, securing, and bracing the assembled panels in position at the site. Erection has two components: falsework and formwork. The *falsework* supports the *formwork,* which, in turn, supports and shapes the concrete work. Occasionally, standard panels have to be modified because of irregularities at the construction site.

Formwork *removal* and *repair* necessitate the careful stripping of formwork from the concrete surfaces and the disassembly of the supporting falsework. Falsework and formwork are inspected for damage, abandoned if necessary or repaired for future reuse. Reuse of formwork is one of the major considerations in the economics of the construction of concrete-framed buildings.

16.2.4 03300 Cast-in-Place Concrete

Cast-in-place concrete consists of a mixture of cement, sand, aggregate, water, and various additives to produce certain characteristics in the finished product. As the title implies, cast-in-place concrete is continuously placed into position in the building in its plastic or fluid state, as distinct from precast concrete, which is installed on a unit basis and in a solid form. The primary issues in concrete work are complete and proper hydration of the cement powder and the correct selection and proportioning of good-quality ingredients. The principal components of cast-in-place concrete work are ingredients, mixing, placing, consolidation, screeding, finishing, and curing, each of which is discussed briefly below.

Concrete *ingredients* include cement powder, sand, aggregates of gravel or other material, water, and additives to improve the quality of the final product, such as powders or liquids to improve air entrainment, alter the rate of set, impart color, waterproof the product, or for some other purpose.

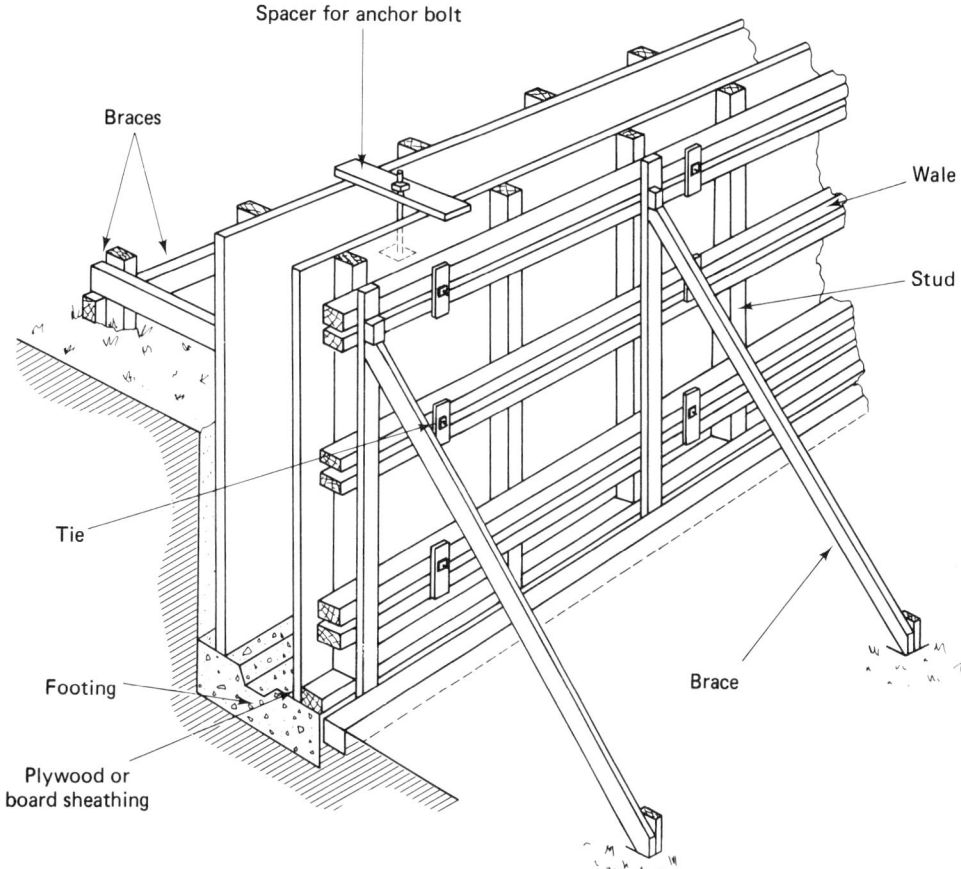

Figure 16.3 Formwork assembly. [From S. W. Nunnally, *Construction Methods and Management* (Englewood Cliffs, N.J.: Prentice-Hall, Inc., 1980).]

Concrete *mixing* involves the careful selection and measurement by weight of all ingredients relative to the proposed strength, durability, and appearance of the final product, and the uniform homogeneous integration of these ingredients in small, large, or continuous batches, either by hand or machine, on-site or off-site.

Concrete *placing* can involve wheeling, pouring, chuting, or pumping the mixture into position, depending on its consistency and final position in the building. The concrete is usually evenly spread around by hand to fill all formwork spaces.

Concrete *consolidation* involves use of hand tampers or mechanical or electrical vibrators to ensure good compaction of the concrete into the forms. Care must be taken not to drop or overwork the concrete in its plastic state and not to disturb related formwork or reinforcing steel already in position.

Concrete *screeding* is the operation that establishes the thickness of horizontal cast-in-place concrete slabs. Screeds are boards or pipes placed at intervals in a grid formation, the tops of which are located at the proposed elevations for the top surfaces of the slabs. Screeding is the act of striking off the plastic concrete at the level of the tops of the screeds.

Concrete *finishing* can involve the use of hand or machine trowels to produce smooth surfaces, hammers or brushes to produce rough or textured surfaces, or a variety of chemicals, such as hardeners or coloring agents, to produce specialty finishes. Power sandblasting and mechanical bush hammering are also used to produce architectural finishes on cast-in-place concrete surfaces.

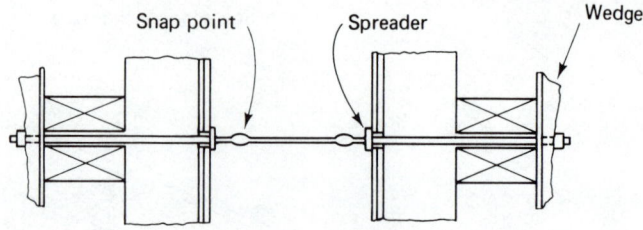

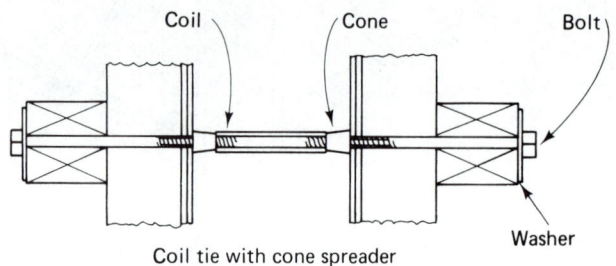

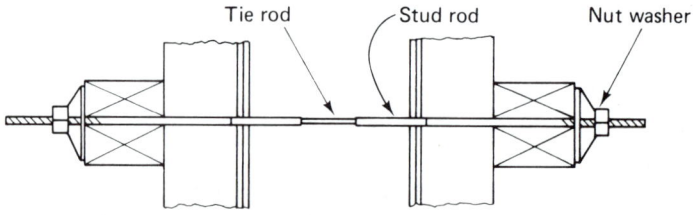

Figure 16.4 Formwork accessories. [From S. W. Nunnally, *Construction Methods and Management* (Englewood Cliffs, N.J.: Prentice-Hall, Inc., 1980).]

Concrete *curing* is probably the most critical step in concrete work, as it is intended to ensure that the design strength, durability, and appearance are achieved in the final product. Proper curing involves careful control of moisture in and around the concrete and can be achieved by the use of curing compounds, covers of plastic or burlap, sprinklers, humidifiers, or prudent use of the formwork.

16.2.5 06100 Rough Carpentry

In general, the term **rough carpentry,** as distinct from "finish carpentry," embraces all permanent *structural* woodwork and related rough hardware, which is normally *concealed* from view, throughout the building (see Figures 16.5 and 16.6). Occasionally, some structural framing is exposed to view, as for example in a post-and-beam house, an open carport, or an industrial warehouse, but such work would still be classified as rough carpentry. On small jobs, glued laminated posts and beams may be part of rough carpentry, whereas on larger projects, such specialized work may be measured and priced separately from the rest of the structural wood frame.

Rough carpentry does not include **finish carpentry,** which consists of permanent woodwork, the function of which is primarily *aesthetic,* such as baseboards,

Pumping concrete. [From S. W. Nunnally, *Construction Methods and Management* (Englewood Cliffs, N.J.: Prentice-Hall, Inc., 1980).]

Hand-floating concrete. [From R. C. Smith, *Principles and Practices of Light Construction,* 3rd ed. (Englewood Cliffs, N.J.: Prentice-Hall, Inc., 1980).]

paneling, glazed partitions, shelving, fine siding, stairs, and related trim, among other items. Furthermore, rough carpentry does not include wooden formwork for concrete work, nor does it usually include millwork, or architectural woodwork, such as cupboards and cabinets, although it may include the *installation* of such items by the general contractor. Wood doors and frames are also excluded from rough carpentry. For further detail on the scope and subdivision of carpentry work, refer to the Masterformat of the Construction Specifications Institute, identified in Appendix B.

Power-floating concrete. [From R. C. Smith, *Principles and Practices of Light Construction,* 3rd ed. (Englewood Cliffs, N.J.: Prentice-Hall, Inc., 1980).]

Rough carpentry consists of many components, of which some of the more important and common are classified below.

1. *Framing.* Wood framing consists of beams, bridging, bucks, cripples, gussets, joists, lintels, posts, purlins, rafters, ridges, studs, wall plates, and related items.
2. *Trusses.* These consist of open-web joist or rafter systems, having top and bottom chords, blockings, braces, fillers, struts, ties, and related items.
3. *Sheathing.* Walls, subfloors, roofs, and underlays are usually sheathed for structural reasons with plywood, particleboard, or some other type of panel or shiplap board.
4. *Decking.* Floors and roofs are often constructed with solid wood decking, about 50 mm or 2 in. thick or greater, usually tongued-and-grooved on edges and ends, and secured in position with large spikes in predrilled holes.
5. *Building paper.* Walls and roofs are usually lined externally with asphaltic paper, and internally with plastic vapor barriers, overlapped at edges and stapled in position.
6. *Blockings.* Rough carpentry also includes all wood bases, bearers, blockings, cants, copings, grounds, sleepers, strapping, strips for nailing, and similar items to support subsequent carpentry or other work.
7. *Insulation.* Soft batt insulation, made of fiberglass, mineral wool, or other material, installed inside wood-framed floors, walls, and ceilings or roofs, is usually included under the rough carpentry classification, together with any vapor barriers in conjunction with the insulation. Note that Masterformat places Insulation in general in Division 07, whereas Rough Carpentry is placed in Division 06.
8. *Labor items.* Cutting and drilling through floors, walls, partitions, ceilings and roofs, installing cabinets and counters, and the like.

Chap. 16 / Introduction

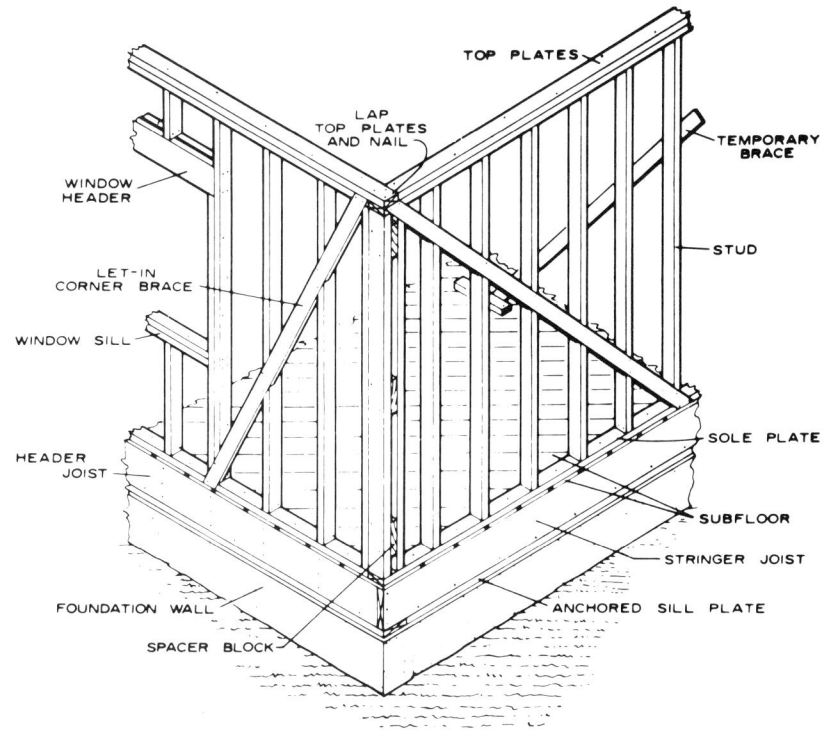

Figure 16.5 Exterior wall framing. [From S. W. Nunnally, *Construction Methods and Management* (Englewood Cliffs, N.J.: Prentice-Hall, Inc., 1980).]

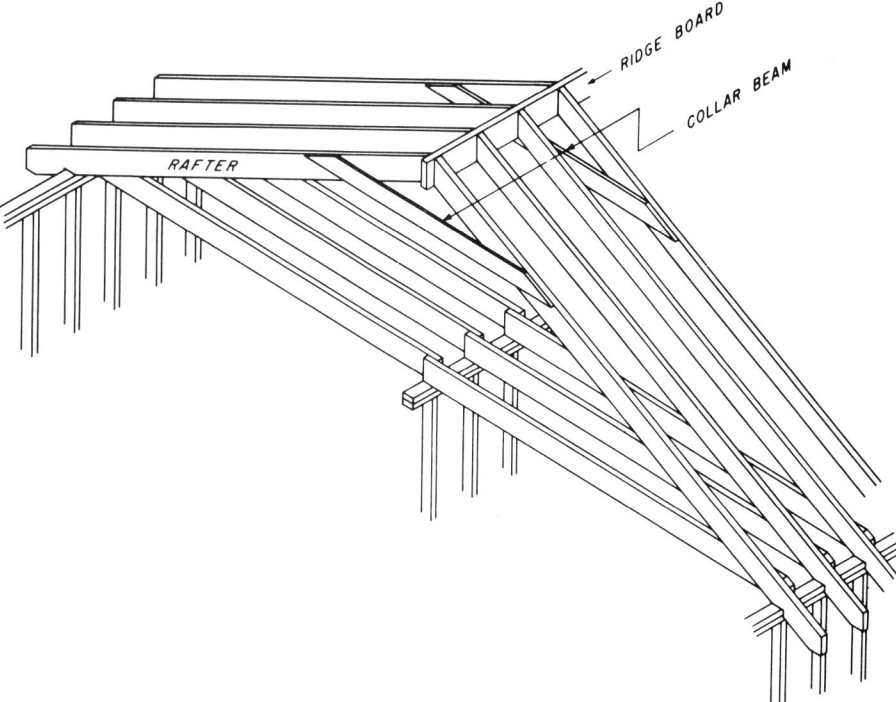

Figure 16.6 Roof and ceiling framing. [From S. W. Nunnally, *Construction Methods and Management* (Englewood Cliffs, N.J.: Prentice-Hall, Inc., 1980).]

16.2.6 09250 Gypsum Drywall

The term **drywall** applies to all categories and types of finishing boards and panels which are attached to the structural frame of the building without the use of so-called "wet" products, such as plaster (see Figure 16.7). Drywall is very common and is encountered on almost every construction project, regardless of building function, type, size, or quality. It is for this reason that drywall has been chosen to illustrate some of the principles of estimating in this book.

Gypsum drywall is one category of drywall, and as with most construction materials, products, or systems, it comes in a variety of types, thicknesses, and sizes. The gypsum that is used to make the drywall is a naturally occurring material, generically known as hydrated or hydrous calcium sulfide, having the chemical symbol $CaSO_4 \cdot 2H_2O$; it is refined and processed for use in construction. It is resistant to vermin, nonhazardous in its manufactured state, provides excellent fire resistance, and is relatively inexpensive and easy to work with, both before and after installation. The chief properties of gypsum drywall board and the main characteristics of its installation are listed briefly below. Reference to the product catalogs of the major manufacturers of gypsum products will provide more detail.

1. *Composition.* Gypsum drywall board has a central core of processed and compressed gypsum, covered on each face and the long edges with kraft or other building paper, sized or unsized. It can be factory-finished with ornamental veneers, such as sheet vinyl, aluminum foil, or other covering, on one or both faces.
2. *Dimensions.* Board thicknesses vary from 6 to 25 mm or 1/4 to 1 in. Widths vary from 60 to 120 cm or 24 to 48 in. Standard lengths vary from 2 to 5 m or 6 to 14 ft. Thicker or larger sizes are available on special order.
3. *Types.* Backing, decorative, fire-rated, insulating, regular, sheathing, and other types are available from various manufacturers. Quality is governed by ASTM, CSA, UL, and other specifications.
4. *Fastenings.* Adhesives, bars, clips, nails, screws, staples, and other devices specially made for this purpose are available.
5. *Finishing.* Cement compounds, joint fillers, kraft tape, skim coats, and other commodities can be utilized, depending on contract requirements.
6. *Configurations.* Gypsum drywall can be installed in single or double layers on a structural frame or substrate, as is often specified for simple wood- or metal-framed buildings. It can also be installed freestanding in

Figure 16.7 Taping and filling joints. [From R. C. Smith, *Principles and Practices of Light Construction,* 3rd ed. (Englewood Cliffs, N.J.: Prentice-Hall, Inc., 1980).]

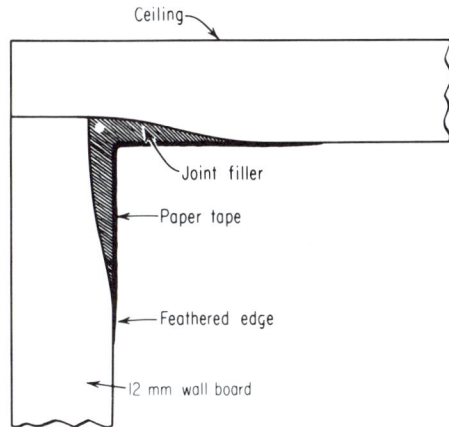

Figure 16.8 Detail: internal corner. [From R. C. Smith, *Principles and Practices of Light Construction*, 3rd ed. (Englewood Cliffs, N.J.: Prentice-Hall, Inc., 1980).]

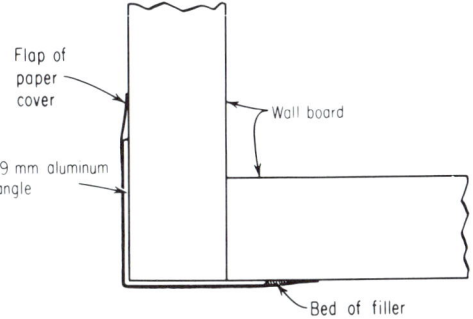

Figure 16.9 Detail: external corner. [From R. C. Smith, *Principles and Practices of Light Construction*, 3rd ed. (Englewood Cliffs, N.J.: Prentice-Hall, Inc., 1980).]

a semisolid, solid, double- or triple-solid configuration, as is occasionally encountered in commercial and institutional buildings. Gypsum boards should be installed horizontally, rather than vertically, wherever possible.

7. *Trim.* External corners, open ends and edges, and other changes of plane or cuttings are usually trimmed with galvanized steel drywall trim, pinned in position. All such trim and joints between adjacent boards are taped and filled, and are usually left ready for subsequent finishing by painting or other covering (Figures 16.8 and 16.9).

QUESTIONS

16.1. Define "mass" or "bulk" excavation. Distinguish such excavation from "trench" and "pit" excavation. Illustrate your answer with a small but clear sketch.

16.2. Identify and briefly comment on the two primary considerations in the design of formwork for concrete.

16.3. Why is hydration of cement of such importance in the mixing of concrete? Relate hydration to water/cement ratio.

16.4. Distinguish between rough carpentry and finish carpentry. Describe and illustrate two examples of each.

16.5. Describe gypsum in chemically correct terminology and give its chemical formula. List the chief physical and dimensional characteristics of standard gypsum board.

CHAPTER 17
MEASUREMENT FOR CONTRACTORS

17.1 GENERAL OBJECTIVES

17.1.1 Preamble

In this chapter, methods that can be used by *contractors* to measure the work of selected *trades* are presented first, followed in Chapter 18 by methods that can be used by *designers* to measure selected *elements* of buildings.

17.1.2 Objectives

The first purpose of this chapter is to illustrate and exemplify the specific methods, techniques, and systems of measurement discussed in Part II. This will be done by showing their application to selected construction trades by means of worked examples, based on building materials, construction methods, and drawing information contained within the book. The second purpose is to permit the student to practice applications on a series of exercises of graduated difficulty, using the illustrations as models against which work and progress can be checked.

17.2 METHODS FOR CONTRACTORS

17.2.1 Preamble

In this section, methods of net measurement of certain structural and architectural trades used by contractors have been selected and adapted from the CIQS Method of Measurement, with permission. The trades selected for measurement study correspond to those selected for the study of materials and methods in Chapter 16, and are again identified as follows:

Masterformat	Title of Trade
02200	Earthworks
03100	Concrete formwork
03300	Cast-in-place concrete
06100	Rough carpentry
09250	Gypsum wallboard

Understanding and mastery of the methods and techniques of measurement of these five trades should lead to competence in measurement of most other normal and traditional trades.

Measurements in the listed trades are stated as being one of the following: **lineal,** in meters or feet; **superficial,** in square meters or square feet; **cubical,** in cubic meters or cubic yards; or **numerical,** by enumerating items or features. A method for measuring other trades not described and discussed could easily be devised by the reader by application of the principles to be observed in the trades that have been detailed.

In general, keep work of different types, specifications, or descriptions **separate.** Measure work net in place; the description presumes that all aspects of delivery, hoisting, setting, fastening, and overlaps and waste will be accounted for in the pricing. Use either the metric or imperial measurement system; do not mix the two systems. State dimensions in order of number, length, width, and depth or height.

17.2.2 02200 Earthworks

The measurement of components of earthwork logically relates to the nature of that work. Start at the existing grade level, with clearing and stripping, and work downward, putting the largest, most important, and most expensive items in proper sequence.

1. *Clearing.* Small trees, shrubs, and bushes are particularly described and *numbered* for removal. An *item* is included for removal of minor debris; the quantity is 1. Alternatively, clearing may be generally described and measured *superficially* for large areas.

2. *Stripping:* measure *superficially,* stating the depth or thickness in the description.

3. *Excavation:* measure *cubically* for most classifications. Isolated postholes or small pits may be *numbered.*

4. *Trenching:* measure *lineally,* with cross-sectional areas stated. Final totals are expressed *cubically.*

5. *Trimming:* measure *superficially,* keeping horizontal, vertical, and sloping work separate, and separating work done by hand or by machine.

6. *Protection:* Shoring is measured *superficially* for the area of supported surfaces; fences and barricades are measured *lineally.*

7. *Dewatering:* measure *numerically,* by including an item with the quantity 1 for each category of this type of work.

8. *Backfill:* measure *cubically,* keeping different materials and types of filling separate.

9. *Grading:* measure *superficially,* stating the depth or thickness of raking or scarifying of the surface.

10. *Finishing:* measure *superficially,* describing the nature of the following or subsequent treatment. Items 9 and 10 are often combined on small jobs.

11. *Disposal:* measure *cubically* or occasionally as a *numbered* item.

Chap. 17 / Measurement For Contractors

In measurement of earthworks, no allowance is made for swelling and shrinkage; this phenomenon is included in the pricing. This simplifies the arithmetic necessary to reconcile and confirm the quantities of stripping, excavation, backfill, and disposal.

When cut and moved, each type of soil and earth adopts its own angle of repose or natural slope as a result of gravity. For example, solid rock can be cut vertically, whereas running sand will adopt a slope of approximately 45 degrees. Slope allowances can be determined for each type of soil, measured cubically, and added to the quantities excavated. Figure 17.1 illustrates the principle and the technique.

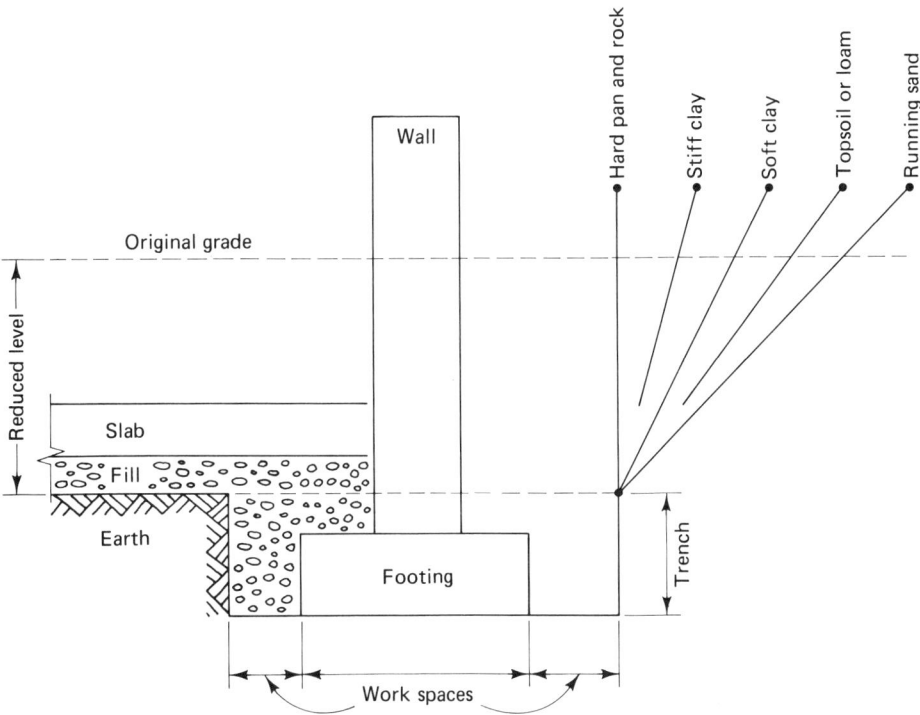

Figure 17.1 Slope allowances.

Notes on Figure 17.1: Slope Allowances

1. It is assumed that the original grade and the final grade are at the same elevation. If they are different, adjustments must be made to allow for the difference.

2. The excavation necessary to get down from the original grade level to the underside of the gravel below the lowest floor slab or skim coat is properly called *reduced-level digging*. It is also known as *mass* or *bulk excavation*, and in some cases, *basement excavation*.

When doing calculations for reduced-level excavations, it is useful to remember that in most cases, the quantity of material excavated *inside* the outside faces of the below-grade external perimeter walls has to be dug out and removed, usually off-site. The quantity of material *outside* the outside faces of such walls has to be dug out and then replaced, either with native excavated material or with different imported material. The calculations usually can be arranged to show these quantities separately for convenience.

3. Slope allowances are best calculated from the reduced-level elevation up to the original grade level at the angle appropriate to the soil type, according to the diagram.

Compacting fill. [From S. W. Nunnally, *Construction Methods and Management* (Englewood Cliffs, N.J.: Prentice-Hall, Inc., 1980).]

4. Trench and pit excavation for foundations and services usually starts at the reduced level and goes down to the bottoms of the trenches or pits. Allowance must be made for suitable working space, to accommodate formwork and workers at the bottoms of the trenches or pits. It is possible (but not usually necessary) to include a slope allowance for trenches and pits, unless they are unusually deep (exceeding 1 m or 4 ft).

5. To calculate the amount of excavation required for simple level rectangular basements, trenches, and pits, it is usually sufficiently accurate to multiply the net length, width, and depth together and add slope allowances, to produce the volume to be excavated.

To calculate the amount of excavation required for trenches and pits in situations other than a simple level rectangular configuration, it may be necessary to use the formula shown for prisms in Figure 11.3. In this case, the slope allowance can be incorporated into the basic formula.

To calculate the amount of excavation required for reduced-level excavation where the original and final grades are not level or are otherwise irregular, it may be necessary to devise a grid on tracing paper with lines at suitable spacings over the areas to be excavated and to determine the differences in "before" and "after" elevations either at the center points or at the four corners of each square of the grid pattern. In places where the terrain is more uneven than others, the axes of the grid can be at more frequent intervals than at other more uniform parts, to produce more accurate results. An example is shown in Figure 17.2.

Chap. 17 / Measurement For Contractors

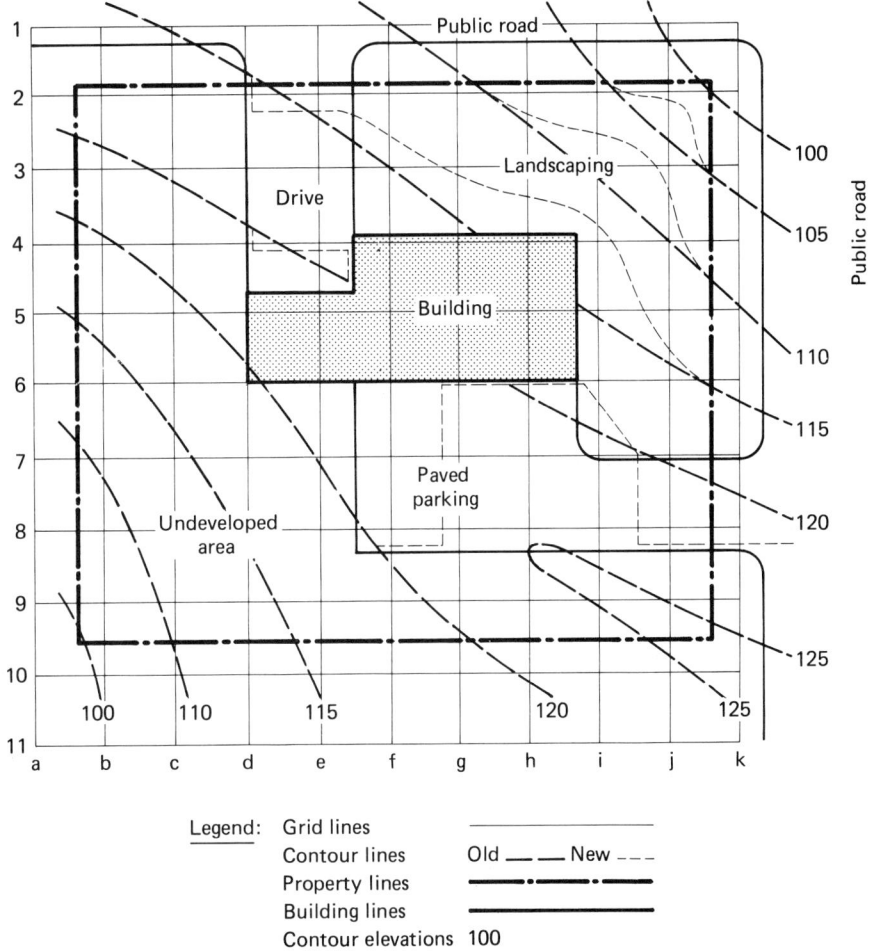

Figure 17.2 Plot plan.

The plot plan shows an example of a grid. If one examines the grid square bounded by the numbers 3 and 4 and the letters i and j, the existing and proposed ground elevations can be determined as follows, assuming that elevations are noted in feet:

Location	Existing	Proposed
3i	109.00	112.00
4i	112.00	115.00
3j	107.00	109.00
4j	110.00	111.00
Averages	109.50	111.75
		−109.50
Amount of fill		= 2.75

If the grid network lines are spaced at 10-yard intervals, the volume of earth to be filled will be 100 sy times 2.75 lf, which equals 91.67 cy for that specific grid square.

17.2.3 03100 Concrete Formwork

In general, items of formwork are measured on the basis of the area of actual surface of concrete work being formed. Start at the foundations in the main frame of the building and work upward, through walls, columns, beams, slabs, and so on.

1. *General.* Formwork is measured *superficially.* Formwork to different parts of the building is kept separate. Heights requiring unusually high falsework are separately stated.
2. *Narrow widths.* Long and narrow areas of formwork (600 mm or 24 in. or less in width) are better measured *lineally,* with the girth or width stated in the description.
3. *Deductions.* Actual voids and wants in the formwork are deducted for the area. Openings in the concrete, such as for normal doors and windows, which would not be cut out of the formwork system are *not deducted* from the area of formwork.
4. *Features.* Small features, such as inserts, holes for pipes, column bases, and capitals, and uncut openings should be *described and numbered* in the measurement for pricing.

17.2.4 03300 Cast-in-Place Concrete

Cast-in-place concrete usually has to be fairly clearly detailed on drawings and in specifications; its measurement is therefore comparatively straightforward and uncomplicated. Start at the lowest foundation level and work upward towards the roof as far as the concrete work extends.

1. *General.* All concrete items are kept separate and are usually taken off *lineally* or *superficially* and then converted to *cubical* measure to express the final total for each item.
2. *Deductions.* All voids and wants are deducted. No deduction is made for amounts of concrete displaced by inserts, such as reinforcing steel or formwork ties, under 5% of the measured volume of concrete; this is because virtually all concrete has a small amount of inserted components, and the pricing will reflect this reality.
3. *Features.* Ornamental, architectural, structural, or other features are usually measured *lineally* if less than 600 mm or 24 in. in girth, and *superficially* if greater. Small openings for pipes, vents, and the like, and inserts such as anchor bolts are usually described and *enumerated.*

Formwork in place.

Formwork removed.

Simple wall framing.

17.2.5 06100 Rough Carpentry

Rough carpentry work can often be complicated, but its measurement is usually fairly simple. Start from the lowest point and work up through the building to the roof levels.

1. *Framing:* Measure *lineally*, then convert totals to *volume*, either in cubic meters or foot board measure.
2. *Trusses:* Measure either *lineally* or *enumerated*, giving a full description.
3. *Sheathing:* Measure *superficially*; keep diagonal or sloping work separate. Deduct voids and wants.
4. *Decking:* Measure as for sheathing.
5. *Building paper:* Measure *superficially*, with overlaps added.
6. *Blockings:* Measure as for framing.
7. *Insulation:* Measure as for building paper.
8. *Labor items:* Cutting and drilling through wood floors, walls, and ceilings or roofs are *described* and *numbered*. Angle, bevel, or circular cutting at edges is measured *lineally*. Installation of cupboards, cabinets, and the like is usually measured *lineally*.

17.2.6 09250 Gypsum Drywall

Gypsum drywall measurement involves examination of the finish schedule for the project, as well as the Drawings and Specifications, to determine which surfaces of the building frame have to be so finished, and how.

1. *General.* Measurement is *superficial*, with work of various types or at higher than normal levels (above 4 m or 12 ft) being separately stated.
2. *Deductions.* There is no deduction made for openings less than 2 m² or 20 sf, such as are encountered at normal doors and windows. Such openings are described and *numbered*.
3. *Features.* Beading, circular cutting, scribing, trim, and other longitudinal features are measured *lineally*. Cutting or drilling small openings for vents, pipes, and the line are described and *enumerated*. Standard holes for electrical switch boxes and power outlets are not measured; their presence is accounted for at the pricing stage.

17.3 WORKED EXAMPLES

17.3.1 Preamble

The remaining pages of this chapter contain worked measurement examples, based on the foregoing selected methods of trade measurement, applied to parts of several sets of the project drawings included in Part VI. In every case, the examples are identified by figure number, project title, section number, and title; they consist of a typed specification checklist and one or two pages of handwritten measurement or take-off, associated with some explanatory notes intended to focus attention on particular issues of format, style, or content. Although each of these worked examples is taken out of context to some extent, reference to Chapter 5 should assist readers to put this part of the estimating process into its proper place.

Measurement exercises are shown in imperial or metric dimensions according to the system used on the respective drawings. In each case, the column rulings on the estimate sheets are slightly different, to accommodate the slight variation in the way imperial or metric dimensions are entered on such sheets. If necessary, either can be used for the other, but it is preferable to use the correct layout for the selected system. Standard metric or imperial quantity or measurement sheets are obtainable through any stationery store or bookstore, or can easily be designed and manufactured by estimating personnel for use in their own companies. Pricing or recapitulation sheets are common to both metric and imperial systems.

As a general comment, note the details inserted at the top of each page of the estimating forms and the care taken to make the measurement readable, understandable, and logical, according to the precepts enunciated elsewhere in this book. Table 17.1 shows factors that some readers may find helpful to convert dimension lumber from nominal imperial to actual metric sizes and to convert inches to decimal parts of inches.

TABLE 17.1

Conversion Factors for Softwood Lumber
Imperial to metric sizes

Nominal thickness (in.)	Actual thickness (in.)	Metric thickness (mm)
2	1½	38
2½	2	51
3	2½	64
3½	3	76
4	3½	89
4½	4	102
5	4½	114
6	5½	140
7	6½	165
8	7½	191
9	8½	216
10	9½	241
12	11½	292
14	13½	343
16	15½	394
18	17½	445
20	19½	495

Thus a 2 in. × 4 in. stud equals 38 mm × 89 mm in size.

TABLE 17.2
Inches to decimal fractions of a foot

Inch	Decimal	Inch	Decimal
1	0.08	7	0.58
2	0.17	8	0.67
3	0.25	9	0.75
4	0.33	10	0.83
5	0.42	11	0.92
6	0.50	12	1.00

17.3.2 Trades Measurement

The specific trades in specific projects which have been measured in this section are identified below for reference.

Project and section titles	Figure no.
Residential Foundation	
IMP-1 Excavation	17.3
MET-1 Concrete and Formwork	17.4
Retaining Wall	
IMP-1 Concrete and Formwork	17.5
Wood-Framed House	
MET-3 Excavation and Drains	17.6
MET-3 Concrete and Formwork	17.7
MET-3 Rough Carpentry	17.8
Post-and-Beam House	
IMP-4 Concrete and Formwork	17.9
IMP-4 Rough Carpentry	17.10
IMP-4 Finishes	17.11

Notes on Figure 17.3: Excavation Work

1. *Preliminary calculations:* It is useful to calculate the mean perimeter of the building, as this figure can be used to establish quantities of work in several items, as can be seen in this example. Also, the basic areas and depths of the primary excavation operations can be determined in advance.

2. Note that in items 1, 2, and 3, areas are determined before volumes are calculated. Extensions are rounded off to the nearest whole number up. It should be remembered that excavation work is not very precise, so some approximations are more acceptable here than in the work of some other trades, such as in glass and glazing, or storefront systems, or architectural woodwork, for example.

3. On such a small, simple job, it is likely that both the stripping of topsoil and the reduced-level digging would be carried through without adjustment for the two small recesses, resulting in some additional backfill. The estimator has to keep the whole of such processes in mind, to avoid forgetting to add back some work. A small note to this effect can be made on the drawings as a memory aid.

4. Note the use made of the mean perimeter for trench and drain lengths. Note the use made of item numbers to identify reflected dimensions.

5. Note the small calculations shown in the description columns to explain entries in the dimensions columns, with the entries made approximately opposite or level with such side calculations. In general, every entry on the form should be explained in words or figures.

6. The owner might wish the surplus topsoil to be left on the site for later use in gardening. If so, the item describing removal of the soil would have to be reworded.

7. The description of item 6 could be expanded to include providing asphalt paper over the joints of the drain tile, if this was specified.

8. The reduced-level digging on this exercise is too small or shallow to make a significant allowance for slopes or protection of cut faces. Instances of such allowances are given in later examples.

```
                    - COMPANY NAME -

SPECIFICATION CHECKLIST FOR ESTIMATING      PAGE 1  OF 1
Project: Residential Foundation      No: IMP-1 Estimator: GMH
Section: Excavation & Drainage       No: 02200 Date: April

EXCAVATION

    Strip top soil, 6" thick
    Reduce level dig, 10" deep
    Trench Excavation, 8" deep
    Trim trench bottoms level
    Temporary shoring to sides

DRAINAGE

    Agric. drain tile, 4" diam.
    Bends & branches to ditto
    Drain gravel around ditto

FILL

    Backfill over drain gravel
    Backfill topsoil
    Remove surplus earth & topsoil
    Clean up

END
```

Figure 17.3 Excavation (IMP-1).

Chap. 17 / Measurement For Contractors

QUANTITY SHEET

PROJECT	Residential Foundation.	ESTIMATOR	GMW	ESTIMATE NO.	IMP-1
LOCATION	Vancouver.	EXTENSIONS	LFW	SHEET NO.	1 of 2
ARCHITECT/ENGINEER	Karen Consultants.	CHECKED	GMW	DATE	April.
CLASSIFICATION	Excavation. 02200				

DESCRIPTION	NO.	DIMENSIONS						ESTIMATED QUANTITY	UNIT

Preliminary Calculations:

```
                              area                    depth
                              11-3              98-6    99-1   grade
mp:    49-0                   37-9              -0-9   -97-9   bed
       23-10                  49-0 x 37-10              1-4
      2)77-10 = 145-8  wkg spa: 2/12"  2-0    2-0       -0-6   top soil
  adj: 4/2/12" = -6-0          51-0   34-10             0-10
       137-8 = mp.
```

EXCAVATION:

1. Remove top soil, 6" thick:

| | area | 51-0 | 34-10 | | 1776 SF (x 6") = 888 cf = | | 33 | CY |

2. Reduced level digging, 10" deep:

| | item 1 | area | | | 1776 SF (x 10") = 1474 cf = | | 55 | CY |

3. Trench Excavation, 8" deep:

```
mp: 137-8   | 1-0
+4/2/06 4-0 | 2-0
    141-8   | 0.6   141-8   3-6     496 SF (x 8") = 332 cf =   12½  CY
```

4. Trim trench bottom level: (area - item 3)   500 SF

5. Temporary shoring at sides:

```
mp: 137-8   | 0.6
+4/2/2-0 16-0 | 0-10
    153-8   | 0-6    153-8   2-0      307
   at top:           153-8   1-0      154
                                      461 SF                   460 SF
```

Figure 17.3 (cont.)

QUANTITY SHEET

PROJECT: Residential Foundation (cont.) **ESTIMATOR:** GNM **ESTIMATE NO.** IMP-1
LOCATION: **EXTENSIONS:** **SHEET NO.** 2 of 2
ARCHITECT/ENGINEER: **CHECKED:** **DATE:** April
CLASSIFICATION: Drainage & Fill 02400.

DESCRIPTION	NO.	DIMENSIONS							ESTIMATED QUANTITY	UNIT
DRAINAGE										
6. 4" diam. agric. drain tile:		137.8								
	4/2/1.6 + 12.0	149.8			150 LF				150	LF
7. Extra for 4" bends to ditto:					10 ea				10	EA
8. Ditto for 4" branches to ditto:					4 ea				4	EA
9. Drain gravel around tile:		137.8								
	4/2/1.3 - 10.0	149.8	1.0	0.8	100 cf					
		127.8	0.6	0.8	43					
					143 cf = (1.58 cy)				2	CY
FILL										
10. Backfill over gravel, 10" deep:		137.8								
	4/2/1.0 8.0	145.8	2.0		292					
	add recesses:	11.3	8.3		93					
	(13-2=11.0)	11.0	4.11		54					
					439 SF (× 10") = 365 cf				14	CY
11. Backfill topsoil outside, 6" thick										
	item 10	area			439 SF (× 6") = 220 cf				8	CY
12. Remove surplus earth:										
	add	item 2			55 cy					
	- ddt	item 10			-14				41	CY
12. Remove surplus top soil:										
	add	item 1			33					
	- ddt	item 11			-8				25	CY
END										

Figure 17.3 (cont.)

Chap. 17 / Measurement For Contractors

Notes on Figure 17.4: Concrete Work

1. *Preliminary calculations:* The basic slab area, wall heights, and perimeter dimensions are established.

2. In item 1, the volume is calculated directly, because the formwork dimensions can easily be extracted later.

3. In item 6, the screed is a device used to determine the thickness of a proposed concrete slab, and usually consists of simple pieces of lumber or pipe sections, set out in a grid, with the upper level at the proposed elevation of the top of the slab.

4. Item 7 could be split into two items of work, one for steel troweling or finishing, and the other for curing the slab. A similar comment applies to item 10, concerning the joint and seal.

5. In item 8, the 15% for laps is not guessed but is worked out from knowing the width of the roll of vapor barrier and the width of the necessary overlaps.

6. In item 9, the compaction of the gravel under the slab is allowed for in the price, not in the measurement.

```
                    - COMPANY NAME -
SPECIFICATION CHECKLIST FOR ESTIMATING      PAGE  1 OF 1
Project: Residential Foundation    No:MET-1  Estimator:  GMH
Section: Concrete Work             No:03300  Date:     April

CONCRETE (18 mPa, 20 mm ag.)
    Footings, 200 thick
    Walls, 200 thick
    Slab on Grade, 100 thick

FORMWORK (Erect & Strip)
    Footings
    Walls
    Slab Screeds, 100 high

MISCELLANEOUS
    Gravel under Slab, 125 thick
    Poly under Slab
    Trowel & Cure Slab
    Separation Joint, 10 thick
    Keyway, 50 x 50

END
```

Figure 17.4 Concrete work (MET-1).

MEASUREMENT SHEET

JOB TITLE: Residential Foundation JOB NO.: MET-1
ESTIMATOR: GMH EXTENSIONS: LFW CHECKER: GMH PAGE NO. 1 OF 2
DATE: April SECTION NO.: 03300 SECTION TITLE: Concrete Work

ITEM	DESCRIPTION	NO.	DIMENSIONS	EXTENSION	EXTENSION		QUANTITY	UNIT
	Prelim. Calcs.			11500	7500	7500	31300	top
				3500	4000	7500	28600	btm
				15000	3500	10000	1700	
	Slab area		15000 10000		15000	15000	- 200	ftg.
	less 2/400		- 800 -800			2 x 25000	1500	ht
			14200 9200		outside peri:	50000		
				less 4/2/300		-2400		
				mean peri:		48600	m	
	Concrete (18 mPa, 20 mm ag.)							
1.	Ftgs., 200 mm thick							
	m.p.		48600 600 200	5.83 m³			5.83	m³
2.	Fnd'n wall, 200 mm thick							
	m.p.		48600 x 1500	72.90 m² (x 200)			14.58	m³
3.	Slab on grade, 100 mm thick							
			14200 9200	130.64 m²				
	ddt. wng.							
	2/400 = 800							
	+ 4000		4800 1500	-7.20				
				123.44 m² (x 100)			12.35	m³

Figure 17.4 (cont.)

MEASUREMENT SHEET

JOB TITLE: Res. Fndn. JOB NO.: MET-1
ESTIMATOR: Gum EXTENSIONS: CHECKER: PAGE NO. 2 OF 2
DATE: SECTION NO.: 03200 SECTION TITLE: Formwork

ITEM	DESCRIPTION	NO.	DIMENSIONS	EXTENSION	EXTENSION		QUANTITY	UNIT

Form Work (Erect & Strip):

4. Ftgs.
   item 1   2 48600 × 200   19.44 m²   19.44 m²

5. Fndn Wall.
   item 2   272900 m²   145.80 m²   145.80 m²

6. Screeds for slab, 100 mm high — (item 3)   123.44 m²

7. Keyway, 50×50 (mean peri.)   48.60 m

   Miscellaneous Work:

8. Steel trowel & cure slab (area)   123.44 m²

9. W/proof membrane u/slab
   area   123.44 m²
   edges   +15%   18.60
                  142.04 m²   142.00 m²

10. Gravel under slab, 125 mm thick
    area   123.44 m² (× 125)   16.50 m³

11. Separation joint and seal.
    nP   48600   48.60 m
    less 4/2 100 − 0.80
                   47.40 m   47.40 m

    END

Figure 17.4 (cont.)

Notes on Figure 17.5: Concrete Work

1. *Preliminary calculations:* The overall length of the wall and its footing are first established. Next, the drops and elevations at the ends and intermediate steps are determined.

2. Note the small side calculations to explain dimensions in item 1, as well as the use of item numbers to identify reflected dimensions elsewhere.

3. Note the use of intermediate extension columns to show subtotals before rounding off final totals, and that the first entry of each extension shows its selected unit.

4. Although all the concrete in this wall is of the same specification, note that several items of essentially different work have been developed in the estimate, such as the wall, the footing, the keys, the buttresses, and so on.

5. In item 10, the abbreviation "pythag" indicates the use of the Pythagorean theorem to determine the sloping length; it would also be possible, but not as good, to scale this length from the drawing.

6. Some estimators would not include a separate item for finishing the top of such a wall, on the assumption that in placing the concrete, the wall could be struck off neatly enough at the top of the forms. It really depends on what the specification for the work requires.

```
              - COMPANY NAME -
SPECIFICATION CHECKLIST FOR ESTIMATING      PAGE  1 OF 1
Project: Retaining Wall          No: IMP-2  Estimator: GMH
Section: Concrete Work           No: 03300  Date: April

CONCRETE

    Main wall, 20" thick
    Main footing, 20" thick
    Main key, 8 x 12 overall
    Buttresses, 20" wide, sloped

FORMWORK

    Main wall
    Main footing
    Lower key
    Keyway
    Buttresses
    Sloping faces

MISCELLANEOUS

    Finish top of wall
    Clean up

END
```

Figure 17.5 Concrete work (IMP-2).

Chap. 17 / Measurement For Contractors

QUANTITY SHEET

PROJECT: Retaining Wall
LOCATION: Vancouver
ARCHITECT/ENGINEER: Lindsay Eng. Inc.
CLASSIFICATION: Concrete Work 03300
ESTIMATOR: GMW
EXTENSIONS: LFH
CHECKED: GMW
ESTIMATE NO. IMP-2
SHEET NO. 1 of 2
DATE: April

Preliminary Calculations:

		drops	elev."	wall	drops
		2·6	upper 9·2	top 15·5	1·8
wall: 49·3		2·6	−drops 6·8	btm 2·6	+2·6
+ 2/10" 1·8		1·8	lower 2·6	ht. 12·11	4·2
ftg length: 50·11		6·8 ✓			+2·6
					6·8 ✓

CONCRETE, (3000 psi, 3/4" agg.)

1. Main Wall, 20" thick:

		49·3 ×	12·11	637 SF	
ddt. top:		6·7	5·7	37	
"	½	26·3	5·7	74	
15·5 18·11 btm:		14·9	1·8	25	
−9·10 −0·10 "		9·10	4·4	43	
5·7 18·1 "		18·1	6·8	121	
				937 SF (×1·8) = 1565 cf	58 CY

2. Main footing, 20" wide × 10" thick:

wall	1	50·11		51 lf	
steps	2	2·6		5	
	1	1·8		2	
				58 lf	
			height: ×1·8	97 sf	
			width: ×5·10	565 cf	21 CY

3. Main key, 8" wide × 12" deep:

item 2		50·11	1·0	51 SF (×8") = 34 cf	1½ CY

4. Buttresses, 20" wide × 60" high:

	4/½	2·6	5·0	25 SF (×1·8) = 42 cf	1½ CY

Figure 17.5 (cont.)

QUANTITY SHEET

PROJECT: Retaining Wall (cont.) **ESTIMATOR:** Gulf. **ESTIMATE NO.** IMP-2
LOCATION: **EXTENSIONS:** **SHEET NO.** 2 of 2
ARCHITECT/ENGINEER: **CHECKED:** **DATE:** April
CLASSIFICATION: FORMWORK 03200

DESCRIPTION	NO.	DIMENSIONS				ESTIMATED QUANTITY	UNIT
FORMWORK (erect and strip.)							
5. Main Wall:							
item 1	2	940.0	SF	1880	SF		
ends		6.0					
		7.4					
		13.4	1.8	23			
				1903	SF	1900	SF
6. Main footing:							
item 2	2	97.0	SF	194	SF		
ends	1+2	5.10	1.8	29			
steps	2	5.10	2.6	29			
				252	SF	250	SF
7. Main lower key:							
item 3	2	51.0	SF	102	SF		
ends	2	1.0	0.8	2			
steps	3	1.0	0.8	2			
				106	SF	106	SF
8. Keyway, 4" wide × 8" deep:							
wall.		49.3	LF.	50	LF	50	LF
9. Buttresses, (sides only.):							
item 4	2	25.0	SF	50	SF	50	SF
10. Extra over for sloping faces:							
pythag.	4	5.7	1.8	38	SF	38	SF
11. Labor only, finishing top of wall:							
wall:		50.0	1.8	84	SF	84	SF
END.							

Figure 17.5 (cont.)

Notes on Figure 17.6: Excavation Work

1. *Preliminary calculations:* In this case, the basic dimension given for the height of the basement has to be modified to determine the actual depth of the main basement bulk excavation.

2. A slope allowance in the ratio of 1:1 has been made in these calculations. As the depth is established at 2.05m, the distance the excavation has to be carried at the top, beyond the necessary working space of 600 mm, will also be 2.05 m at each side of the building.

3. The basic dimension for stripping topsoil will be the net area of the building at the grade level, plus 100 mm to clear the footings, plus the additional distance necessary to allow for a 600 mm working space, plus a 2.05 m slope allowance (at 1:1) for a minimum 2.75 m all around. This has been raised and rounded off to 3.00 m, as shown in item 1.

4. In item 2, net measurement of excavation less the slope allowance could have been used, with the allowance handled in a separate item, either as volume of earth to be dug or area of shoring to be provided. The price of each option could then have been calculated to see which is cheaper.

5. Note the side calculations in item 3, used to explain the entries in the dimension columns.

6. In items 6 and 7, the costs of the fittings will be priced for their value extra over the cost of the basic drain tile pipe. This is a common way of handling such specialty items which avoids complicating the measurement of the basic item of work—the pipe itself.

7. The drain rock in item 9 could be priced on the basis of actual measured quantity (5.73 m^3), but the material is not expensive and there will certainly be some spillage, so the total is entered as 6.00 m^3. The pricing of this item would make allowance for some spillage as well.

8. In item 10, note the use of item numbers for reflected dimensions. Also notice that dimensions are in millimeters, extensions are in meters to two decimal places, and final totals are in whole numbers, which are also used for the final balancing of totals at the end.

```
                - COMPANY NAME -
SPECIFICATION CHECKLIST FOR EXTIMATING      PAGE  1 OF 1
Project: Wood Frame House        No: MET-3  Estimator: GMH
Section: Excavation & Drainage   No: 02200  Date: April
```

EXCAVATION

    Strip topsoil, 150 thick
    Basement excavation
    Trench excavation, ftgs, drns.
    Backfill & trimming
    Sand bedding in trenches
    Remove & dispose of excess soil

DRAINAGE

    Ag. drain tile, 100 diam.
    Fittings to ditto
    Drain gravel around ditto
    Drain gravel at window wells
    Ashpalt felt over tile joints, 2.5 kg/m^2

NOTES

    Allow slopes 1:1 in excavation

END

Figure 17.6 Excavation (MET-3).

Chap. 17 / Measurement For Contractors

MEASUREMENT SHEET

JOB TITLE: Wood Frame House   JOB NO.: MET-3
ESTIMATOR: Gmrt   EXTENSIONS: LFH   CHECKER: Gmrt   PAGE NO. 1 OF 3
DATE: April   SECTION NO.: 02200   SECTION TITLE: Excavation

ITEM	DESCRIPTION	NO.	DIMENSIONS	EXTENSION	EXTENSION		QUANTITY	UNIT
	Prelim Calcs		*depth*				*length width*	
							10300 7500	
	basic	2400	1825	wt. spec	2/600		1200 1200	
	jst	184	slab 75	one side			11500 8700	
	plate	36	blma 25	other side			11500 8700	
	space	200	grvl 125	slopes	2/2050		4100 4100	
	top soil	150	2050	total			27100 24500	
		1825 (say 1825) ✓		average			13550 10750 ✓	
	EXCAVATION							
1.	Strip top soil; stock pile on site.							
	7500 10300							
	2/3000 6000 6000		16300 13500	220.05 m²				
			+150	33.01 m³			33.00	m³
2.	Bulk exc'n; incl 1:1 slopes.							
	bsmt		13550 10750	145.66 m²				
			×2050	298.61 m³			300.00	m³
3.	Trench exc'n; footings							
	10300 10300							
	7500−2/400							
	2/17800 9500	9500 500	4.75 m²					
	35600							
	+4/2/100							
	34800		39800 820	27.94				
				32.59 m²				
			×50	1.63 m³			2.00	m³
	(continued)							

Figure 17.6 (cont.)

MEASUREMENT SHEET

JOB TITLE: W.F. House JOB NO.: MET-3
ESTIMATOR: GMH EXTENSIONS: CHECKER: PAGE NO. 2 OF 3
DATE: SECTION NO.: 02400 SECTION TITLE: Drainage.

ITEM	DESCRIPTION	NO.	DIMENSIONS	EXTENSION	EXTENSION	QUANTITY	UNIT
4.	Ditto; plumbing service						
	2100						
	7000		2 5000 500	5.00 m²			
	5100 ÷ 2		× 2550	12.75 m³		13.00 m³	
	DRAINAGE						
5.	Ag. tile perimeter drain, 100 mm ⌀						
	m0		10800				
			7500				
		2	17800 − 35600				
	adjust 4/2	150	1000				
			37600				
	to street	1 5000	5000				
			42600	42.60 m		43.00 m	
6.	Ditto; bends (2 × 4 corners)			8 ea		8 ea	
7.	Ditto; branches (1 × 4 R.W.)			4 ea		4 ea	
8.	Drain rock around tile, 300 mm deep.						
	drns		37600 400 300	4.51 m³			
	windws	4	1200 500 200	0.48			
	shafts	4	350 350 1500	0.74			
				5.73 m³		6.00 m³	
9.	Hand trim bottoms of exc.ⁿ						
	bsmt		item 2	145.56 m²			
	trnch		" 3	37.59			
	plmbg		" 4	5.00			
				187.95 m²		184.00 m²	

Figure 17.6 (cont.)

Chap. 17 / Measurement For Contractors

MEASUREMENT SHEET

JOB TITLE: W.F. House (cont.) JOB NO.: MET-3
ESTIMATOR: GMcH. EXTENSIONS: CHECKER: PAGE NO. 3 OF 3
DATE: SECTION NO.: 02200 SECTION TITLE: Backfill

ITEM	DESCRIPTION	NO.	DIMENSIONS	EXTENSION	EXTENSION	QUANTITY	UNIT

BACKFILL

10. Backfill native material in trenches.
    add: bsmt   item 2   298.61 m³
         trenches   " 4   12.75
                                    311.36   311.36
    ddt: house   10300 7500 2050   158.36
         entry   1560 900 1400   1.89
         drain   item 9   5.73
         windows. 4 1200 400 400   0.68 – 172.81
                                                    138.57   139.00 m³

11. Sand fill in trenches:
         2 5000 500 500   2.50 m³   3.00 m³

12. Remove excess material off-site.
    add: total   item 2   300 m³
         trench   " 3   2
          "   " 4   13
                                 315
    ddt: fill   " 10   – 139
                                 176 m³   176.00 m³

13. Remove excess top soil off-site.
         total   item 1   33.05 m³
    ddt: hse   10300 7500 150 – 11.59
                                 21.42   21.00 m³

    END.

Figure 17.6 (cont.)

Notes on Figure 17.7: Concrete Work

1. *Preliminary calculations:* It is useful to calculate the center line perimeter and either the outside or inside perimeters.
2. The listing of items usually proceeds from the lowest level up through the structure to the upper levels. Note the standard practice of side-noting all extension entries, and the care taken to enter figures in their proper columns and to underline significant totals.
3. Note the practice of associating like items with like items: for example, main footings with entry footings, and main walls with entry walls.
4. It is common to include compacted gravel under a concrete slab as part of the concrete work in Division 3, although some estimators might include such an item under Division 2, Siteworks. The trimming of trench bottoms is also often included with this section of work, although it can be as easily measured in the excavation section, as shown in this book. The only difficulty with such allocations may arise out of attempts to arrange subcontracts for one or another portion of the work, if due consideration is not given to omissions or duplications caused by such practices.

Figure 17.7 Concrete work (MET-3).

```
           - COMPANY NAME -
SPECIFICATION CHECKLIST FOR EXTIMATING      PAGE  1 OF 1
Project: Wood Frame House          No: MET-3  Estimator: GMH
Section: Concrete Work             No: 03300  Date: April

CONCRETE

    Footings, various dimensions
    Stair footing 350 x 150
    Foundation wall, 200 thick
    Slab on Grade, 75 thick

FORMWORK

    Footings, continuous
    Fndtn Walls & Openings
    Keyways
    Slab edges

MISCELLANEOUS

    PC Step @ entry
    Gravel under slab, 125 thick
    Blinding, 25 thick
    Vapor barrier under slab
    Trowel & Cure slabs
    Clean exposed surfaces
    Asphalt paint on basement walls

DRAINAGE

    (include with Excavation)

END
```

MEASUREMENT SHEET

JOB TITLE: Wood Frame House JOB NO.: MET-3
ESTIMATOR: Gmw EXTENSIONS: LFH CHECKER: Gmw PAGE NO. 1 OF 4
DATE: April SECTION NO.: 03300 SECTION TITLE: Concrete work.

ITEM	DESCRIPTION	NO.	DIMENSIONS			EXTENSION	EXTENSION	QUANTITY	UNIT
	<u>Prelim. calcs.</u>		hght	2400		10300			
			- 1st	184		7500			
			- plt	38		17800 ×2 =	35600		o.p.
			+ flr.	75		less 4/2/100 =	800		
				2253			34800		m.p.
1.	Ftgs, contin.					area	vol		
	mp	34800	400	200		6.96	1.78		m³
	10300 - 2/300 =	9700	300	150		1.46	0.44		
						8.42	3.22 m³	3.22	m³
2.	Ditto, isolated.								
	entry	1700							
		650							
		2350	350	150		0.35	0.13 m³	0.13	m³
3.	Main Walls; 200 thick								
	area	34800		2253		78.41 m²			
	Ddt: windows	4	1000	500		-2.00			
						76.41			
	Volume		×200				15.30 m³	15.30	m³
4.	Stair Walls; 150 thick								
	area	1200							
		2 900							
		3000	150	1500		4.50	0.68 m³	0.68	m³

Figure 17.7 (cont.)

MEASUREMENT SHEET

JOB TITLE: W.F. House JOB NO.: MET-3
ESTIMATOR: Gnsh EXTENSIONS: LFH CHECKER: PAGE NO. 2 OF 4
DATE: April SECTION NO.: 03300 / 03100 SECTION TITLE: Concrete / Formwork

ITEM	DESCRIPTION	NO.	DIMENSIONS	EXTENSION	EXTENSION	QUANTITY	UNIT
5.	Slab on grade; 75 mm thick						
			10300 − 2/200 9900 7100	70.29 m^2			
			7500 − 2/200 ×75		5.27 m^3	5.27	m^3
6.	Ditto; 100 mm thick						
	entry		1500 1000 ×100	1.50 m^2	0.15 m^3	0.15	m^3
	FORMWORK (Erect & Strip)						
7.	Ftgs. contin.						
	item 1	2	8420	16.84 m^2		16.84	m^2
8.	Ftgs. isolated						
	item 2	2/	1200 = 2400				
		2/	1000 = 4000				
			6400 150	0.96 m^2		0.96	m^2
9.	Walls, main fndn.						
	item 3	2/	78410 (gross)	156.82 m^2		156.82	m^2
10.	Walls, entry						
	item 4	2/	4500	9.00 m^2		9.00	m^2
11.	Window jambs & sill; 200 mm wide						
		4/	1000	4.00			
		4/2	500	4.00		8.00	m

Figure 17.7 (cont.)

MEASUREMENT SHEET

JOB TITLE: W.F. House (cont'n) JOB NO.: MET 3
ESTIMATOR: Gary EXTENSIONS: CHECKER: PAGE NO. 3 OF 4
DATE: SECTION NO.: 03100 / 03300 SECTION TITLE: Formwork miscellaneous

ITEM	DESCRIPTION	NO.	DIMENSIONS	EXTENSION	EXTENSION	QUANTITY	UNIT
12.	Form entry slab edges; 100 mm high						
	1200 + 2/150		1700	1.70			
		2	900	1.80			
				3.50 m		3.50	m
13.	Form key in ftgs (38 × 89)						
	mp		34800	34.80			
	center		9700	9.70			
	stair		1200	1.20			
		2	900	1.80			
				47.50		47.50	m
14.	Slab screeds, 75 mm high						
	item 5			70.24 m²			
	" 6			1.50			
				71.74		71.74	m²
	MISCELLANEOUS WORK						
15.	P.C. Concrete step at stair; 1000 × 1000 × 200.					1	No
16.	Cut back ties and clean fndn walls						
	main			156.82 m²			
	entry			9.00			
				165.82 m²		165.82	m²
17.	2 cts asphalt paint; below grade walls						
	item 3		(net area)	76.41 m²		76.41	m²

Figure 17.7 (cont.)

MEASUREMENT SHEET

JOB TITLE: WF House (Cowin)   JOB NO.: MET 3
ESTIMATOR: GMB   EXTENSIONS: CHECKER: PAGE NO. 4 OF 4
DATE: SECTION NO.: 03300   SECTION TITLE: Miscellaneous

ITEM	DESCRIPTION	NO.	DIMENSIONS	EXTENSION	EXTENSION		QUANTITY	UNIT
18	Gravel under S.O.G.; 125 mm thick							
	area		item 5		70.29 m²			
	add. ftgs	33600	200	6.72	9.63			
		9700	300	2.91	60.66			
	add. entry	1200	750		0.90			
					61.56 m²			
			× 125		7.70 m³		7.70	m³
19	Sand blinding over gravel; 25 mm thick							
	area		item 18		60.66 m²			
			× 25		1.52 m³		1.52	m³
20	Poly vapor barrier under SOG; 0.15 mm thick							
	area		item 5		70.29 m²			
	laps		+10%		7.03			
					77.32 m²		77.32	m²
21	Steel trowel basement slab (item 5)						70.29	m²
22	Broom finish entry slab (item 6)						1.50	m²
23	Sack rub finish ext walls above grade							
	OP	35600	200	7.12 m²				
	add windows 4	1000	200	0.80				
				6.32 m²			6.32	m²
24	Anchor bolts; 12 mm ⌀ at 2400 centers (200 long)						20	ea
	END							

Figure 17.7 (cont.)

Notes on Figure 17.8: Carpentry Work

1. *Preliminary calculations:* There is little need for many such calculations in carpentry work. Most calculations, such as the basic perimeters and story heights, can usually be derived from earlier parts of the estimate, such as the concrete work. In this particular case, the height (or length) of the basic wall stud is determined to be 2102 mm, which is useful to know, as the standard length of such studs is usually 2310 mm. Therefore, 2310-mm studs will be used, and there will be an allowance for waste on material, when the pricing is developed.

2. Again note that each entry in the extension columns is justified by description or calculation in the side-note columns.

3. The original measurement of rough carpentry work for this building extended to about 12 handwritten pages. Only the first page has been included in this example. Using the checklist and the first page as a guide, the student should develop the rest of this measurement as an exercise in format, style, and content.

4. To assist students to check their results, the following total quantities of materials are presented as an approximate guide. Slight variations in quantities as established by different estimators are to be expected, but most acceptable results will fall within about 5% of the following totals.

Dimension lumber:	10 m^3
Sheathing plywood:	300 m^2
Sheathing paper:	110 m^2
Number of items (+ or −):	45

```
           - COMPANY NAME -
SPECIFICATION CHECKLIST FOR ESTIMATING      PAGE  1 OF  1
Project: Wood Frame House        No: MET-3  Estimator: GMH
Section: Rough Carpentry         No: 06100  Date: April
```

WALLS

    Plates @ base & top
    Studs @ 400 mm centers, 600 exterior
    Lintels
    Rough Bucks

FLOORS

    Sill plates & bolts
    Joists @ 400 mm centers
    Headers @ openings
    Cross bridging; ribbon bridging

ROOF FRAME

    Trusses @ 600 mm centers
    Gable ladder frame
    Gable ends @ 400 mm centers
    Fascia backer and board
    Soffit frame; roof vent frames

SHEATHING

    Flooring: t&g select plywood
    Walls & roof: sheathing grade ditto
    Soffits; roof vents
    Building Paper, stapled

MISCELLANEOUS

    Stair framing
    Basement strapping
    Drywall backing
    Backing & boxing for cabinets
    Girts, blockings, backings for fixtures
    Joist Hangers
    Rough hardware

END

Figure 17.8 Carpentry work (MET-3).

Chap. 17 / Measurement For Contractors

	MEASUREMENT SHEET						
JOB TITLE: Wood Frame House						JOB NO.: MET-3	
ESTIMATOR: GMH	EXTENSIONS: LFH		CHECKER: GmH			PAGE NO. 1 OF 12 *	
DATE: April	SECTION NO.: 06100		SECTION TITLE: Rough Carpentry				
ITEM	DESCRIPTION	NO.	DIMENSIONS	EXTENSION	EXTENSION		QUANTITY	UNIT

```
                                                              2400
   NOTE: All dimension lumber:-       Prelim. Calcs.          - 184 jst
         Douglas Fir, const'd grade.       3/78               - 114 plt
                                                              2102
   ─────────────────────────────────────────────────────────

   WALLS - BASEMENT.

1. Sill plates 38×89, ramset to concrete. (random lengths)
     N.S.   10300
     -2/200   400         9900              9.90 m
     2700+400+270         3470              3.41
     EW      1000
     +(960-100)           1860              1.86
      1000-100             900              0.90
                                           16.07 m                0.06 m³

2. Top plates, ditto, ditto.
       item 1    2/6070                    32.14 m                0.11 m³

3. Precut studs, ditto (2310 long) at 400 centers.
     16070÷400    41                        41    No.
       ends        3                         3
       doors      4/2                        8
       corners    4/2                        8
                                            60    No.
       studs              60×2310          138.60 m                0.47 m³

4. Lintels at openings 38×140 (2440 long)
             2/2440                         4.88 m                 0.03 m³

* Continued (see notes in text regarding Figure 15-7)
```

Figure 17.8 (cont.)

Notes on Figure 17.9: Concrete Work

1. *Preliminary calculations:* The basic building dimensions are established in a few key figures.
2. Note the technique to show half-units in total quantities; care must be taken not to write ambiguous figures in such cases.
3. As before, the troweling and curing of the concrete slab could be separated into two items, but the areas of each will remain the same.
4. In item 9, note that the quantity of gravel is net and rounded off; the compaction is allowed for in the pricing.
5. The reinforcing steel is shown measured and calculated here, simply because it appears on the drawings and because in such a small and simple project, the general contractor would, in all probability, purchase and place such an item directly, without a subcontractor becoming involved.

- COMPANY NAME -

SPECIFICATION CHECKLIST FOR ESTIMATING: PAGE 1 OF 1
Project: Post & Beam House   No: IMP-4   Estimator: GMH
Section: Concrete Work   No: 03300   Date: April

CONCRETE

    Foundation Wall, 8" thick, 2500 psi
    Slab on Grade, 4" thick, 3000 psi
    Thickening 8" thick, (ditto)

FORMWORK

    Foundation Walls
    Form recess
    Slab Edges, 4" high
    Ditto, 8" high

MISCELLANEOUS

    Steel trowel & cure slab
    Gravel under slab, 4" thick, compressed
    Slab Screeds, 4" high

REINFORCEMENT

    #5 bars in foundations
    6"x6"x6/6 WWF in slab

END

Figure 17.9 Concrete work (IMP-4).

Chap. 17 / Measurement For Contractors

QUANTITY SHEET

PROJECT	Post and Beam House	ESTIMATOR	GMH	ESTIMATE NO.	IMP-4
LOCATION	Blaine, WA.	EXTENSIONS	LFH	SHEET NO.	1 of 2
ARCHITECT/ENGINEER	Byles/Weston	CHECKED	GMH	DATE	April
CLASSIFICATION	Concrete & Formwork — 03200, 03300				

DESCRIPTION	NO.	ℓ	w	d				ESTIMATED QUANTITY	UNIT
Preliminary Calculations				64-0			_Wall Heights_		
				16-0					
		4-0	1-0	2/ 80-0	160-0		3-0		
− 2/2"		0-4	0-4	2/ 3-8	7-4		0-1½		
		3-8	0-8		0-8		0-4		
					168-0		3-5½ ≈ 3-6		
CONCRETE									
1. 8" fnd. wall (2500 psi)									
		168-0	×	3-6	588 SF (×8")	394 CF		14½	CY
2. 4" slab on grade (3000 psi)									
slab		63-6	15-6		984 SF				
entry		7-6	3-6		26				
hrth.		3-6	2-3		8				
					1018 SF (×4")	336 CF		12½	CY
3. 8" thickening (ditto)									
chmy		3-6	2-3		8 SF				
		3-4	1-0		3				
					11 SF (×8")	7 CF		7	CF
FORMWORK									
4. 8" fnd. walls (erect + strip)									
wall area	2	588-0	SF		1176 SF				
bulkhead	3	0-8	3-6		7				
					1183 SF			1183	SF
5. 2"×5½" recess					160 LF			160	LF
						(cont.)			

Figure 17.9 (cont.)

QUANTITY SHEET

PROJECT: P. and B. House
ESTIMATOR: GWW
ESTIMATE NO.: 1MP.4
LOCATION:
EXTENSIONS:
SHEET NO.: 2 of 2
ARCHITECT/ENGINEER:
CHECKED:
DATE: April
CLASSIFICATION: 03300 (cont.)

DESCRIPTION	NO.	l	w	d					ESTIMATED QUANTITY	UNIT
6. 4" slab edges, braced.										
	7-0	4-6		14-6	15 LF					
	1-6	3-6		8-0	8					
					23 LF				23	LF
7. 8" ditto, ditto.										
	2	6-9			14 LF				14	LF
MISCELLANEOUS WORK.										
8. Steel trowel and cure slab										
area		1020-0 SF			—				1020	SF
9. 4" gravel under slab (10% compaction)										
area		—			1018 SF					
add. wnt.		157-0	0-2		26					
					992 SF (×4")	328 CF			12	CY
10. Slab screeds, 4" high					area				1020	SF
REBAR.										
11. #5 str. in fnd. wlls.										
	2	170-0			340 LF (×1.043 #)				355	lbs
12. 6"×6"× 6/6 welded wire fabric.										
slab		area			1018 SF					
laps		5%			52				1070	SF
END.										

Figure 17.9 (cont.)

Notes on Figure 17.10: Rough Carpentry

1. *Preliminary calculations:* There is little point in preparing extensive preliminary calculations for this project, as there are few dimensions that relate to many parts of this particular section of work.

2. Note that the final quantities of dimension lumber are given in board feet (BF), which is a volume measure. Most reference books used by estimators show labor productivity for carpentry work in volume units, and most lumberyards quote prices for lumber in volume units, either in board feet or cubic meters.

To convert lineal feet of lumber to board feet, divide the cross sectional area of the piece in nominal inches by 12 and multiply the result by the length of the piece in feet. For example, a piece of fir joist, 2 in. thick, 10 in. deep, and 18 ft. long contains ($2 \times 10 = 20/12 = 1.67 \times 18$) 30 board feet. Similarly, twenty 2×4 studs each 12 ft. long contain ($2 \times 4 = 8 / 12 = 0.67 \times 12 = 8 \times 20$) 160 board feet.

3. In item 9, note the special use of extension columns for separating additions from deductions.

4. In item 16, the material for such blockings will almost certainly come from off-cuts from other items. Some estimators may prefer that every item with a material component should have a material quantity shown, on the basis or precept that one thing should not stand for another in estimating.

Figure 17.10 Rough carpentry work (IMP-4).

```
            - COMPANY NAME -

SPECIFICATION CHECKLIST FOR ESTIMATING      PAGE 1  OF 1
Project: Post & Beam House           No: IMP-4  Estimator: GMH
Section: Rough Carpentry             No: 06100  Date: April

All Dimension Lumber: Construction Grade Fir

EXTERIOR WALLS:

North, South: 2 x 6 framed
East, West:   2 x 4 framed
3/4" Standard Sheathing Plywood
Building Paper stapled with 4" laps
3" Mineral Rock Wool Insulation, with vapor barrier

INTERIOR WALLS:

2 x 4 Plates & Studs
Blockings at Fittings

MISCELLANEOUS:

2 x 6 Sills : take with Finish Carpentry
2 ply x 45 lb Asphalt Felt DPC under sills
Allow for cutting Sheathing to angle
Allow for rough hardware (nails, spikes)

END
```

QUANTITY SHEET

PROJECT: Post and Beam House
LOCATION: Blaine, WA.
ARCHITECT/ENGINEER: Byles/Wesson
CLASSIFICATION: Rough Carpentry Work 06100
ESTIMATOR: Gmt
EXTENSIONS: LFH
CHECKED: Gmt
ESTIMATE NO.: IMP-4
SHEET NO.: 1 of 3
DATE: April

DESCRIPTION	NO.	DIMENSIONS			ESTIMATED QUANTITY	UNIT

EXTERIOR WALLS.

1. 2×6 Wall plates, bolted at 5' o.c.:

   65·6
   −3·0
   ───
   62·6   N.   62·6   63 lf
           S.   56·6   57
   64·6   ───
   −8·0   120 lf   120 BF

2. 2×4 ditto, ditto:

   16·0
    4·0
    0·3
   ───
    20·3   E,W   2   20·3   41 lf
           add W. 1   7·0   7
                                ───
                                 34   25 BF

3. 2×6 top plates:

           N.   62·6   63 lf
           S (partial) 16·0   16
                                ───
                                 79 lf   80 BF

4. 2×4 ditto:

           singles   2   20·3   41 lf   30 BF

5. 2×6 studs in framed walls, 16" o.c. (out of 10')

   62·6 = 48+4
   1·4   52   5·0   260 lf   260 BF

6. ditto, ditto (out of 8'0")

   16·0 = 13+0
   1·4   13   4·0   52 lf
           upper: 5   3·6   18
                     2   5·0   10
                                ───
                                 80 lf   80 BF

7. 2×4 studs, ditto, ditto.

   7·6   20·3 = 17
   −·3   1·4   +4
   ───   ──
   7·3   21   2/21 7·3   305 lf   205 BF

Figure 17.10 (cont.)

Chap. 17 / Measurement For Contractors

QUANTITY SHEET

PROJECT: P & B House
LOCATION: Blaine, Wa.
ARCHITECT/ENGINEER:
CLASSIFICATION: 06100 (Cont.)
ESTIMATOR: Gmw
EXTENSIONS:
CHECKED:
ESTIMATE NO. IMP-4
SHEET NO. 2 of 3
DATE: April

DESCRIPTION	NO.	l	w	h	add	ddt.			ESTIMATED QUANTITY	UNIT
EXTERIOR (Cont)										
8. 2x4 girts fitted between studs.										
item 2.		as before			34 lf				25	BF
9. ¾" sheathing plywood on studs.										
N.		62·6	x	5·0	313 sf					
7·6 S.		16·0		4·0	64					
1·0										
0·4 E.W.	2	20·3		8·10	358					
8·10 ends	3	0·6		8·10	13					
ddt dr.		3·0		7·6	—	22				
returns	2	4·0		8·6	68					
		1·0		8·6	9					
					825					
					−22	22				
					803				800	SF
10. Asphalt Building paper on plywood.									800	SF
11. 3" Rockwool Batt insulation:										
		area			800 SF					
less projⁿ.	2	4·0	x	8·0	−64					
					736				740	SF
12. Raking cutting & waste on sheathing.										
	2	20·6			41 lf.					
	2	4·0			8					
					49				50	LF

Figure 17.10 (cont.)

QUANTITY SHEET

PROJECT: P. & B. House (cont.)
ESTIMATOR: GWW
ESTIMATE NO. IMP-4
LOCATION:
EXTENSIONS:
SHEET NO. 3 of 3
ARCHITECT/ENGINEER:
CHECKED:
DATE: April
CLASSIFICATION: 06100 (cont.)

DESCRIPTION	NO.	DIMENSIONS						ESTIMATED QUANTITY	UNIT
INTERIOR PARTITIONS									
13. 2×4 Bottom plates, ramset to conc. @ 48" o.c.									
N,S	3	10·0			30	lf			
	3	5·0			15				
		2·6			3				
EW		20·6			20				
	1·3	2·0			8				
					76			50	BF
14. Ditto top plates					76	lf		50	BF
15. Ditto studs at 16" o.c. (cut at 10'-0")									
76·0/1·4 = 58	58	8·6			493	lf			
space 14	77	8·6			655				
ends 12					1148	lf		770	BF
corners 15									
junct. 36									
77									
2×4 in flat gable frames									
3/8·6 = 25·6									
3/2·0 6·0	4	32·0			128	lf		85	BF
31·6									
16. Labor Only:									
Blockings for cabinets					35	lf			
Ditto for fixtures					22				
					57	lf		60	LF
17. 2 ply × 45 lb asphalt felt damp proof									
6" wide: (2×6 plate)					120	lf		120	LF
4" wide: (2×4 plate)					110	lf		110	LF
END.									

Figure 17.10 (cont.)

Chap. 17 / Measurement For Contractors

Notes on Figure 17.11: Finishes

1. *Preliminary calculations:* The basic wall height must be clearly established. Some small reminders are also included at this point, such as the note about the use of the Keene's cement in the kitchen and bathrooms.

2. Notice that every entry in the extension columns is explained with a side note in the description columns.

3. The original drawings for this building (prepared in 1950) showed a lath-and-plaster finish, so this finish has been measured here. Today, it is more likely that a gypsum drywall board finish would be incorporated in such a project. The only significant change would be in the item description. The quantity of work would remain the same.

```
                - COMPANY NAME -

SPECIFICATION CHECKLIST FOR ESTIMATING      PAGE 1 OF 1
Project:  Post & Beam House         No: IMP-4 Estimator: GMH
Section:  Finishes - Stucco/Plaster No: 09200 Date: April

EXTERIOR WORK

2 cts Stucco on metal lath,
   wi. ¼" white quartz dash
26 g. galv. beads and casings

INTERIOR WORK

Gypsum Lath & Plaster on Walls:
   Keene's on Kit. & Bath
    Plain on remainder
Keene's on Bath Ceiling only
26 g. galv. beads & casings
END
```

Figure 17.11. Finishes (IMP-4).

QUANTITY SHEET

PROJECT: Post & Beam House.
LOCATION: Blaine WA.
ARCHITECT/ENGINEER: Byles/Weston
CLASSIFICATION: Finishes — Stucco/Plaster   09200
ESTIMATOR: GMW
EXTENSIONS: LFH.
CHECKED: GMW
ESTIMATE NO.: IMP-4
SHEET NO.: 1 of 4
DATE: April.

DESCRIPTION	NO.	DIMENSIONS			ESTIMATED QUANTITY	UNIT
Preliminary Calcs. Wall Ht. :		5-0			north wall 5-0	
		2-6			kit. closet 6-0	
		1-0			recesses: K. and B.	
wedge :		0-2				
		8-8			(no deducts under 40 SF)	
EXTERIOR – STUCCO						
1. 2 cts stucco on metal lath, with ¼" white quartz dash.						
64.0 / +0.4 N.		64.4	×	4.8	300 SF	
4.0 panel		4.0		2.6	10	
−0.2 S.		16.0		3.10	62	
16.0 W.E.	2	20.3		6.9	355	
4.0 / 0.3 return	2	4.7		8.6	76	
5.0 / 2.6 / 8.6 N.E.	2	0.10		8.6	15	
1.0 / 0.3 −4.8 N. ends	2	0.7		3.10	5	
					825 SF	92 SY
2. 26 ga. galvanized steel casing beads						
N.	2	64.4			129 LF	
panel		9.0			–	
dr.		17.0			26	
S.	2	20.0			40	
E. dr		17.0			–	
wndw		9.4			26	
f. plc	2	5.0			10	
rec.	3	8.6			26	
					257 LF	260 LF
3. Ditto, ditto corner bead.						
all	7	8.6			60 LF	60 LF

Figure 17.11 (cont.)

Chap. 17 / Measurement For Contractors

QUANTITY SHEET

PROJECT: P & B House (Cont.) ESTIMATOR: Grant ESTIMATE NO. IMP-4
LOCATION: EXTENSIONS: SHEET NO. 2 of 4
ARCHITECT/ENGINEER: CHECKED: DATE: April
CLASSIFICATION: 09200 (Cont.)

DESCRIPTION	NO.	L	W	H	Add.	Add.			ESTIMATED QUANTITY	UNIT
INTERIOR - PLASTERWORK.										
4. 3/8" gypsum Lath, 1/4" plasterbase										
1/8" gypsum plaster finish, on walls.										
32·0 / 3·0 ① LW. N.		29·0	×	5·0		145	SF			
29·0 E.		15·8		8·8		136				
2·0 ddt. f/pl.		4·0		8·8	34	—				
2·5 S.			nil		—	—				
1·8										
1·0 closet W.		9·4		6·0		56				
2·8 lobby		6·0		8·7		52				
9·4 ② Hall N.		15·8		5·0		78				
5·2 panel		3·0		2·6		8				
0·4 closets	2/2	2·3		8·7		77				
0·6 E.W.	2	8·4		8·8		145				
1·0 Furn.	2	1·10		8·8		49				
12·6 S.		15·8		1·8		136				
8·6 ③ Bed N.		11·8		8·8		102				
12·0 E.W.	2	10·0		8·8		173				
-0·4 closet	2	2·3		1·8		39				
11·8 S.			nil			—				
④ Bed N.		15·8		8·8		136				
E.W.	2	15·8		8·8		272				
② ③ S.		4·0		8·8		34				
Closet	2	2·3		8·7		39				
						1676				
					34	—34				
						1642 SF			180	SY
5. Extra finishing ends of partitions										
exposed at fan-lights. (5" wide × 12" high)						8			8	ea

Figure 17.11 (cont.)

QUANTITY SHEET

PROJECT: P. & B. House (cont.) ESTIMATOR: Groff ESTIMATE NO. IMP-4
LOCATION: EXTENSIONS: SHEET NO. 3 of 4
ARCHITECT/ENGINEER: CHECKED: DATE: April
CLASSIFICATION: 09200 (cont.)

Description	No.	l	w	h	Add.					Estimated Quantity	Unit
PLASTERWORK (cont.)											
6. 3/4" gypsum lath, 1/2" plaster base, 1/8" keene's cement, on WALLS.											
@ Kit. W.		9-4	×	8-8	81						
wash	2	1-10		8-8	32						
Closet		5-4		6-0	32						
"	2/2	1-8		6-0	40						
S.		9-8		8-6	82						
					267 sf					30	SY
7. Ditto, on CEILINGS.											
@ Bath.		5-6	×	7-2	39 sf					5	SY
8. 26 g. galvanized steel corner beads.											
①	8	6-0			48						
②	4	8-8			35						
③	8	8-8			70						
④		nil			—						
⑤, ⑥	5	8-8			43						
					196 lf					200	LF
9. 26 g. galvanized steel casing bead.											
① N.	2	23-0			46 lf						
E.	2	24-5			49						
t/r.	2	8-0			16						
wdw		9-6			10						
dr.		21-0			21						
S.W.		nil			—						
lobby		8-6			9						
dr.		21-0			21						
(carried forw'd):					172					—	

Figure 17.11 (cont.)

QUANTITY SHEET

PROJECT: P. & B. House (Cont).   ESTIMATOR: Gown   ESTIMATE NO. MP-4
LOCATION: EXTENSIONS: SHEET NO. 4 of 4
ARCHITECT/ENGINEER: CHECKED: DATE: April
CLASSIFICATION: 09200 (Cont)

DESCRIPTION	NO.	DIMENSIONS				ESTIMATED QUANTITY	UNIT
PLASTERWORK (Cont).							
10. 26g. casing beads (cont) b/f:				172 lf.			
① w.dw.		12·0		12			
② S.	2	15·8		32			
panel	2	2·6		5			
"		3·0		3			
drs	4	21·0		14			
④ clg	2	12·8		25			
⑤ dv.		21·0		21			
parts	2	8·6		17			
⑥ dv.		21·0		21			
parts	3	8·6		26			
				418 lf.		420	LF
END							

17.4 ADDITIONAL TRADES

17.4.1 Preamble

The construction of any building usually involves the work of a number of separate or discrete trades. In a simple job, such as a small storage or dwelling building, there may be as few as 8 or 10 trades involved in the structure, the services, and the finishes. In a large complex building, such as a hospital or scientific laboratory, there could be as many as 50 trades. Although theoretically there may be no upper limit to the possible number of trades, experience suggests that there are between 65 and 70 distinct titles which may be utilized from time to time to identify the work of various trades.

In Chapter 16 and in the foregoing sections of this chapter, five commonly encountered trades were specially selected as vehicles to describe in detail aspects of their general subject matter and their particular methods of measurement. These trades were as follows:

1. 02200   Earthworks
2. 03100   Concrete formwork
3. 03300   Cast-in-place concrete
4. 06100   Rough carpentry
5. 09250   Gypsum wallboard

As was stated, these five were specially selected because they permit examination of measurement techniques in numerical, lineal, superficial, volumetric, and mass modes. If the issues raised for discussion and examination in each of these five examples have been properly understood, the general principles of measurement applicable to the remaining 60 to 65 trades likely to be encountered should be able to be derived with little difficulty.

However, it is acknowledged that some readers of this book in remote geographical areas may not have access to standard methods of measurement for all trades, beyond the ones detailed herein. Furthermore, some students and instructors find it helpful to have some guidelines by which to measure their knowledge and application of what has been learned. Therefore, recommended measurement techniques for an additional 10 commonly encountered construction trades have been included in this section in synoptic or abbreviated form for ready reference and actual practice.

There will of course be differences of opinion as to which 10 of the possible 65 or 70 trades should have been selected for review. These 10 were selected as being typical, regular, common, and representing Divisions 02 through 09 of Masterformat, which deal with normal structural and architectural aspects of buildings. Each selection is identified by its current Masterformat number, a proper title, and appears in two parts: (1) a brief description of the main features of the work usually supplied and installed by each trade, and (2) a guide to the significant measurement items and techniques.

In general, measure all classes, types, and grades of work **separately.** If the height of any work above grade is a significant factor, make sure that an appropriate statement is included in the estimate and an allowance made for scaffolding or hoisting. Allow for interim and final cleanup, preparation of shop drawings where necessary, temporary storage of materials and tools, protection of work against damage, submission of test samples, written guarantees, or other data, and for the provision of necessary temporary services relative to the work of any specific trade, among other things.

17.4.2 Selected Trades

04200 Unit Masonry

Main Features: Unit masonry work includes all concrete and cinder blockwork, clay brick and tile work, and stone bearing and veneer work, whether glazed or unglazed, such as is commonly found in bearing and partition walls, chimneys, walkways, wall facings, and the like. This work also includes items built into masonry, such as ties, plugs, flashings, dampproof courses, and expansion joints. The work may also include supply and installation of internal insulation, concrete fill, masonry caulking, and mortar.

Measurement technique: Initially, measure masonry work **superficially.** Do not deduct small openings less than 1 square meter or 10 square feet. Separate exterior or interior walls, load- or non-load-bearing partitions, glazed or nonglazed veneers, backings or facings, surface treatments, and labor items. Convert brickwork to units of 1000 bricks; convert block and other work by stating the number of block or other units; convert reinforcement to units of length or mass; do not convert stone work. Express mortar in **volume.** Measure corners, jambs, sills, lintels, string courses, and the like **lineally.** Measure work on surfaces of masonry **superficially.**

05100 Metal Framing

Main Features: Metal framing includes all preformed structural steel beams, bracing, brackets, columns, joists, lintels, plates, posts, struts, ties, trusses, and related engineered connections. The work also embraces related fasteners, such as bolts, nuts, rivets, and welds, and bracing cables where required. Open-web long-span steel joists are frequently encountered and usually included in this work.

Measurement Technique: Initially, measure structural steelwork **lineally** from centerline to centerline, and convert to weight or mass. Separate vertical, horizontal, sloping, and curved work. Measure composite or built-up trusses **lineally** along the span between the load points. Describe girts, base plates, sag rods, purlins, spacers, braces, fastenings, and other similar components in detail. Allow for threading or machining at ends of components as necessary. Measure each type of protective paint or galvanized finish **superficially** over the surface actually treated.

05500 Metal Fabrications

Main Features: Metal fabrication includes custom-made building components, such as access doors, channel and angle frames, cover plates, lintels, protection guards, shelf angles, railings, stairs, threshold plates, and so on, fabricated from steel angles, bars, channels, plates, rods, sheets, and tubing. Fastenings, fixings, and rustproofing are usually included in this work. Ornamental metalwork is usually excluded from this work.

Measurement Technique: Initially, measure components of fabricated items in **detail,** and express totals in mass or weight. Separate and subdivide composite items, such as staircases, under one heading, stating quantities for each composite part. Describe and enumerate all fixing and fastening components. Measure minor items **lineally,** such as moldings, trim, rails, and the like. Protective coatings, such as paint or galvanizing, can be described to be included for pricing with each item, or can be measured independently for separate pricing.

06200 Finish Carpentry

Main Features: Finish carpentry work includes standard wood products, such as prefinished solid wood, particleboard, or plywood sheets, and preprofiled boards and trim, such as exterior siding, fascia boards, bead and baseboards, and interior wood paneling or partitions and related trim, essentially intended for aesthetic purposes. It also includes all fastenings, adhesives, and other accessories necessary to secure such carpentry work in position. Finish carpentry specifications frequently require the installation of cabinetwork, such as counters and cupboards used in kitchens.

Measurement Technique: In general, measure work less than 300 mm or 12 in. in girth **lineally,** and work greater than these dimensions **superficially.** Measure running trim **lineally,** making no deduction for minor interruptions, such as at door or window openings. Measure stair components in **detail,** under one heading. Include for all fastenings normally associated with finish carpentry work, such as glues, pins, and finishing nails or screws, but exclude all rough carpentry items. Separate interior from exterior work. Measure installation of cabinets **lineally.**

07500 Membrane Roofing

Main Features: Membrane roofing includes felt and asphalt (tar and gravel) built-up roof systems, composition roofing such as asphalt shingles, vapor barriers on roof decks, as well as a variety of synthetic layered systems made with neoprene or butyl rubber materials. The work includes preparation of wood, concrete, steel, or other deck materials, using primers or patent adhesives. Insulation in conjunction with roofing is usually included with the work of this trade. Performance bonds and testing of completed work is usually a feature of specifications governing such installation.

Measurement Technique: Measure roofing systems **superficially;** ignore minor deductions under 1 square meter or 10 square feet in area. Fully describe the components of each system. Specify the height above grade and the substrate on which work will proceed. Separate roof insulation from roof membranes or vapor barriers, even though the quantities are identical, and separate level from sloping or curved work. Describe and **enumerate** special work around vents, skylights, hoppers, and the like. Include items for testing and bonding requirements.

07600 Flashing and Sheet Metal

Main Features: Flashing and sheet metal work includes a number of relatively thin gage or lightweight sheet metals, such as aluminum, copper, galvanized steel, lead, and zinc, installed on buildings in wide or narrow strips around the edges of roof membranes to produce weatherproof perimeter joints. The work includes fastening devices such as clips, nails, reglets, and rivets, as well as sealing products. Copings, gravel stops, gutters, pitch pockets, and scuppers are also considered to be part of the work of this trade.

Measurement Technique: Generally, measure flashing systems at cants, hips, valleys, eaves, and roof edges **lineally,** stating the various gages and girths of each system. Include fastenings devices, such as nails, screws, clips, or reglets, in the description of the item to be priced. Measure soldering or calking of joints **lineally,** if required. Describe and **enumerate** scupper boxes, hoppers, gum pockets, and the like. Coordinate this work with roofing membrane system work.

08400 Entrances and Storefronts

Main Features: Entrances and storefront work includes main frames, consisting of tubular aluminum extrusions complete with anchors, fastening and reinforcing devices, integral doors with frames, complete with stops, thresholds, and hardware, integral opening vented or jalousie units, spandrel panels, insect screens, casings, beads, and trim. This work may or may not include glass and glazing work, depending on the specification.

Measurement Technique: Measure sills, mullions, transomes, and other framing, reinforcing, or finishing members **lineally.** Enumerate and fully describe opening portions, such as doors or vents, and state their sizes. **Enumerate** anchors, joints, and fastenings. Fully describe the characteristics of each member or component, with regard to metal and finish types, cross-sectional dimensions, and accessories, such as sealants, beads, and trim. Measure spandrel, insulated, or screened panels **superficially.** Allow for cleaning. If necessary, measure glazing work as described in the next section.

08800 Glass and Glazing

Main Features: Glass and glazing work includes glass of all types, glazing materials and accessories, such as clips, pins, putty, and neoprene, or other types of beadings, mirrors and shelves made of glass, and work done on glass, such as beveling and cutting of edges, and acid etching or sandblasting of surfaces. Much of this work is now done at the window-frame factory, not at the construction job site.

Measurement Technique: Measure glass panels and labor on surfaces of same **superficially,** in rectangular panels in increments of 20 mm or 8 in. along each side. Measure irregular shapes to nearest overall incremental rectangular dimensions. Measure perimeter work such as beveling, cutting, and glazing **lineally.** Enumerate sealed insulated units, stating various dimensions. Enumerate and describe drilled or cut holes. Allow for cleaning.

09650 Resilient Flooring

Main Features: Resilient flooring work includes sheet and tile materials made of vinyl, rubber, asphalt, cork, and linoleum, including preparation of substrates, primers, adhesives and glues, vinyl and rubber base and trim, and cleaning, polishing, and waxing (if necessary) of finished surfaces.

Measurement Technique: Measure sheet, tile, and finishing work greater than 300 mm or 12 in. in girth **superficially;** measure narrow work (less than 300 mm) and trim **lineally.** Ignore deductions less than 1 square meter or 10 square feet in areas. Keep nonstandard work, such as special designs or ornamental features worked in the flooring materials, separate. Measure stair work **lineally** (across the width of the stair), stating widths, heights, breadths, and numbers of treads and risers; the total length should be the sum of all the tread widths.

09900 Painting

Main Features: Painting work includes so-called "wet" finishes, primarily required for aesthetic reasons, such as paints of many types, stains, varnishes, mastics, lacquers, and gold or silver leaf. The work usually includes preparation and priming of concrete, metal, wood and other surfaces for these finishes as well as protection and cleanup afterward. It does not include factory-applied coatings, such as are found on prefabricated items such as electrical panels, kitchen cabinets, toilet partitions, and school lockers, nor does it normally include strictly protective coatings such as are found on structural steelwork or mechanical or plumbing items.

Measurement Technique: Measure work greater than 300 mm or 12 in. girth **superficially.** Measure narrow work (less than 300 mm) and trim **lineally.** Ignore deductions for openings less than 4 square meters or 40 square feet (about the area of two sides of a normal pass door or a typical residential window). Separate all different treatments or paint systems with respect to substrate, system type, location, paint formula, finish, texture, and the like.

QUESTIONS

17.1. In excavation work, verbally explain the reason for making a slope allowance at the faces of cut surfaces. Illustrate graphically and numerically how this allowance is made, using a brief example.

17.2. State two good reasons why formwork for concrete is best measured superficially instead of measuring quantities of all of the components that comprise the formwork system.

17.3. What is the recommended sequence for measuring (or taking-off) items of concrete work from project drawings? Justify or support your answer with some simple examples.

17.4. How would you measure labor items in rough carpentry so that they can be properly priced? How is the waste on materials, which such labor items usually generate, accounted for?

17.5. Explain why in drywall construction, openings if less than 2 square meters or 20 square feet are not deducted from the gross area. Mention one possible pitfall for the estimator because of the foregoing instance of nondeduction.

CHAPTER 18
MEASUREMENT FOR DESIGNERS

18.1 GENERAL OBJECTIVES

18.1.1 Preamble

In this chapter, methods that can be used by designers to measure the work of selected *elements* are presented for review and discussion. Each selection is presented by title, inclusions, exclusions, and measurement. As in the rest of the book, the work of measuring the mechanical and electrical service elements is not detailed, as such work is normally farmed out to specialist consultants by designers. A method for measuring the service elements could easily be devised by the reader by application of the principles to be observed in the elements that have been detailed.

18.1.2 Objectives

The objectives of this chapter are first, to describe actual methods of measurement of various elements of buildings, and second, to illustrate these methods by actual exemplification, using one of the sets of project drawings included in Part VI. A third objective is to permit the student to practice measurement of elements of buildings other than those exemplified, using the illustrations as a guide.

As a general comment on technique, keep different types of elements separate; for example, if 30% of the area of the roof of a building was supported by a glued laminated structural system, and the remaining 70% was supported by a long-span steel joist system, the Roof Construction Element 2.c would be measured and priced as two items.

18.1.3 List of Elements

The following list of building elements was developed and published by the Canadian Institute of Quantity Surveyors (CIQS) and is reproduced here with permission. Although it has some limitations and indeed some omissions, the list is certainly a workable basis and an acceptable standard on which to begin to develop probable costs of any proposed building. The list is included here as a matter of convenience for quick reference; the measurement of each element in the list is explained in detail immediately following the end of the list.

1. Substructure
   a. Normal foundations
   b. Basement
   c. Special foundations
2. Structure
   a. Lowest-floor construction
   b. Upper-floor construction
   c. Roof construction
3. Exterior Cladding
   a. Roof finish
   b. Walls below ground floor
   c. Walls above ground floor
   d. Windows
   e. Exterior doors and screens
   f. Balconies and projections
4. Interior Partitions and Doors
   a. Permanent partitions and doors
   b. Movable partitions and doors
   c. Glazed partitions and doors
5. Vertical Movement
   a. Stairs
   b. Elevators and escalators
6. Interior Finishes
   a. Floor finishes
   b. Wall finishes
   c. Ceiling finishes
7. Fittings and Equipment
   a. Fittings and fixtures
   b. Equipment
8. Services
   a. Electrical
   b. Plumbing and drains
   c. Heating, ventilating, and air-conditioning
9. Site Development
   a. General
   b. Services
   c. Alterations
   d. Demolition
10. Overhead and Profit
    a. Site overhead
    b. Head-office overhead and profit
11. Contingencies
    a. Design contingency
    b. Escalation contingency
    c. Post-contract contingency

Chap. 18 / Measurement For Designers

18.2 METHODS FOR DESIGNERS

18.2.1 Substructure

a. Normal Foundations

Inclusions: Foundation walls, wall and column footings, column caps, grade beams, base plates, anchor bolts, dampproofing, and normal shoring or protection up to the top of the lowest-floor construction, including storm drain tile and excavation and backfill for such items.

Exclusions: Lowest-floor construction, excavation for floors and basements, basement walls and waterproofing thereto, and special foundations.

Measurement: Superficial. Gross horizontal area of the lowest floor to the outside faces of perimeter walls, or to the outer extent of foundations if there are columns or similar foundations isolated beyond the walls.

b. Basement

Inclusions: Mass excavation and backfill to permit construction of a basement.

Exclusions: Work included in normal foundations below basement floors, and basement walls and waterproofing to same, which are measured later.

Measurement: Cubical. Content of the basement volume to the outside faces of perimeter walls and from the underside of the lowest-floor construction up to grade level.

c. Special Foundations

Inclusions: Rock work, special shoring, engineered rafts, piling, caissons, and dewatering.

Exclusions: Minor abnormalities in normal foundations described above.

Measurement: Rockwork: *cubical;* shoring and rafts: *superficial* area of system; piling and caissons: *lineally* vertically; dewatering in one *numbered* item.

18.2.2 Structure

a. Lowest-Floor Construction

Inclusions: Slab on grade, including fill beneath and vapor or water barrier, suspended slab or subfloor over crawl space, including supporting columns, beams, and joists and dampproof skim or sealer coat on grade below, together with small sumps, pits, and joints normally associated with such floors.

Exclusions: Pony or foundation walls below such floors, all finishes on top of such floors, and machine and equipment bases.

Measurement: Superficial. Gross horizontal area of the floor as in Section 1.a.

b. Upper-Floor Construction

Inclusions: Suspended slabs and subfloors above lowest floor, including supporting beams, columns, fireproofing, joists, and recessed balconies.

Exclusions: Floor and ceiling finishes. All columns, beams, or other supports for roof systems above topmost floor. Projecting balconies.

Measurement: Superficial. Gross horizontal area of the floor as in Section 1.a. Make no deduction for openings.

c. Roof Construction

Inclusions: Beams, boards, columns, joists, purlins, trusses, slabs, sheathing, and other structural components.

Exclusions: Barge boards, cant strips, drains, fascias, flashing, gutters, insulation, rainwater leaders, roof finishes, skylights, and internal or external soffit finishes.

Measurement: Superficial. Gross horizontal or sloping area of the roof system, measured to its extremities.

18.2.3 Exterior Cladding

a. Roof Finish

Inclusions: Barge boards, cant strips, flashing, fascias, gutters, insulation, roof finishes, integral skylights, and external soffit finishes under eaves. Parapets and copings above the general roof level can be included or stated separately.

Exclusions: Drains, gutters, rainwater leaders, and internal soffit finishes.

Measurement: Superficial. Gross area of the roof as in Section 2.c.

b. Walls below Ground Floor

Inclusions: Perimeter basement walls of every type, including insulation, external parging or cement render coats, waterproofing, and integral pilasters or columns.

Exclusions: Excavation, internal finishes.

Measurement: Superficial. On outside face of wall, from top of lowest-floor construction above foundations to ground or grade level.

c. Walls above Ground Floor

Inclusions: Perimeter exterior walls of every type, including insulation, vapor barriers, construction joints, and exterior applied finishes.

Exclusions: Internal finishes, and parapets or copings above roof levels included with roof finishes.

Measurement: Superficial. On outside face of wall, from top of foundation or perimeter walls measured below ground level to top of wall for sloped roofs and top of roof for flat roofs. Deduct areas of windows and doors, except in curtain wall systems. Ignore isolated projections such as columns or sills and recessed at doors and windows.

d. Windows

Inclusions: Calking, dampcourses, finishing frames, glass, glazing, hardware, mullions, sash, screens, sills, transoms, and any other necessary components. Also concrete, stone, or other surrounds.

Exclusions: Curtain wall systems, glazed partitions, and storefront systems, where these are clearly distinguishable from windows.

Measurement: Superficial. Compare with deductions from walls, and also show window area as a percentage of wall area.

Chap. 18 / Measurement For Designers

e. Exterior Doors and Screens

Inclusions: Calking, dampcourses, hollow or solid doors (revolving, sliding, swinging), finishes, frames, hardware, operating devices, roll-ups, screens, sills, shutters, storefront systems, and subframes; also, concrete, stone, or other surrounds.

Exclusions: Curtain wall, internal glazed partitions.

Measurement: Superficial. Overall sizes in the vertical plane.

f. Balconies and Projections

Inclusions: Structure and finishes of balconies, canopies, overhangs, sunshades.

Exclusions: Structural components of balconies or overhangs included with other elements.

Measurement: Superficial. Measure the most significant plane of the element.

18.2.4 Interior Partitions and Doors

a. Permanent Partitions and Doors

Inclusions: Permanent dividing walls and partitions of every type, toilet partitions, and finished doors, frames, and hardware in walls and partitions.

Exclusions: Applied finishes to permanent walls and partitions.

Measurement: Superficial. Gross area of walls, partitions, and doors in the vertical plane. State area of doors separately.

b. Movable Partitions and Doors

Inclusions: Demountable, folding, or movable partitions of every type, and finished doors, frames, and hardware in such partitions. Related permanent structural components.

Exclusions: None.

Measurement: Superficial. Gross area of system in the vertical plane. State area of doors separately.

c. Glazed Partitions and Doors

Inclusions: Permanent interior glazed partitions of every type, internal storefront systems, borrowed light systems, and glazed vestibules, including finished doors, frames, and hardware in such partitions.

Exclusions: External doors and storefront systems adjacent to entrance vestibules.

Measurement: Superficial. Gross area of system in the vertical plane. State area of doors separately.

18.2.5 Vertical Movement

a. Stairs

Inclusions: Balustrades, finishes, floor framing, handrails, landings, risers, soffits, stringers, supports, and treads.

Exclusions: Wall enclosures. Landings included as part of floor elements.

Measurement: Lineal. Total length of center line of system, measured up the incline across risers and horizontally across integral intermediate landings.

b. Elevators and Escalators

Inclusions: Conveyors, dumbwaiters, elevators, and hoists, including entrance cabs or cars, cables and counterweights, doors and frames, guiderails, escalator treads and moving balustrades, operating machinery and enclosures, controls, and sumps, bases, pits, and related structural components.

Exclusions: Wall enclosures. Landings included as parts of floor elements. Power supply.

Measurement: Numerical. State type, overall size, total height of run, number of floors served, speed, degree of control, quality, power characteristics, and other significant features of complete installation.

18.2.6 Interior Finishes

a. Floor Finishes

Inclusions: Every type of applied finish (such as carpeting, coloring, hardeners, paints, resilient flooring, terrazzo, and specialty finishes), including floating, topping, and troweling, bases and curbs, and related features, such as inset mats and frames, cover plates, and the like.

Exclusions: Integral finishes already included with the structural element. Unfinished structural subfloors left exposed. Loose rugs and area carpets, not permanently installed.

Measurement: Superficial. Gross area in the horizontal plane.

b. Wall Finishes

Inclusions: Every type of applied finish (such as paint, plaster, stain, stucco, tile, terrazzo, and specialty finishes).

Exclusions: Integral finishes already included in the structural element (such as vinyl-covered wallboard). Unfinished structural walls left exposed. Unfinished portions of walls above ceiling lines. Loose drapes and curtains, not permanently installed.

Measurement: Superficial. Gross area in the vertical plane. State area of doors separately.

c. Ceiling Finishes

Inclusions: Every type of applied or suspended finish (as for walls), including acoustical systems, ornamental cornices, bulkheads, and the like.

Exclusions: Integral finishes already included in the structural element. Unfinished structural ceilings left exposed. Specialty illuminated electrical, mechanical, or combined installations.

Measurement: Superficial area in the horizontal plane of the building. Keep sloped, domed, dished, or vertical areas separate. Measure cornices lineally.

18.2.7 Fittings and Equipment

a. Fittings and Fixtures

Inclusions: Wood, metal, or plastic built-in cabinets, counters, cupboards, shelf units, closet trim, and similar items, appropriate to the building type.

Exclusions: Operable equipment.

Measurement: Superficial. Gross area of floor space. If sufficient detail is available, measure *lineally*.

b. Equipment

Inclusions: Permanently installed operable equipment, machinery, or devices, used in gymnasia, hospitals, kitchens, laboratories, laundries, pools, and the like.

Exclusions: Roughing-in of electrical, mechanical, plumbing, or other services for such equipment.

Measurement: Superficial. Gross area of floor space. If sufficient detail is available, *enumerate* items.

18.2.8 Services

a. Electrical

b. Plumbing and Drains

c. Heating, Ventilating, and Air-Conditioning

Note: See Section 18.1.1.

18.2.9 Site Development

a. General

Inclusions: Group 1: grading, landscaping, paving, planting, ramps, reflecting pools, seeding, sodding, terraces, and the like. *Group 2:* bumpers, canopies, curbs, fences, hedges, railings, retaining walls, stairs and steps, and the like. *Group 3:* benches, parking meters, poles, seats, signs, and the like.

Exclusions: Services in Section 9.b.

Measurement: Group 1: *superficial.* Group 2: *lineal.* Group 3: *numerical.*

b. Services

Inclusions: Provision of electric, drainage, gas, phone, water, and other services or utilities, including necessary and normally related items such as sumps, tanks, and pits with covers. Ornamental lighting and fountains.

Exclusions: General items identified in Section 9.a.

Measurement: Superficial. Gross area of the entire site.

c. Alterations

Inclusions: Minor changes to an existing building because of proposed adjacent new construction.

Exclusions: Major changes to an existing building because of proposed integral new construction.

Measurement: Measure minor alterations in *detail.* Relate cost to the gross floor area of the new construction. Include major changes with measurement of respective new elements.

d. Demolition

Inclusions: Actual destruction or removal of existing buildings, preparatory for construction of new buildings.

Exclusions: Alterations or partial dismantling of buildings to accommodate new construction work.

Measurement: Numerical. State building size and type.

18.2.10 Overhead and Profit

a. Site Overhead

Inclusions: Indirect on-site costs, such as bonding and insurance costs, cleaning, cutting and patching, ancillary and administrative equipment such as hoists, first aid and protection, excess labor expenses, permits and fees, progress photos, temporary services, site offices and access, supervision, and similar items, related to the building construction process.

Exclusions: Direct costs, such as for normal labor, materials, and production equipment already allowed for in other construction elements.

Measurement: By *percentage* of cost of the other elements.

b. Head-Office Overhead and Profit

Inclusions: Indirect off-site costs, such as business operations insurance, pension and incentive plans, public relations, head-office rental costs, employee salaries, dividends, and all related expenses and costs. Profit markup.

Exclusions: Costs attributable to actual construction work or to site overhead.

Measurement: By *percentage* of cost of the other elements. State overhead and profit separately (one represents cost, the other return on investment).

18.2.11 Contingencies

a. Design Contingency

Measurement: A *percentage* included to permit necessary design changes to be made as the project develops.

b. Escalation Contingency

Measurement: A *percentage* included to cover increases in cost between the date of the estimate and the date of the award of a contract to build.

c. Post-Contract Contingency

Measurement: A *percentage* included to permit changes to be made during construction, due to unforeseen circumstances.

18.3 WORKED EXAMPLES

18.3.1 Preamble

The following pages contain worked examples and notes, based on the foregoing selected methods of elements measurement, applied to parts of one set of the project drawings included in Part VI.

Chap. 18 / Measurement For Designers

18.3.2 Elements Measurement

The calculations in the five examples given below indicate applications of the recommended measurement techniques to selected elements of the Wood-Framed House (MET-3) to determine the quantity of such elements. The calculated quantities are later shown inserted into the Element Summary (Figure 20.1) for this project, and in Figure 20.2, where they have also been priced.

EXAMPLE 1: Element 1.b—Basement

Area:	$10300 - 2 \times 200$	$= 9900$ mm
	$7500 - 2 \times 200$	$= 7100$
	9900×7100	$= 70.29$ m^2
Depth:	$2400 - 200 + 225$	$= 2425$ mm
Volume:	70.29×2.425	$= \underline{170.18 \text{ m}^3}$

EXAMPLE 2: Element 2.a—Lowest Floor

Area:	10300×7500 mm	$= \underline{77.25 \text{ m}^2}$

EXAMPLE 3: Element 3.a—Roof Finish

Width:	$7500 + 2 \times 600$	$= 8700$ mm
Half:	8700×0.5	$= 4350$
Rise:	4350×0.33	$= 1450$
Slope:	(Pythagoras)	$= 4585$
Length:	$10300 + 2 \times 200$	$= 10700$
Area:	10700×4585	$= 49.06$ m^2
Roof:	49.06×2	$= \underline{98.12 \text{ m}^2}$

EXAMPLE 4: Element 3.c—Walls above Ground

Length:	$10300 + 7500$	$= 17800$ mm
Total:	17800×2	$= 35600$
Height:	$2400 + 184$	$= 2584$
Area:	35600×2584	$= \underline{91.99 \text{ m}^2}$

EXAMPLE 5: Element 6.a—Trowel Floor

Area:	(as element 2.a)	$= \underline{77.25 \text{ m}^2}$

Students may attempt to calculate the remaining elements in a similar manner, and to compare their results with those given in the Element Summary. Some minor discrepancies in totals may be encountered, but calculated quantities to be acceptable, should be within about 5% of those given.

18.4 SELECTED PARAMETERS

18.4.1 Preamble

Because of the widespread use of area and volume parameters to establish estimated costs for proposed buildings, some guidelines for the proper application of measurement methods and techniques are appropriate in this book. It should be understood that the fundamental weakness in utilizing such methods lies not in the difficulty of measurement, but rather in the virtual impossibility of developing a truly appropriate and reliable unit price to apply to the measured quantities. Synopses of the two methods are shown below. Areas should be expressed in square meters or square feet; volumes should be expressed in cubic meters or cubic feet.

18.4.2 Area Method

General: Establish the gross area of the building by measuring each floor to the significant plane of the outside faces of the perimeter walls, and adding the area of full-height columns extending beyond that plane. Ignore minor external projections such as exterior balconies, string courses or canopies, minor internal voids for normal stairs, elevators, and the like, and space occupied by internal walls, columns, piers, and the like.

Deductions: Where a major void extends through two or more floors, such as in a foyer two or more storys in height, include the largest area of the feature at one level only and deduct the void area at other levels.

Inclusions: Roughed-in areas, such as crawl spaces, unfinished basements or upper floors, accessible tunnels and trenches, and *finished areas,* such as enclosed penthouses, porches, usable attic areas, machine rooms, garages, exterior stairs and fire escapes, and the like.

Exclusions: Unfinished areas, such as crawl spaces, inaccessible tunnels and trenches, unusable attic areas, and *finished areas,* such as unenclosed walkways, open porches, carports, isolated chimneys, open courtyards and terraces, roof overhangs, and the like.

18.4.3 Volume Method

General: Establish the gross volume of the building by measuring the horizontal area as described in the preceding section, and multiplying that area by the vertical height measured from the underside of fill at the lowest floor levels to the upper side of the roof levels. Where roofs are not flat, add the true volume of the roof spaces above the upper side of the ceiling system.

Deductions: None.

Inclusions: Essentially the same as for the area method.

Exclusions: Essentially the same as for the area method.

18.4.4 Additional Items

Those items listed as exclusions above, as well as other sitework, such as landscaping, patios, exterior stairs, ornamental fountains, flagpoles, and so on, can be measured and priced *in detail,* and their costs separately stated, for addition to the basic building construction cost established in the estimate.

Chap. 18 / Measurement For Designers

QUESTIONS

18.1. If supporting beams and columns beneath a suspended concrete slab are specified to be included in the costs of the floor system, why are ceiling finishes below the same system excluded from such costs? Where are such excluded costs included in an elemental estimate?

18.2. Briefly explain how it is recommended to measure passenger elevators in elemental analysis. What attributes of such elevators should be stated in the item? Identify at least three sources of cost data for passenger elevators.

18.3. If the roof construction of a small building consisted of factory-made wood trusses, 30 ft long, 6 ft high, and set at 24 in. on centers, explain how you would measure this feature in an elemental analysis (calculation is not required for your answer).

18.4. Explain how you would measure a dogleg, wood-framed staircase having one square landing halfway between a lower floor and an upper floor served by the complete stair, for pricing in an elemental estimate. Specifically comment on how you might handle the landing.

18.5. How are allowances for overhead costs included in elemental analysis estimates? How are the magnitudes of such allowances established?

CHAPTER 19
PRICING EXAMPLES SELECTED TRADES

19.1 GENERAL OBJECTIVES

19.1.1 Preamble

In this chapter, techniques that can be used by contractors to price the work of selected trades are presented first. In Chapter 20, techniques that can be used by designers to price selected elements of buildings are explained.

19.1.2 Objectives

The first purpose of this chapter is to illustrate and exemplify the specific methods, techniques, and systems of pricing discussed in Part III. This will be done showing by their application to selected construction trades, by means of worked examples, based on building materials, construction methods, and drawing information contained within the book. The second purpose is to permit the student to practice applications on a series of exercises of graduated difficulty, using the illustrations as models against which format and content can be checked. Although it is a stated objective of this book not to digress into all possibilities of materials and methods, a number of useful exercises could be developed to consider the pricing of items of work other than the ones exemplified in the text, and to consider methods of doing work differently to the methods selected for illustration purposes.

19.2 METHODS FOR CONTRACTORS

19.2.1 Preamble

In this section, the five trades selected for pricing study correspond to those selected for the study of materials and methods in Chapter 18, again identified as follows:

Masterformat	Title of trade
02200	Earthworks
03100	Concrete formwork
03300	Cast-in-place concrete
06100	Rough carpentry
09250	Gypsum wallboard

Understanding and mastery of the methods and techniques of pricing of these five should lead to competence in measurement of most other normal and traditional trades.

List Prices. In this book, costs for labor are calculated using the following unit prices:

Skilled tradesmen or journeymen	$20.00 per hour
Unskilled laborers or helpers	$15.00 per hour

Costs for materials and equipment have been assumed at realistic levels relative to the date of publication of the book. Percentages for overhead costs and profit, wherever used, have been assumed at median levels, common in the industry.

19.2.2 02200 Earthworks

General Issues. The pricing of earthwork (specifically excavation and backfilling of earth) logically relates to the nature of that work. The equipment available for doing earthwork is, of course, very varied and extensive, and consists of machines and tools for *digging,* such as augers, backhoes, shovels of different kinds, and trenching machines; for *grading,* such as bulldozers, rakes, scarifiers, and scrapers; and for *compaction,* such as tampers, sheepsfoot drums, and various types and weights of rollers. Other machines such as cultivators and harrows, are often used in landscaping operations.

The key factor is *productivity* of the equipment proposed for use. Productivity, in turn, consists of a number of elements, such as fixed time, variable time, and efficiency of the machine, as well as the quality and quantity of the material to be excavated. **Fixed time** is the time it takes for a machine, such as a backhoe, to do its basic work—digging out enough earth material to more or less fill the bucket. **Variable time** is the time it takes for the backhoe to maneuver around to unload the excavated material into a truck or onto a dump and return to the digging operation. The combination of fixed and variable time is called the **cycle time.** Efficiency is related to the ability of the machine to move around the site; in any given time cycle, a wheeled vehicle may be 75% efficient, whereas a tracked vehicle may achieve 85% efficiency, resulting in reciprocal adjustment factors of 1.33 and 1.18, respectively. Most manufacturers provide simple formulas and typical data to permit reasonably accurate and useful calculations to be made of the foregoing elements.

Material Quality. With respect to the quality of the material to be dug or placed, an allowance must be made for the twin phenomena of bulkage and shrinkage. Any given volume of every type of sand, soil, hardpan, and clay swells to some extent on digging, and can be shrunk to some extent by compaction after backfilling. Factors can be determined to allow for this variable in the pricing process for each type of material; some typical data are given in Table 19.1. The effect of the factors means that for every 1 cubic meter or cubic yard of (say) hardpan to be dug

TABLE 19.1
Bulkage and Shrinkage Factors

Material type	Bulkage (%)	Shrinkage (%)
Running sand	+14	−12
Clean gravel	+16	−14
Loam or topsoil	+20	−17
Clay and hardpan	+25	−20
Dense clay	+33	−25
Solid rock	+50	−30

TABLE 19.2
Adjustment Factors

Material type	Factor
Sand and gravel	+10%
Loam or topsoil	Nil
Dense clay	−10%

up, 1.25 m³ or cy has to be carried away. Conversely, for every 1 cubic meter or yard to be placed and compacted, 1.20 m³ or cy will have to be brought in to the site. It might be noted that the factors in the two columns are not the same.

An allowance can also be made for the comparative ease or difficulty of digging materials of various types. It is easier to dig up topsoil than to dig up stiff clay; to put it another way, it will take a little longer to dig the clay than the soil. Some representative factors are listed in Table 19.2 which can be used to adjust machine outputs; additional factors can be developed for other types of excavation materials. With respect to the quantity of material, in general the normal economic law applies that the greater the total amount, the lesser the resultant unit price.

Metric Example. Assume that a backhoe (Figure 19.1) is selected to dig a level trench in stiff clay, 700.00 m long, 0.75 m wide, and 2.00 m deep, with no allowance for sloping sides, as it is proposed to use close-sheet shoring to support the sides. Also assume that the backhoe is fitted with tracks and a 0.75-m³ bucket.

Quantity: The net quantity of clay to be dug is (700.00 × 0.75 × 2.00) 1050.0 m³. The gross quantity will total (1050 + 25%) 1313 m³, allowing for 25% bulkage. The estimated output of the machine selected is approximately 60 m³ per hour for

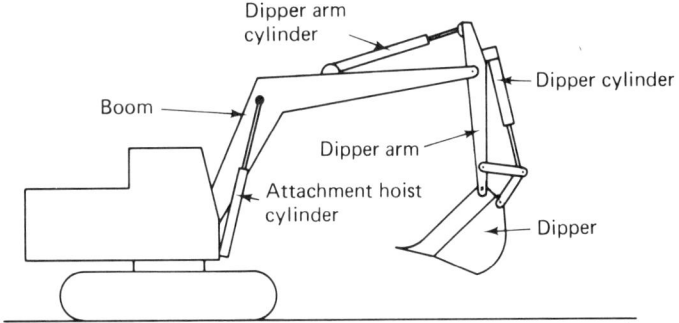

Figure 19.1 Backhoe components. [From S. W. Nunnally, *Construction Methods and Management*, (Englewood Cliffs, N.J.: Prentice-Hall, Inc., 1980).]

Typical backhoe.

work in trenches, taken from manufacturer's data tables, and the soil quality factor for clay is −10%. Probable machine output is therefore (60 less 10%) 54 m³ per hour to dig and dump, without moving around the site.

Fixed Cost: The chargeable cost of the machine, calculated as described in an earlier chapter, is assumed to be $100.00 per hour. The cost of the composite crew to operate such a machine (including a foreman, a driver, and a helper) is assumed to total $60.00 per hour. The combined rate per hour is therefore $160.00.

To find the operating cost, first divide the quantity by the output, in this case (1313 gross ÷ 54) 24.31 hours. Another way of saying the same thing is to use the net quantity of 1050 m³ and divide it by the basic output factor reduced by the reciprocal of the bulkage factor (54 less 20%) to 43.2 m³ per hour, resulting in (1050 net ÷ 43.2) 24.31 hours, as before.

Next, round the time up to the nearest 4 hours, in this case 28 hours, because most such machines, and more particularly crews, are not usually available in units of less than half a day. Multiply the time by the combined rate per hour, in this case (28 × 160) to determine the operating cost at $4480.00 for 1050 m³ net excavation.

To the foregoing operating cost allowances must be added for:

1. Mobilization costs (say, 2 hr × $100 per hour = $200)
2. Overhead costs on the labor component (say, 28 hr × $60 × 50% = $840)
3. Profit of (say) 10% on the whole

To determine the unit price for digging, these costs and calculations can now be summarized as follows:

Item		Costs
Equipment charges		$4480.00
Mobilization costs		200.00
Overhead on labor		840.00
Subtotal	(5520)	
Profit on subtotal		552.00
Total cost for 1313 m³		$6072.00
Cost per m³ (6072/1050) = $5.78		

Chap. 19 / Pricing Examples—Selected Trades

Variable Cost: In the foregoing example it was assumed that the backhoe would dig the material out and dump it at the side of the trench, ready for backfilling. Other possibilities exist: the backhoe could move around the site under its own power to dump the material farther away from the trench, or a selection could be made from a variety of trucks to carry the excavated materials away to a more remote dump.

1. *Self-propelled haulage.* In the first case, the efficiency of the backhoe to move around the site has to be determined. This is a function of the speeds of which the machine is capable, in both loaded and unloaded modes, and in forward and reverse gears. The formula to calculate this factor is as follows:

$$Vt = \frac{Hd}{Hs} + \frac{Rd}{Rs}$$

where Vt = variable time
Hd = haul distance
Hs = haul speed
Rd = return distance
Rs = return speed

If the distance is expressed in meters and the speed in km/h, a factor of (1000/60) 16.67 is applied to convert kilometers per hour to meters per minute. If the distance is expressed in feet and the speed in mph, a factor of (5280/60) 88 is applied to convert miles per hour to feet per minute.

In the foregoing example, the variable time taken to move the backhoe a distance of (say) 30 m and back again might be calculated as follows, using speeds of 3 km/h loaded in forward gear and 5 km/h unloaded in reverse gear:

$$Vt = \frac{30 \text{ m}}{3 \text{ km/h} \times 16.67} + \frac{30 \text{ m}}{5 \text{ km/h} \times 16.67}$$

$$= 0.60 + 0.34 = 0.94 \text{ minute (say 1 minute)}$$

Other speeds and distances will naturally produce other results.

Adding the allowance for inefficiency of a tracked vehicle at 85% (or its reciprocal factor of 1.18), the variable time will increase to 1.18 minutes, in the given situation. It was already calculated that the selected machine could probably remove about 54 m³ per hour without moving around. Using a 0.75-m³ bucket, this output translates to about (54/0.75) 72 cycles per hour, which is the same as 1.20 cycles per minute or 0.83 minute per cycle.

However, we now have to add 1.18 minutes per cycle, to allow for the 30-m run to the dump and back, for a total time of (0.83 + 1.18) 2.01 minutes. So the productivity of the digging procedure will decline to (60/2.01) or approximately 30 cycles per hour. This can now be converted back to output expressed in cubic meters, if required. If 72 cycles could remove 54 m³ in 1 hour, 30 cycles will only remove 22.50 m³ in 1 hour. The time required to excavate, move, and dump 1313 m³ will therefore now be (1313/22.50) 58.34 hours, rounded up to 60 hours, with the cost calculated as before.

2. *Haulage by dump truck.* In the second case, where the excavated material may have to be hauled away from the trench site and dumped at a considerable distance elsewhere, calculation of hauling costs will have to include consideration of the size, speed, and type of trucks, the distance to the dump, the road or traffic conditions likely to be encountered, and any fees involved at the dumping grounds.

In general, the size or cubic capacity of truck selected to dispose of material excavated is recommended by hauling companies to be about four to five times the size of the bucket doing the digging. In the example given above, where the backhoe has a 0.75-m^3 bucket, the minimum size of truck will be (4 × 0.75) 3-m^3 capacity. The selected backhoe will dig and dump each bucket-full in about ⅓ minute (i.e., 0.75 m^3 every 20 seconds), so the 3-m^3 truck will be filled to capacity in about [(3/0.75) × 20] 80 seconds.

As long as allowance is made for bulkage in the excavation calculations, no further allowance need be made for bulkage at this point, because the presumption is that the quantities in both the bucket and the truck are struck (or net capacity) quantities; alternatively, one might say that because the bulkage in both the bucket and the truck is about the same, one makes compensation for the other. This is one of the few exceptions to the general estimating rule that it is bad form, and poorer practice, to let one thing stand in place of another in an estimate.

It takes an additional period of about 1⅓ minutes (80 seconds) to get each loaded truck (of the size selected) out of the way and an empty one into position beside the backhoe. This maneuver is called "spotting" in the trade. At this rate, it would be necessary to spot 45 of the selected trucks every hour (45 × 1.33 = 60 minutes) to keep the backhoe in continuous operation. Depending on the distance the material has to be hauled to the dump, the speed of the truck, and the road and traffic conditions encountered along the way, a time can be estimated for each truck to load, dump, and return.

If, in this example, one assumes a period of 20 minutes per round trip per truck per load, each truck can be involved only 3 times per hour or once every 20 minutes at the site. We have already established that it takes about 80 seconds to load each truck and another 80 seconds for spotting, for a total of 160 seconds or about 3 minutes. It would therefore require a minimum of (20/3) 7 trucks of the selected size to keep the backhoe fully occupied for each hour of operation under optimum conditions, which seldom occur.

If a larger and faster type of truck is selected, say one able to hold 5 m^3 and make the return trip in 15 minutes, they may take longer to spot and load, but fewer trucks may be necessary. A number of similar calculations could be made to determine the optimum number, relative to the size of the job, the rental rates of the trucks, the costs of the backhoe, and so on, trading off one factor for another in the calculations to achieve a balanced result.

It is also possible to divide the gross quantity of material to be removed by the capacity of the trucks selected to do the haulage and so determine the likely total number of truck loads of earth to be moved from the site to the dump. Using this figure and having established how many trucks of a given size are necessary to keep the backhoe optimally operating on an hourly basis, the minimum time required to rent or utilize these trucks can easily be calculated. It is recommended that the student perform such a calculation for practice.

One should bear in mind, however, that all such calculations have a certain imprecision built into them, as it is clearly impossible to absolutely predict the precise soil quality, the exact digging and loading times, or spotting and cycling operations, with mathematical certainty. In the foregoing example, a more reasonable figure would probably be about 5 or 6 trucks. The slight loss of production that almost always occurs in such inexact work would probably be more than offset by the possible savings in rentals. One could also use detachable skips or truck boxes, which could be deposited at the site while the motorized truck takes filled skips off to the dump for unloading. Other possibilities may come to mind, but they all require direct confirmation by field observation and cost accounting, to relate estimated outputs and costs to reality.

The foregoing example could be reworked in imperial units, as a practice exercise.

19.2.3   03100   Concrete Formwork

In general, the simpler an item of construction work is to measure, the more complex it is to price, and vice versa. To some extent, this precept was encountered in the measurement and pricing of excavation work, but it is nowhere more clear than in its application to concrete formwork. The measurement of formwork is simply the surface area of the formwork that actually touches the surface of the concrete to be formed. As has been shown, the quantities of formwork are usually generated by the estimator easily and directly from the concrete calculations. However, the pricing of formwork involves considerable knowledge of a large variety of discrete but closely related aspects of materials and labor.

Before discussing the specifics of formwork materials, systems, and labor, it should be stated that from an estimating point of view, concrete formwork is best considered as an aspect of construction equipment, not as an aspect of materials. The reason for this view lies in the fact that although formwork is necessary work, it is not part of the permanent structure, any more than a concrete mixer or a pickup truck is part of the structure, although both are also necessary for construction to proceed. Also, in many cases, formwork that is supplied to one site will be moved back to the builder's yard for storage and for subsequent use at another site.

Not every estimator subscribes to this view, however. Many estimators view the materials in formwork as part of the permanent structure, because in many cases, much of these materials are in fact used first for formwork to concrete, and later for permanent carpentry framing or sheathing elsewhere in the structure of the building. As was stated earlier, though, it is not recommended to assume that one item of work will offset another; each should be given its own weight and attention in the estimate.

The key to reliable formwork pricing is sound knowledge of two fundamental issues:

1. Selection and reuse of materials
2. The type and amount of labor

Materials

Selection: Formwork consists of some type of sheet material to mold the concrete into shape, some framing to support the sheeting, some hardware to hold the sheets and frames together, and some shores or props to hold the assembly in position.

The sheet materials may be of shiplap lumber, plywood, particle- or plasterboard, sheet steel or fiberglass, among other products. Linings or inserts may be made of wood, metal, rubber, fiberglass, or plaster molds to produce ornamental work. The surface of the formwork that will touch the concrete is usually treated in some way, either by spreading a thin coat of a special releasing oil or by selecting form materials that have a specific type of resin or plastic coating on one surface.

The framing is usually of dimension lumber, although special steel shapes and aluminum bracing are also available for this purpose. The hardware consists of nails (both common and double-headed), form ties in a variety of configurations, wedges to secure the ends of the ties, as well as a variety of shims and other metal braces and attachments, such as are used for column forms. The shores and braces are often made of common lumber, although special hollow steel and aluminum tubes with threaded telescoping inserts are also widely used because of their ease and accuracy of handling, as well as their comparatively low cost and reliability. Open-web long-span joist systems, consisting of wooden chords with tubular aluminum webs, are also available from some suppliers to support formwork for soffits of suspended slabs, column drops, band beams, and the like.

Reuse: The single largest economic factor in the pricing of formwork systems is the number of times the components of the selected system are going to be used. There are three things to note about reuse:

1. Reuse applies primarily to the material components.
2. Each of the material components may be subject to a different reuse factor.
3. The number of reuses to be applied to each component in each formwork item of work is a matter of judgment and experience combined.

The foregoing points may seem to be statements of the obvious, but it is not uncommon for estimators to make the mistake of calculating the total labor and material cost for any one formwork system and then dividing the *total* cost by an estimated number of reuses. A moment's reflection will disclose the fallacy of this technique with respect to the labor costs, a fallacy that is explained in detail in the next section.

Common sense also suggests a possibility, borne out by experience, that formwork made of rough boards or dimension lumber is good for only about two or three uses, whereas formwork made from plywood specially manufactured and treated for this purpose may be good for 20 to 30 uses, and formwork made from fiberglass or steel can be used between 200 and 300 times, or even better if care is taken in its handling. Furthermore, some components of the formwork system may not be able to be reused as often as others, and allowance can be made for such variation.

For example, plywood panels have to be drilled to accommodate the form ties; after a few uses, there may be simply too many holes in the wrong places in the plywood for it to be satisfactorily used again. However, the lumber and bracing used to support and secure the plywood may still be good for many more uses. At the other end of the scale, most common makes of form ties can be used only once, although there are some expensive types with threaded inserts intended for multiple reuse. And the common nails used for assembly of the formwork panels and boxes probably could be recovered and straightened out for reuse, but the labor cost to do this would probably more than offset the material cost to supply new nails in every case.

Labor

If one were to analyze the costs involved in the construction of the frame of a typical reinforced-concrete building, such as an office block or a parking garage, it is likely that the costs of the formwork would comprise about 50% of the total structural cost, with the concrete and the steel making up the other half. Furthermore, about two-thirds of the cost of the formwork half (or one-third of the whole) will be consumed by the cost of the labor components attributable to the formwork.

It therefore follows that if one wishes to make significant savings in the cost of a concrete-framed building, the first thing to examine would be the labor component of the formwork. The object would be to simplify the design shapes of the concrete and therefore of the formwork, as distinct, say, from changing the mix of the concrete or cutting down on the quality or amount of the reinforcing steel, much of which is governed by building codes that do not permit much latitude in any event.

A typical formwork crew consists of a foreman, some journeyman carpenters, and some semiskilled helpers. The main elements of formwork design, fabrication, erection, and removal were dealt with in Chapter 14. The design and fabrication of the units or panels is exclusively done by the carpenters; in some localities, the design must be approved by an engineer. The erection and stripping of the forms, including the placement and removal of the form ties, is done by a combination of carpenters and helpers, with proportionally more carpenters involved in the erection and more helpers in the stripping. The cleanup of the stripped forms and their movement about the site is done exclusively by the helpers. Repairing damaged forms is usually a job for the carpenters.

It should be noted that the labor cost for design and fabrication can be incorporated into the calculations for reuse of materials, as these costs can be spread over the number of uses. The labor cost for erection, stripping, cleaning, and moving, once expended, cannot be reused, any more than time can be reversed. The worked examples show the technique to make such allowance in the correct manner where reuse becomes a factor.

It might also be mentioned that some estimators like to include in the formwork costs the additional cost to patch the tie holes left in the concrete surfaces after the forms have been removed. Application of the basic estimating precept of keeping everything separate would suggest exclusion of this item of work from the formwork.

Table 19.3 gives some representative factors for work of this type, based on a crew of six or seven carpenters and two or three laborers. Depending on circumstances, the factors in the table could increase or decrease by as much as 50%.

TABLE 19.3
Factors for Formwork Labor[a]

Position	Assemble	Erect	Strip	Repair
Footings	4.00	3.00	2.00	2.00
Walls	6.00	2.50	1.50	1.50
Columns	9.00	4.00	2.00	2.00
Slabs[b]	5.00	2.00	1.50	1.50
Beams	10.00	4.50	2.50	2.00
Stairs	15.00	5.00	3.00	2.50

[a]Man-hours per 10 m^2 or 100 sf of formwork.
[b]Soffits of suspended slabs.

Worked Examples (Imperial). Two worked examples are given next, the first being a simple vertical lineal form to establish the perimeter of a slab, either on grade or at the edge of a suspended slab, and the second being a typical formwork panel system used in a foundation wall of a simple commercial or residential building. The abbreviations which are used in the examples are explained in Table 19.4.

TABLE 19.4
Abbreviations

lf	= lineal feet	bf	= board feet
fbm	= foot board measure	mbf	= 1000 board feet
lb	= pounds weight	msf	= 1000 square feet
o.c.	= on centers	ft	= feet

1. Perimeter Form: Such work would normally be done with solid dimension lumber, braced at frequent intervals (about 6 lf) with simple wooden or metal vertical supports at the sides, about 2 ft long. On a small job, these forms would be installed once for the entire job; on a larger project, two or three reuses might be anticipated. Assume one use only in this example.

Material cost (per 100 lf):

$$
\begin{aligned}
&2 \times 8 \text{ dimension lumber: } 100 \text{ lf} \times 1.33 \text{ bf} &&= 133 \text{ fbm} \\
&2 \times 4 \text{ braces @ 6 ft o.c. } 100/6 \times 2.00 \text{ lf} &&= 23 \text{ fbm} \\
&\text{Add for waste on } (133 + 23) \text{ 156 fbm @ 3\%} &&= \underline{5 \text{ fbm}} \\
&\text{Total quantity of form lumber} &&= 161 \text{ fbm}
\end{aligned}
$$

$$
\begin{aligned}
&\text{Total cost of lumber: 161 fbm @ \$250/mbf} &&= \$40.25 \\
&\text{Add hardware and waste: 6 lb @ \$2.50/lb} &&= \underline{15.00} \\
&\text{Material cost for 100 lf} &&= \$55.25
\end{aligned}
$$

Material unit price (55.25/100 lf) = $0.55/lf

Labor cost (per 100 lf):

$$
\begin{aligned}
&\text{Carpenter: 4.00 hr @ \$20.00} &&= \$ \ 80.00 \\
&\text{Laborer: \ \ 3.00 hr @ \ \ 15.00} &&= \underline{45.00} \\
&\text{Labor cost for 100 lf} &&= \$125.00
\end{aligned}
$$

Labor unit price (125.00/100 lf) = $1.25/lf

A similar calculation would be made for the formwork used to make a simple concrete footing, where boards are arranged along the two sides of the proposed footing, braced on the outsides, and held in position with spreaders across the top. Allowance would have to be made to include for the materials along both sides of the footing if the unit price is to represent the price per lineal foot of the footing, as distinct from a price per lineal foot of the formwork. Also, plywood could have been used instead of dimension lumber in the foregoing system.

2. Foundation Walls: In this example, assume that the formwork system consists of normal prefabricated plywood panels, with dimension lumber supports and standard hardware. Also assume eight uses after the first use.

Material cost (100 sf)	First use	Reuse
100 sf formply @ $350/msf	$35.00	$ nil
Allow 10% waste	3.50	3.50
150 fbm lumber @ $250/mbf	37.50	nil
Allow 05% waste	1.88	1.88
7 lb hardware @ $2.50/lb	17.50	nil
1 lb hardware (reuse only)	nil	2.50
Allow 10% waste	1.75	2.00
Total material cost	$97.13	$9.88

The total proposed number of uses is 9. The total cost is therefore (1 × $97.13) plus (8 × $9.88) = $176.17. The average cost per use for 100 sf will be ($176.17/9) $19.58. The material unit price for this system at these assumed material supply prices will therefore be ($19.58/100) $0.20 per square foot.

The minimum amount of support lumber necessary to make up reusable form panels can easily be determined by considering a typical panel of plywood, 32 sf in area, having 2 in. × 4 in. dimension lumber all around the edges and at 16 in. on centers lengthwise in the field of the panel. This would result in a total length of 40 lf of lumber attached to the plywood, plus another 16 lf for wales and (say) 16 lf for bracing, giving a total of 72 lf or 48 bf per panel. Three such panels cover

Chap. 19 / Pricing Examples—Selected Trades

96 sf and require 144 bf of lumber; these data can be rounded off at 100 sf and 150 bf, respectively.

If it was proposed to use shiplap lumber instead of plywood for the sheeting, an allowance must be made for the additional material in the overlap. In lumber having nominal dimensions of 1 in. × 8 in., the overlap would amount to 12.5%, so to provide 100 sf of coverage, 112.5 sf of shiplap would be required. It might be noted in passing that in imperial units, the area in square feet of lumber which has a nominal thickness of 1 in. will be the quantity of that lumber in board feet. Thus 112.5 sf of 1 in. × 8 in. shiplap is 112.5 bf.

The general formula for accommodating all factors is shown below, where n equals the number of proposed uses:

$$\text{unit price} = \frac{(\text{first-use \$}) + [(n-1) \times \text{reuse \$}]}{n \times \text{unit area}}$$

$$= \frac{\$97.13 + (8 \times \$9.88)}{9 \times 100 \text{ sf}}$$

$$= \$0.1958 \text{ (or 20 cents per square foot)}$$

A similar approach is adopted to calculate the total labor price for this system, with subsequent application of the foregoing formula to determine the labor unit price:

Labor cost (100 sf)	First use	Reuse
Assembly (carpenter)		
First use: 6 hr @ $20	$120.00	$ nil
Second use: 1 hr 20	nil	20.00
Erection (two workers)		
Carpenter: 3 hr @ $20	60.00	60.00
Helper: 1 hr 15	15.00	15.00
Stripping (two workers)		
Carpenter: 1 hr @ $20	20.00	20.00
Helper: 2 hr 15	30.00	30.00
Moving and repairing		
Helpers: 2 hr @ $15	30.00	30.00
Total labor cost	$275.00	$175.00

Applying the formula given above, the labor unit price is now calculated as follows:

$$\text{unit price} = \frac{\$275.00 + (8 \times \$175.00)}{9 \times 100}$$

$$= \$1.86 \text{ per square foot}$$

In the section of the calculation dealing with moving and repairing, it might be argued that carpenters should be used for the repairs, with the helpers only doing the moving. If necessary, this or any other part of the calculations can be broken down into smaller parts to permit more refined consideration of specific detail, depending on circumstances.

The two foregoing examples could be reworked in metric units as practice exercises.

19.2.4 03300 Cast-in-Place Concrete

In general, the pricing of cast-in-place concrete involves consideration of a fairly large number of closely related components, most of which can be subsumed under three primary elements or headings:

1. The supply of the concrete
2. The provision of equipment
3. The labor to place the mix

The Concrete. Elements of cost for concrete materials include (but are not limited to) the items in the following list:

1. Size and quality of aggregate
2. Type and amount of cement
3. Type and amount of additives
4. Heating of ingredients, if necessary
5. Amount of calcium chloride, if necessary
6. Method of delivery
7. Delivery distance
8. Quantity required
9. Cash, trade, or volume discounts
10. Local and national taxes
11. Testing and inspection
12. Proportioning of ingredients
13. Water/cement ratio and slump

Ready-mix concrete is widely used now for all but the smallest of projects. The companies that supply ready-mix concrete usually quote unit prices for delivering concrete of various specified qualities, quantities, and distances. The methods by which such prices are determined are beyond the scope of this book. However, a brief review of methods used to establish concrete costs using small site mixers will give some insight into the larger aspects of ready-mix costs while providing a useful methodology for the small practitioner. The complexities of modern concrete technology make for an interesting study, highly recommended to anyone who aspires to be competent in this field.

The smallest practical concrete mixing machine is called the *one-sack mixer,* so named because its capacity is such as to accommodate one sack of cement along with the aggregates and water necessary to make one batch of concrete. A typical concrete mix will require one part cement, two parts sand, and four parts aggregate. It is customary to measure all parts by weight, including the water. One sack of cement weighs 40 kg, the amount of sand would weigh 80 kg, and the coarse gravel aggregate would weigh 160 kg. Water would be added to the mix according to empirical (or experiential) tables to produce appropriate hydration. The amount of water might be in the region of 20 liters, which weigh 20 kg. The costs of the separate ingredients can be confirmed from local suppliers. (In imperial units, a sack of cement weighs either 80 or 87.5 pounds, depending on the manufacturing company, and has a volume of about 1 cubic foot. The other ingredients can be converted in simple proportion: 1 kg = 2.2 lb.)

The entire batch would therefore weigh about 300 kg, and would take about 2.5 minutes to be mixed by the specified machine ready for placing. The machine will therefore produce about 25 batches per hour or up to ($25 \times 7.5 = 188$), say,

200 batches per working day. Each batch has a volume of approximately 0.25 m^3, so the optimum output of the machine will be around (200 × 0.25) 50 m^3 per day.

The work crew required to keep such a machine in full production might consist of a foreman or journeyman operating the machine and handling the water, two or three helpers moving the loose cement, sand, and gravel around in wheelbarrows, and another two or three helpers moving the mixed concrete from the machine to the chute, crane, or hoist being used to place the material in its final position. This crew of six or seven will therefore produce about 6 or 7 m^3 per hour or 1 m^3 per man-hour, under ideal conditions. The labor hours might safely be estimated at about 2.0 per m^3 under average conditions in such an operation, with the estimate being confirmed later by actual observation in the field. The hourly cost of a specific crew can readily be determined by adding up each of their respective hourly rates. The hourly cost of the machine must also be included in the calculation.

The Equipment. Elements of cost for equipment used to transport and place concrete at the site can include any or all of the following:

1. Chutes
2. Cranes
3. Pumps
4. Vibrators
5. Buggies
6. Hoists
7. Conveyors
8. Other

Under "other," one might consider more esoteric pieces of equipment, such as helicopters, special monorails, and the like, now being used to deliver and place concrete in particular situations. Also, specialty work, such as underwater work, high-pressure work, or steam autoclaving, will naturally require more specialized consideration. The costs of all such equipment are determined in the manner described in an earlier chapter; such costs can be confirmed by checking rental rates charged by local companies.

The Labor. Elements of cost for labor to place concrete include the following:

1. General type and location of building project
2. Size, shape, and complexity of the job
3. Time of year and time available for work
4. Quality and quantity of available labor
5. Amount and degree of rigor of inspections
6. Appropriateness of available equipment

It is theoretically possible to develop factors to allow for each or all of the foregoing attributes, giving each a factor of 1.00 and increasing or decreasing the factor as appropriate. For example, extremely poor weather may warrant a factor of 1.05 for element 3, whereas an exceptionally productive crew might rate a 0.95 for element 4. It might be an interesting exercise to develop such a table of factors and a methodology for their application. Generally, an overall allowance is made to the work as a whole, based largely on experience and "gut feeling" about the project and its prospects, and usually expressed as a percentage increase or decrease on the developed unit prices.

Table 19.5 shows some approximations of labor hours for placing ready-mix concrete in a variety of common configurations. To convert the following factors from cubic yards to cubic meters, increase all values by about 10% and round off to the nearest tenth of an hour. In every case, such data can and should be confirmed by actual site observation.

TABLE 19.5
Approximate Labor Hours for Placing Ready-Mix Concrete

Concrete position	Method for placing per cubic yard			
	By hand	Buggy	Crane	Pump
Foundations	1.25	0.80	0.40	0.30
Columns and piers	1.50	1.25	0.60	0.45
Slabs on grade	1.00	0.75	0.30	0.25
Suspended slabs	1.10	0.80	0.50	0.35
Walls above grade	1.25	1.00	0.55	0.40
Stairs and landings	1.75	1.50	1.00	0.75

Concrete placed above ground level will require some equipment, such as a hoist, pump, or a crane, to lift it to the required elevation. Typical heights and outputs for concrete pumping equipment can be obtained from any local equipment rental agency. An additional allowance of up to 0.10 labor hour can be added to include for the time of a person to be in charge of the hoisting of the concrete from the ground level to the suspended floor level. As a general guide, one can allow about 1 man-hour per cubic yard of concrete for manual placement in normal or typical situations.

To show the general principles of pricing of concrete work, a worked example for a typical item of work in a concrete-framed commercial or institutional building is shown next. The approach to pricing most other items of concrete work involves only minor modification of the techniques shown and discussed in the example. The example is worked in metric units; students could rework the example in imperial units as a practice exercise.

Worked (Metric) Example. Calculate the cost for supplying and placing (but not finishing) concrete for a suspended floor slab, 30 m long by 20 m wide by 150 mm thick, 3.00 m above grade, using a crane and bucket.

Materials:

Concrete required (30.00 × 20.00 × 0.15) = 90.00 m^3

Material cost:

Supply of ready-mix concrete (say)		$35.00/m^3
Add for heating during winter		0.50/m^3
Subtotal	(35.50)	
Allow for waste (say, 3% on 35.50)		1.07/m^3
Unit cost per cubic meter	=	$36.57

Total material cost: 90 m^3 × $36.57 = $3291.30

Labor: From Table 19.5, assume that the concrete will be placed by crane and bucket. Labor hours for suspended slabs are estimated at 0.50 per cy or 0.55 per m^3. Another 0.05 labor hour for hoisting could be added for an estimated total of 0.60 hour per m^3. The quantity of concrete to be placed is 90 m^3, so the estimated total time will be (90 × 0.60) 54 labor hours. Crew size and configuration would probably consist of a foreman, two helpers charging and discharging the concrete bucket, four helpers spreading the placed concrete, and a cement worker operating a vibrator, for a total of eight people. The total job should therefore take about (54/8) 7 hours maximum, from a "labor" point of view.

Using an appropriately sized crane and bucket to raise ready-mix concrete delivered to the site in trucks, about 140 m³ of concrete can be placed in one 7.5-hour day under optimal conditions, that is, about 19 m³ in 1 hour or 1 m³ in 3.16 minutes. In this example, there is 90 m³ to pour, which will take a minimum of (90 × 3.16) 284 minutes, or 4.75 hours, without interruption or delay. It is more likely that it will take between 5.0 and 5.5 hours to place the 90 m³ of concrete, from an "equipment" point of view, as there will always be some delays in waiting for and spotting the ready-mix trucks, in adjusting the crane and bucket, in spreading the dumped concrete, for coffee breaks, and for many other reasons, such as poor weather, difficulty of access, and so on.

One must now try to reconcile the probable minimum equipment time of 5.5 hours with the probable maximum labor time of 7.0 hours, by adjusting proposed machine capacities, ready-mix truck sizes, and crew configurations to find the optimum balance between the two, namely, where the two factors more or less coincide. In this example, slightly larger equipment and crew sizes would probably achieve a balance around 6.5 to 7.0 hours. As a result, the entire 7.5-hour day for the crew and machine would probably be charged to this item of work, to allow for some machine downtime and for crew setup and cleanup time.

Labor cost:

Foreman:	1 @ $22.00/hr	= $ 22.00
Journeyman:	1 @ 20.00/hr	= 20.00
Helpers:	6 @ 15.00/hr	= 90.00
Vibrator:	1 @ 25.00/hr	= 25.00
Labor cost per hour		= $157.00

Placing cost:

7.5 hr @ $157.00 = $1177.50 for 90 m³

Therefore,

Unit price to place = $13.08/m³

It is also possible to calculate an average labor cost per man-hour. For example, if a crew consisted of one cement mason at $20 per hour and two laborers at $15 each, the average labor cost would be [(20 + 30)/3] $16.67 per man-hour. The costs of the crane and bucket are reckoned as direct overhead costs, either relative to the concrete work or to the site as a whole, and are therefore excluded from the calculation for this particular item of work.

Use of other pieces of equipment, such as power buggies for transportation of the concrete from the crane bucket to the deposit location, will of course change the calculations. Using such buggies, a crew of 10 men should be able to place about 25 m³ per hour, that is, 2.5 m³ per man per hour, or 0.40 labor hour per 1 m³ (or about 0.35 labor hour per 1 cy). The cost of using the buggies must also be included at some point in the calculations.

Finishing: Assuming that a separate fine cement topping is not required on the surface of the structural slab, the finishing of the foregoing specified concrete slab would involve about four cement masons, one operating a bull-float machine, two handling straightedges or wood floats, and one working around the edges of the

Concrete power buggy. [From R. C. Smith, *Principles and Practices of Light Construction,* 3rd ed. (Englewood Cliffs, N.J.: Prentice-Hall, Inc., 1980).]

slab and around columns, openings, or other interruptions in the slab. One of these masons would probably act as a foreman.

The foreman and one of the workers would be required to power float the slab about 4 hours after its initial placement, work that would almost certainly involve overtime payment, for which allowance should be made. One cement mason with a helper can manually finish about 15 m² in about 1 hour; using a mechanical float or trowel, output can be tripled, to about 45 m² per hour. The specified crew of four, using a mechanical float, should be able to finish up to 100 m² per hour under optimum conditions.

Labor cost:
Standard time (at flat rates):

    Foreman: 1 @ $22.00/hr   = $22.00
    Cement masons: 3 @   20.00/hr   = 60.00 = $82.00/hr

Overtime (at double time):

$$\$82.00 \times 2 = \$164.00/\text{hr}$$

Cost of finishing:

600 m² @ 100 m²/hr: 6 hr @ $164.00 = $ 984.00
Bull-float rental: 6 hr @   30.00 =   180.00
   Total labor cost for 600 m²   = $1,164.00

Unit price = 1164/600 = $1.94/m²

An allowance for the use of hand tools can be included in the labor rates used to determine the labor costs. The labor and material costs of any additives to the surface finish, in the form of color or nonslip granules, would have to be also figured into the foregoing calculations. The foregoing examples could be reworked in imperial units as a practice exercise.

19.2.5 06100 Rough Carpentry

As noted earlier, rough carpentry includes the construction of a very wide variety of building components, such as floors, walls, ceilings, and roof systems of many types and configurations. It is impractical, if not impossible, to address every possible aspect of rough carpentry, so attention will be focused on standard horizontal floor and vertical wall framing systems in this section, to exemplify the pricing approach that might be adopted with appropriate modifications for other aspects of carpentry.

List Prices. Some worked examples are developed in detail later in this section, to illustrate the pricing techniques for carpentry work. The prices in Table 19.6 have been adopted for this purpose; local current prices should be substituted as necessary.

Labor Productivity. The amount of work that can be done by carpenters and their helpers in any given time period is naturally going to vary from time to time and from job to job, because of the large number of variables involved in the processes. The ability of each individual worker, the configuration of any particular work crew, the nature of the work to be done, the availability of the proper tools and materials, the time of year, the weather, the quality of the management, and so on, will all have their effects on the output.

Notwithstanding the foregoing generalities, it is possible to give some reasonable approximations or ranges within which qualified and experienced workers should be able to perform work of certain specific types. Some typical data are shown in Table 19.7 as a general guide and for use in practice pricing exercises. Units are expressed in SI metric and in imperial units: productivity is shown in man-hours for a qualified tradesman and a laborer as helper. In imperial units, it is customary to express productivity for rough carpentry work involving dimension

TABLE 19.6

Price List for Carpentry Work

Material item	List price
Dimension lumber	$250.00/mbf
½-in. plywood sheathing	$200.00/msf
½-in. T&G sheathing	$250.00/msf
Common nails	$30.00/keg
Carpenter	$20.00/hr
Helper	$15.00/hr

TABLE 19.7

Carpentry Productivity in Labor Hours

	Units		Productivity	
Item	SI	Imp.	Trade	Labor
Setting wall plates	30 m	100 lf	6–8	1–2
Exterior wall studs	300 m	1000 lf	22–44	6–8
Interior wall studs	300 m	1000 lf	20–30	6–8
Framing floor joists	300 m	1000 lf	14–15	4–5
Install cross-bridging	100 sets		7–9	1–2
Plywood flooring	100 m²	1000 sf	10–15	3–5

lumber in labor hours per 1000 board feet (mbf) rather than in lineal feet or square feet. This convention is adopted in the tables and examples that follow.

Horizontal Joists: To give more detailed consideration to joists, most wood-framed buildings require floor joists, ceiling joists, and roof joists. Joists are probably the least complicated of carpentry items to make and install, and productivity factors related to joists are relatively accurately established, as shown in Table 19.8.

It will be observed that as the cross-sectional dimension of the joist increases, the carpenter's productivity appears to increase (it takes less time to install the same quantity of lumber), because there will be correspondingly fewer pieces of larger dimension in the given 1000-board foot measure. There will also be a slight decrease in productivity because of the added weight of each piece of the larger sizes of lumber. Any corresponding change in the helper's time, although theoretically existing, is negligible and can be ignored. The foregoing productivity figures can be constructively compared to the corresponding data given in Table 19.9 for rafters in a complicated roof frame, having hips and valleys and other interruptions. One can see the significantly higher values and the larger spread in values in Table 19.9 compared to Table 19.8.

To show how labor price is determined, consider the installation of simple 2 × 8 floor joists (Figure 19.2), using values at the low end of the scale:

```
Carpenters:              9 hr @ $20.00   = $180.00
Helpers:                 4 hr @  15.00   =   60.00
  Total labor cost for 1000 bf           = $240.00
```

Unit price per board foot (240/1000) = $0.24/bf

Compare this with the cost of installing 2 × 8 rafters in a complicated roof system, using values at the high end of the scale:

```
Carpenters:             31 hr @ $20.00   = $620.00
Helpers:                 8 hr @  15.00   =  120.00
  Total labor cost for 1000 bf           = $740.00
```

Unit price per board foot (740/1000) = $0.74/bf

It would appear that the second item of work is worth about three times the first item. There are, of course, many reasons for the large difference. Simple floor

TABLE 19.8
Productivity Factors for Installing Joists

Joist size (in.)	Carpenters	Helpers
2 × 6	10–13	4–5
2 × 8	9–12	4–5
2 × 10	8–11	4–5
2 × 12	7–10	4–5

TABLE 19.9
Productivity Factors for Installing Rafters

Rafter size (in.)	Carpenters	Helpers
2 × 4	30–35	8–9
2 × 6	28–33	8–9
2 × 8	26–31	7–8
2 × 10	24–29	6–7

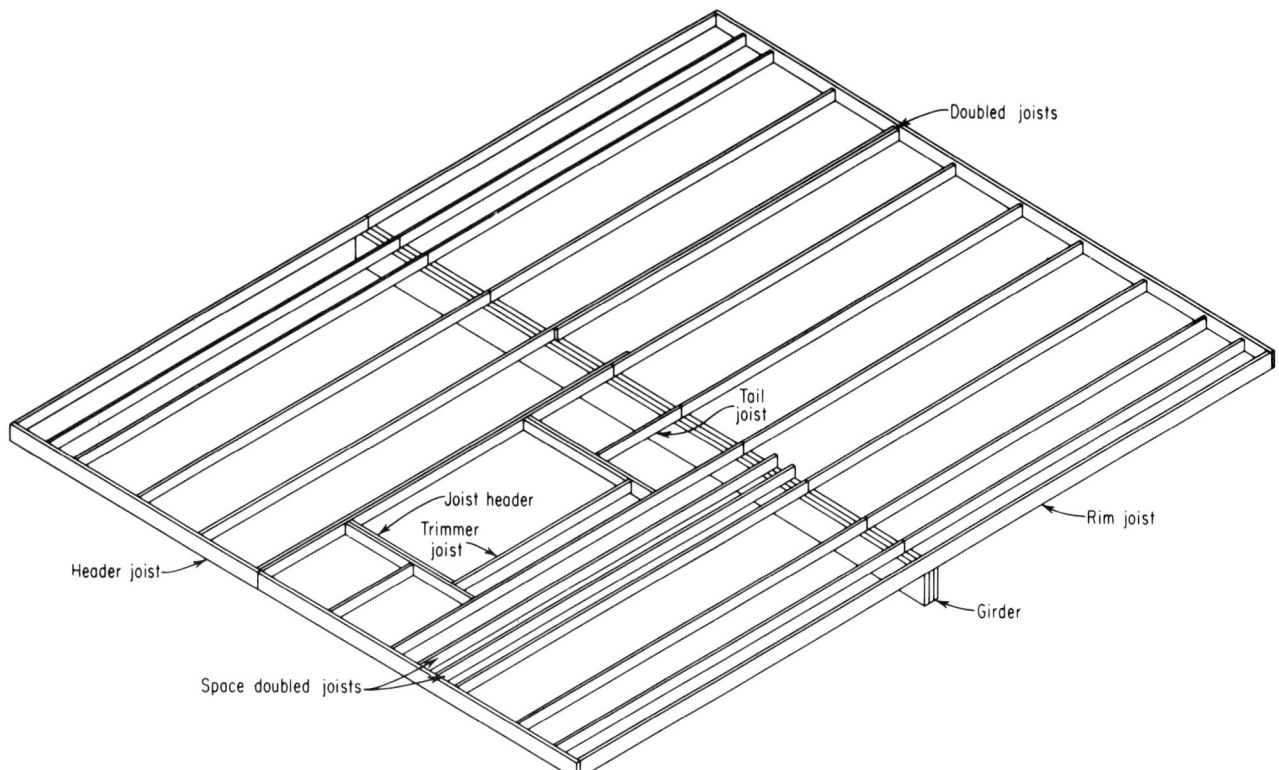

Figure 19.2 Typical floor framing. [From R. C. Smith, *Principles and Practices of Light Construction,* 3rd ed. (Englewood Cliffs, N.J.: Prentice-Hall, Inc., 1980).]

joists seldom have to be cut to length, whereas roof rafters are almost always measured and cut, involving more handling of material. Floor joists can be lifted in bundles to their final positions, whereas most roof rafters require individual attention for their final positioning. Some floor joists go in at lower levels, whereas all roof rafters have to be hoisted to the usually higher roof levels. If the joists require to have rows of cross-braces installed between them, a separate calculation will be necessary to figure out the cost of installing this item, at the maximum rate of about 10 sets of braces installed per carpenter labor hour.

It is also possible to calculate labor hours for joists in terms of area covered. Consider a standard area of 100 sf, with joists 10'-0" long at 16 in. on centers. In a typical width of 10'-0", there are [(10 × 12)/16] 7.5 spaces between the joists; therefore, 8 joists each 10'-0" long will be required, for a total length of 80 lf. From Table 19.8 it takes about 15 hours to install 1000 lf of joists, so the productivity for carpenter labor hours for 100 sf will be about [(80/1000) × 15] 1.20 hours.

In metric units, for 10 m² with joists 3 m long spaced at 400 mm, there will be [(10/3 = 3.3) and (3.3/0.4 = 8.33)] 9 joists each 3 m long for a total of 27 m. From the table it takes about 15 hours to install 300 m of joists, so it will take about [(27/300) × 15] 1.35 hours to install 10 m². The volume of lumber in the standard area is determined by multiplying the foregoing lengths by the nominal cross-sectional areas of the joists selected, and adding a small factor for waste.

Vertical Stud Walls: The external walls of wood-framed buildings consist of horizontal wall plates supporting vertical studs, stiffened with diagonal braces at corners and with girts or firestoppings placed between the studs at about the midpoint. Although stud walls are slightly more complicated than joisted floors or ceilings, productivity factors have been fairly well established at the magnitudes in labor hours per 1000 bf shown in Table 19.10.

TABLE 19.10
Productivity Factors for Installing Studs

Stud size (in.)	Spacing (in.)	Carpenters	Helpers
2 × 4	12	20–25	5–6
	16	22–26	5–6
	24	24–27	5–6
2 × 6	12	19–24	5–6
	16	21–25	5–6
	24	23–26	5–6

To show how labor price is determined, consider the installation of a wall consisting of 2 × 6 studs at 16 in. on centers (Figure 19.3), using values at the low end of the scale:

Carpenters:	21 hr @ $20.00	= $420.00
Helpers:	5 hr 15.00	= 75.00
Total labor cost for 1000 bf		= $495.00

Unit price per board foot (495/1000) = $0.50/bf

For comparison, recalculate the unit price, using values at the high end of the scale:

Carpenters:	25 hr @ $20.00	= $500.00
Helpers:	6 hr 15.00	= 90.00
Total labor cost per 1000 bf		= $590.00

Unit price per board foot (590/1000) = $0.59/bf

Note that the second unit price is about 18% higher than the first one.

It is also possible to express productivity for building framed walls in terms of labor hours per square feet of wall area. If one considers a standard area of 100 sf, with a typical height of 8'-0", the length will be (100/8) 12'-6". In such a portion of wall, there would be three horizontal plates (one at the bottom and two at the top) each 12'-6" long, and at 16-in. centers there will be [(12.5 × 12)/16] 9.38 or 10 studs each 8'-0" long, for a total length of 117'-6". From Table 19.7, it takes an average of 25 labor hours to frame 1000 lf of lumber in interior partitions; it will therefore take [(117.5/1000) × 25] 2.94 or about 3 labor hours to frame up 100 sf of such work, on the average. In simple work, this figure will decline to about 2.5 hours and in more complicated work it may rise as high as 3.5 hours.

Sheathing: Sheathing applied to horizontal floor joists and vertical wall framing is very common, indeed virtually essential, in most wood-framed buildings. In this section, two of the most common types of sheathing will be examined: diagonal shiplap sheathing on floor joists, and plywood or particle board sheathing on wall studs. Table 19.11 gives labor hours per 1000 bf, as before.

TABLE 19.11
Productivity Factors for Installing Sheathing

Boards (in.)	Joist space (in.)	Carpenters	Helpers
1 × 6	12	13–14	5–6
	16	12–13	5–6
	24	11–12	5–6
1 × 8	12	12–13	5–6
	16	11–12	5–6
	24	10–11	5–6

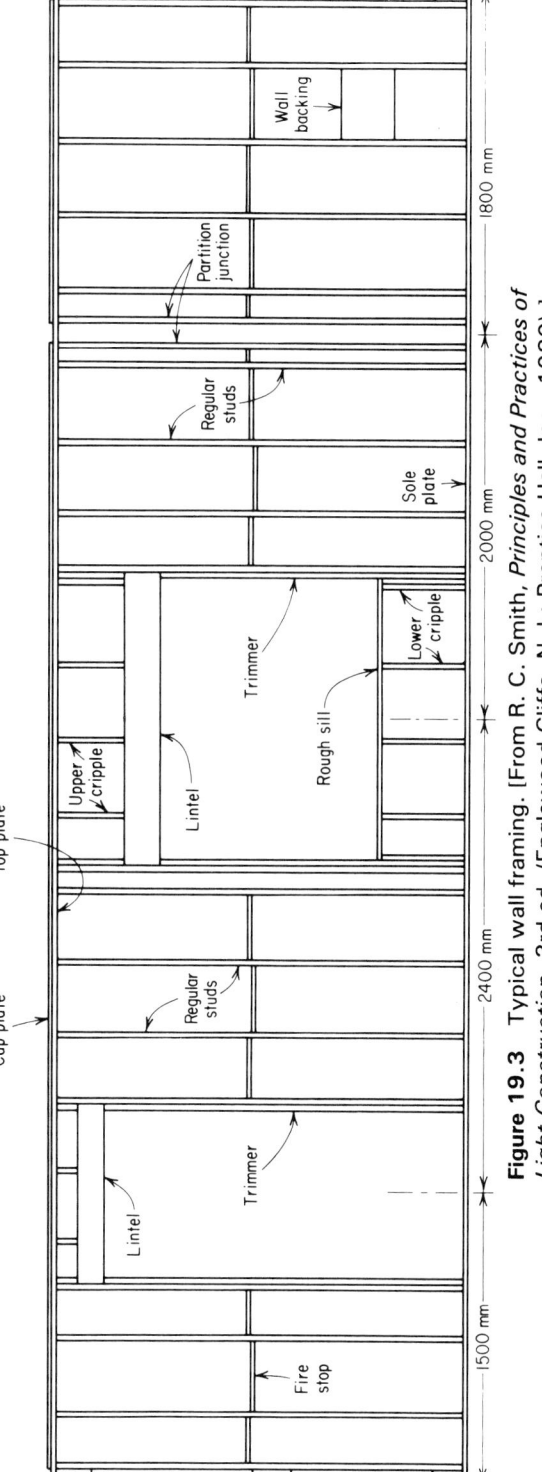

Figure 19.3 Typical wall framing. [From R. C. Smith, *Principles and Practices of Light Construction*, 3rd ed. (Englewood Cliffs, N.J.: Prentice-Hall, Inc., 1980).]

Raising wall sections. [From R. C. Smith, *Principles and Practices of Light Construction,* 3rd ed. (Englewood Cliffs, N.J.: Prentice-Hall, Inc., 1980).]

From the table, the installation cost for a flooring specification calling for 1 × 8 shiplap laid diagonally on joists at 24 in. on centers would be calculated as follows:

$$\begin{array}{lrl}
\text{Carpenter:} & 10 \text{ hr @ } \$20.00 & = \$200.00 \\
\text{Helpers:} & 5 \text{ hr} \quad 15.00 & = 75.00 \\
\text{Total labor cost for 1000 bf} & & = \$275.00
\end{array}$$

Unit price per board foot (275/1000) = $0.28/bf

The productivity factors for 1000 sf of rough plywood sheathing on stud walls are about 8 to 10 hours for carpenter's time and about 4 to 5 hours for helpers. Using such data, the unit price is calculated as follows:

$$\begin{array}{lrl}
\text{Carpenter:} & 8 \text{ hr @ } \$20.00 & = \$160.00 \\
\text{Helper:} & 4 \text{ hr} \quad 15.00 & = 60.00 \\
\text{Total labor cost for 1000 sf} & & = \$220.00
\end{array}$$

Unit price per square foot (220/1000) = $0.22/sf

Conclusion: The labor price for installation of roof sheathing, floor and roof decking, soffits, bearers, blockings, grounds, furring, and other rough carpentry items can be calculated in a manner similar to the examples detailed above. The central and most difficult issue is the correct determination of productivity factors. Such factors can be established by reference to texts, by observation, and to some extent by intuitive thought.

The majority of other carpentry framing items will find their cost levels somewhere between the extremes given in the foregoing examples. It might be a useful

19.2.6 09250 Gypsum Drywall

In typical single-layer construction, gypsum drywall systems consist of regular boards 13 mm or ½ in. thick, or fire-rated boards 16 mm or ⅝ in. thick, secured to wood or steel studs or joists at 400 mm or 16 in. on centers, with annular-ring nails having 6-mm or ¼-in. heads and 32-mm or 1¼-in. shanks, and placed at 200 mm or 8 in. on centers along supports. A few common nails are also used to temporarily secure the boards in position.

In double-layer construction, gypsum backing board, 9 mm or ⅜ in. thick, is first secured to the framing at right angles to the second layer, which is adhered to the backing with joint compound spread with a notched trowel. The board costs approximately \$1.65 per m^2 or \$1.50 per sy. The nails cost approximately 35 cents per kg or 15 cents per lb.

The joints between the boards are then filled with compound and taped with kraft paper tape, available in rolls about 50 mm wide by 75 m long or 2 in. wide by 80 ly long. The indented nail heads are also filled with matching compound. The tape costs about \$1.50 per roll, and the filler compound costs about 50 cents per kg or 25 cents per lb. Assume that about 5 kg or 10 lb or compound is used with each roll of tape. The quantity of tape and filler translates into approximately 1 m per m^2 or 1 ly per sy of drywall to be finished.

In large unobstructed areas, an experienced tradesman with a helper should be able to install about 9 m^2 or 10 sy of single-layer gypsum drywall per hour. In small or confined areas, this productivity factor will decline by about 50%. Taping and filling by hand should achieve about 9 m or 10 ly per hour; it may be noted that machine finishing should improve this output by a factor of 4 or 5. All such productivity factors should be confirmed by actual observation and modified as necessary.

In gypsum drywall and plaster finishes, the basic unit used to price the work is a hypothetical interior partition wall, finished on two sides, 40 m long and 2.5 m high (100 m^2) in metric units, or 100 ft long and 9′-0″ high (100 sy) in imperial units. A separate unit can be calculated to determine unit prices to finish ceilings and other soffits. Prices for the supply and installation of metal trim at typical corners and open edges of drywall can also be worked out.

Imperial Example. Assume 1800 sf of ½-in. boards on two sides of wood studs, taped and filled.

Item	Quantity	Cost	Price	Total
Gypsum board	1800 sf	0.20	$360.00	
Annular nails	15 lb	0.15	2.25	
Common nails	2 lb	0.10	0.20	
Kraft tape	2.5 rolls	1.50	3.75	
Compound	25 lb	0.25	6.25	
(Basic cost)		(372.25)		
Taxes on cost	$372.25	@5%	18.65	
Waste on cost	$392.00	@5%	19.60	
Total material cost				$ 411.60

Item	Quantity	Cost	Price	Total
Tradesmen				411.60
Installing	18 hr	20.00	$360.00	
Finishing	24 hr	20.00	480.00	
Helper	10 hr	15.00	150.00	
Total labor cost				$ 990.00
Small tools	$990.00	@3%	$ 29.70	
Equipment	(allowance)	item	30.00	
Total equipment cost				$ 59.70
Total installed cost for 1800 sf of system:				$1461.30
Cost per square foot ($1461.30/1800 sf) = $0.82/sf				

An allowance of approximately 20% (12% for overhead and 8% for profit) still has to be added to the unit price calculated above, either to each unit price or to the estimate as a whole at the end. This will bring the unit price up to a total of about $1.00 per sf or $10.00 per m² for the composite supply and installation of the specified system. For small jobs, with less than 100 m² or 100 sy, an inefficiency factor up to 25% may be included in the calculations. A simple statistical curve can be developed to relate efficiency to area for ready reckoning.

19.3 CONCLUSIONS

The benefit of choosing a fairly large basic unit of work (such as 100 m² or 100 sy, or as in the example above, 1800 sf) on which to calculate the final unit price is that more refined judgments can be applied to the quantity and unit cost of each element in the buildup of the total cost, and less significance is given to each element when the grand total is divided by the number of units in the original base. If one were to try to calculate a price per square meter or square yard for a system such as drywall using such small units for the elements, any minor misjudgment of quantity or cost or even rounding-off dollars and cents would have proportionately larger effects on the validity of the final unit price. In other words, it is better to divide large costs by large quantities to produce small but accurate unit prices than to calculate small and possibly inaccurate prices which may subsequently be multiplied by large quantities to produce doubtful results.

QUESTIONS

19.1. In equipment costs, distinguish between fixed and variable costs by defining each of these terms and by giving two examples of each type of cost.

19.2. Give three reasons why it is not possible (nor necessary) to calculate precise unit prices for excavation work.

19.3. Why is formwork for concrete best considered by some estimators as "equipment" rather than as "materials"?

19.4. Identify the primary economic factor involved in the calculation of unit prices for formwork. List three secondary factors related to the primary factor.

19.5. Identify and briefly describe four methods by which concrete can be placed in its final position at the site. State one limitation of each method selected.

19.6. Briefly describe the operation of finishing a monolithic structural slab by commenting on the crew and equipment involved. Give an approximation of man-hours to establish probable productivity, using either metric or imperial measurement units.

Chap. 19 / Pricing Examples—Selected Trades

19.7. Explain why carpentry productivity in joist installation appears to increase in proportion to the cross-sectional dimensions of the joists. Also explain why the productivity might actually decrease.

19.8. If a drywall installer has to work in confined areas, such as small closets in an apartment building, by how much might the productivity be expected to decline, expressed as a percentage? Give two practical reasons to support your answer.

19.9. What is the primary benefit of choosing a fairly large basic unit of work on which to calculate final unit prices? What is the risk if such basic units are too small? Show one brief example of each choice to support your contention.

CHAPTER 20
PRICING EXAMPLES SELECTED ELEMENTS

20.1 GENERAL OBJECTIVES

20.1.1 Preamble

In Chapter 19, pricing techniques of interest to contractors were illustrated and explained. In this chapter, pricing techniques of interest to designers are introduced for study and discussion. It will be seen that the basic approach to pricing of elements is not all that different to the techniques explained for trades in the previous chapter. The main difference lies in the amount of detail included, and to some extent, on the quality of the data sources.

20.1.2 Objectives

The first objective in this chapter is to show the entire elemental analysis process in sequence, to put the various parts of that sequence into context for the reader. The second objective is to illustrate the development of unit prices for selected elements for the building in question, which is the Wood-Framed House shown in Part VI. A third objective is to present readers with opportunities to try to develop unit prices of their own, following the general approach illustrated in the examples given, and then to compare their results with the results shown.

20.2 METHODS FOR DESIGNERS

20.2.1 Preamble

Before the preliminary pricing of any proposed design begins, an outline specification should be prepared, first to cause the designer to really consider each of the various parts and the makeup of the whole, and second, to form a definite basis for accurate pricing. It is actually possible to start such deliberations even before any drawings or rough sketches of the owner's requirements are committed to paper or to electronic screen by the designer. However, in most cases, reasonably detailed sketches usually exist in one form or another, before such budget studies get under way.

The outline specification is intended to be used as a base to develop (and to record) material and cost information to be included in the project. Later, as the working drawings and detailed specifications are developed and the budget aspects refined, often by technicians or technologists, they can be kept in line with the original intentions of the owners and the designer by reference to the outline.

The example that follows is based on the Wood-Framed House, for which a set of fully detailed working drawings is included in Part VI. The pricing of the elements is consistent with the methods of measurement described in Chapter 18. Although not commonly done at so late a stage in the development of documentation (nor in such detail for such a small and simple building), it is of course possible and indeed relatively easy to base an outline specification and budget on such detailed information. It really depends on how important the budget is to the owner and the designer, and at what stages they wish to start to develop such data to be carried into the contract negotiation stages of the general design and development of the project. In this particular case, however, it is not possible (or important) for the reader to know whether the outline preceded the drawing details, or vice versa.

20.2.2 Element Identification

The outline shown in Figure 20.1 is based on the List of Elements recommended by the Canadian Institute of Quantity Surveyors, and shown in Section 18.1.3.

ELEMENT OUTLINE FOR THE METRIC HOUSE

1. *Substructure*
a. Normal foundations
Soil conditions: undisturbed clay, densely packed, with a thin layer of top soil
Disposal of excavated material: off-site, except topsoil
Backfill: imported gravel or crushed stone
Type of foundation: continuous concrete footings
    At perimeter: 200 × 400 mm
    At center: 200 × 300 mm
Drain tile: 200 mm agricultural tile at perimeter
b. Basement
Average depth below grade: 1978 mm
Feature: load-bearing frame wall at center
c. Special foundations
Sheet piles: corrugated steel @ window wells (4-off)
Porch footing: strip footing supporting foundation wall
    Footing: 150 × 350 mm
    Foundation: 150 mm thick

2. *Structure*
a. Lowest floor
Concrete: slab on grade, 75 mm thick, 20 mPa, 20-mm aggregate
Dampproofing: polysheet, 4 mil thick
Blinding: skim coat concrete, 25 mm thick
Base: pit-run gravel, 125 mm thick
Perimeter: expansion joint, asphalt-impregnated fiberboard

Figure 20.1 Element outline.

Chap. 20 / Pricing Examples—Selected Elements

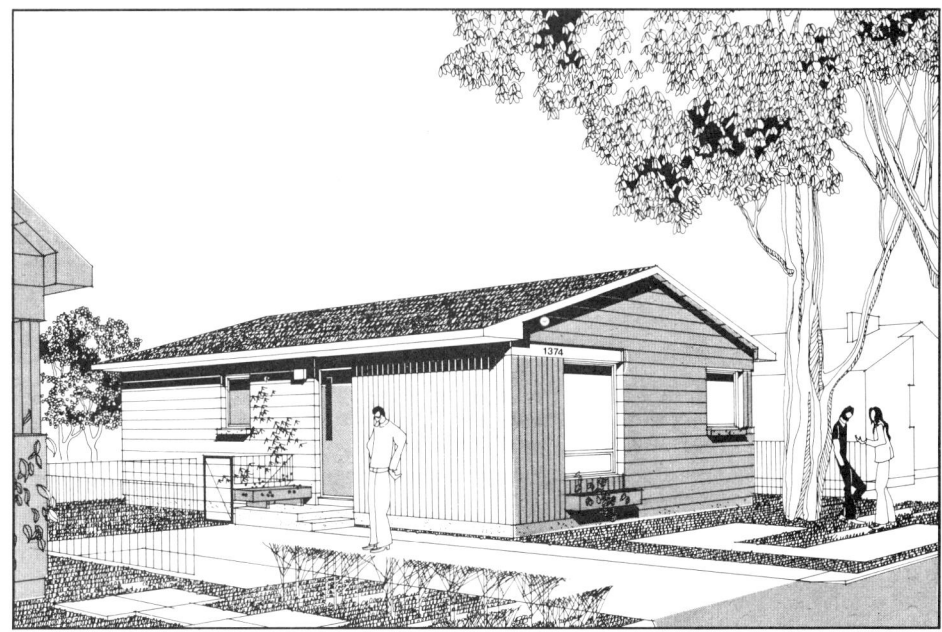

Figure 20.1 (a) Metric House—Perspective.

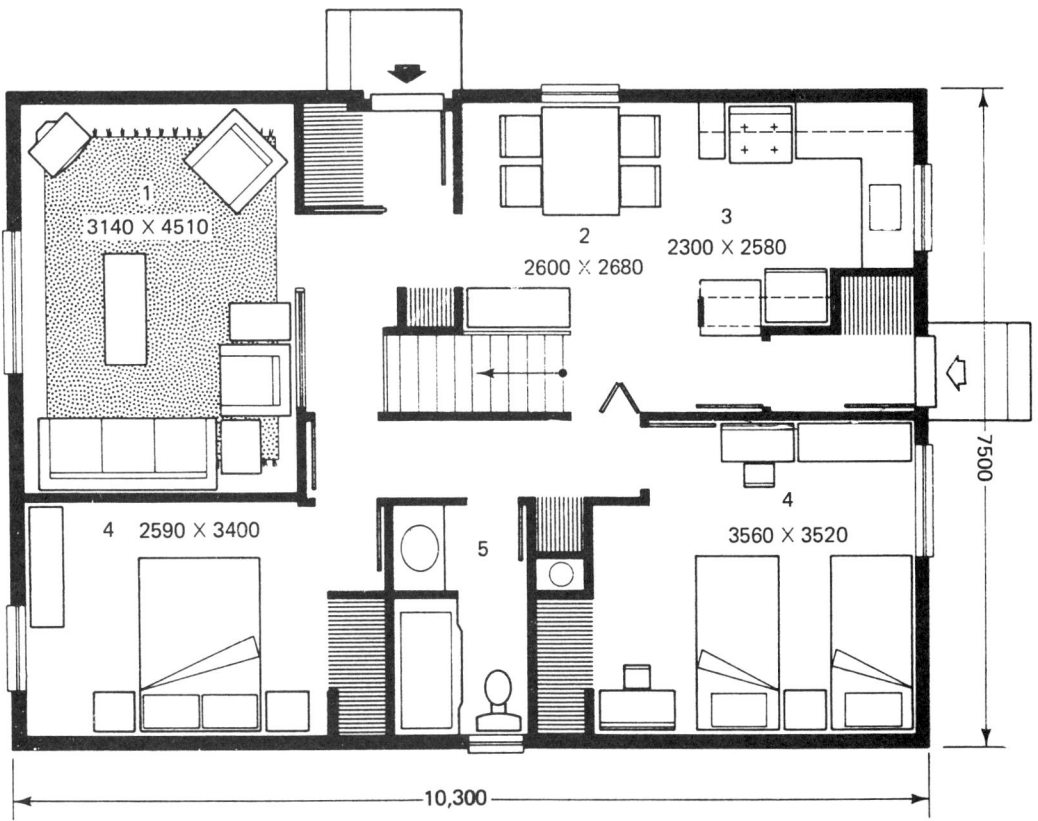

Figure 20.1 (b) Metric House—Floor Plan.

b. Upper floor
Joists: No. 1 spruce, 38 × 184 @ 400 mm o.c. (3750 span)
Hangers: galavanized steel
Sheathing: T&G plywood, 16 mm thick
Braces: ribbon cross-bridging, 19 × 89
Firestopping: to code requirements
c. Roof frame
Type: Douglas fir fink trusses @ 600 mm o.c.
Design: span, 7300 mm; slope, 1:3; load, 1.52 kN/m²
Sheathing: plywood, 13 mm thick
Soffit frame: 38 × 38 fir @ 600 mm o.c.

3. *Exterior Cladding*
a. Roof finish
Cover: asphalt shingles, granule coated, 2.5 kg/m²
Insulation: fiberglass batt, value R-20, 140 mm thick
Vapor barrier: polysheet, 4 mil thick
Flashing: galvanized steel @ ridge, gables, chimney, and vents
Fascia: dressed clear cedar, 19 × 140 mm
Soffit: fir plywood, 9 mm thick, vented
Eave trough: continuous aluminum, 100 mm wide
RWL: sectional aluminum, 50 × 50 mm square
b. Walls below ground
Foundation walls: reinforced concrete, 200 mm thick
Concrete strength: 20 mPa, 30-mm aggregate
Reinforcement: 10 m bars, 2 @ top, 2 @ bottom
c. Walls above ground
Type: western frame, 38 × 140 @ 600 mm o.c.
Material: construction-grade Douglas fir
Dampcourse: asphalt strip, 2.5 kg/m²
Sheathing: fir plywood, 9 mm thick, with building paper
Insulation: fiberglass batt, value R-20, 140 mm thick
Vapor barrier: polysheet, 4 mil thick
Finish: patent vinyl siding, 13 mm thick, with trim
d. Windows
Type: 1. Top hung open out
      2. Side hung and sliding
      3. Fixed sash
Material: residential aluminum, thermal break
Finish: anodized (brown)
Glass: float, double, 13-mm air gap
Hardware: industry standard
e. Exterior doors
Type: top half double-glazed, bottom half solid pine
Frames: No. 1 Douglas fir, one piece, painted
Hardware: pair and half butts, deadbolt, latchkey
f. Balconies and projections—None

4. *Interior Partitions and Doors*
a. Permanent partitions
Type: western frame, 38 × 89 @ 400 mm o.c.
Material: construction-grade Douglas fir
Heights: 1. Main floor: 2400 mm
         2. Basement: 2216 mm
b. Movable doors
Type: flush panel, hollowcore, mahogany veneer
Quality: standard residential, prehung
Finish: stain, 2 coats polyurethane

Figure 20.1 (cont.)

Frames: No. 1 Douglas fir, one piece, stained as doors
Hardware: 1. Swing: Pair of butts
       Lock knobs on master and bath
       Regular latch knobs on rest
    2. Slide: Overhead track on bifold and pocket
       Standard door pulls on all sliders
c. Glazed partition
Type: wood frame, with clear fir trim, painted
Glass: float, 3 mm thick, one piece, fixed

5. *Vertical Movement*
a. Stairs
Type: closed tread and riser, straight run
Material: fir or spruce
Rise, 2400 mm; run, 2510 mm; length, 3320 mm
Components: Stringers: 38×235 mm = 0.06 m^3
     Risers: 16×200 mm = 12 each
     Treads: 19×210 mm = 12 each
Finish: stain, 2 coats polyurethane
Handrail: Steel rail one side, full length
    Steel balustrade other, at top only
Finish: primer and acrylic paint on steel

6. *Interior Finishes*
a. Floors
Basement: concrete, steel trowel, no paint
Kitchen, bathroom: vinyl asbestos tile, with rubber base
Remainder: broadloom carpet and underlay, with wood trim
b. Walls

Basement

Exterior upper: 39×39 fir strapping @ 400 mm o.c.; rigid Styrofoam insulation, 39 mm thick; gypsum drywall, 12 mm thick; 2 coats latex
Exterior lower: concrete, cleaned, 2 coats latex
Interior total: gypsum drywall, 12 mm thick, 2 coats latex

Main Floor

Teak veneer: Living room north wall, 6 mm thick
Ceramic tile: bathroom tub enclosure, 25 mm^2 mosaic
Gypsum drywall: all other walls, 12 mm thick
Stain and seal: living room teak veneer
Primer and 2 coats oil-base enamel: kitchen, dining room, and bathroom
2 coats latex: all other drywall, including closets
c. Ceilings
Exposed frame: laundry, utility, storage
Gypsum drywall: all other ceilings, 12 mm thick
Primer and 2 coats oil-base enamel: kitchen, dining room, and bathroom
Plastic spraytex: living room, bedrooms, hall
2 coats latex: all other drywall, including closets

7. *Fittings and Equipment*
a. Fittings
Kitchen: Base cabinets = 3.3 m prefab units
     Wall cabinets = 3.0 m prefab units
     Countertop   = 3.3 m prefab units
Bathroom: Vanity counter = 1.0 m prefab unit

Figure 20.1 (cont.)

```
            Closets:    Shelving units       = 6 units
                        Coatracks            = 4 units
            Hardware:   Coat hooks           = 6 units
                        Toilet accessories   = 1 set
                        Medicine cabinet     = 1 unit
                        Mirrors              = 1 unit
         b. Equipment
         Included in Section 8

         8. Services
         a. Electrical and fixtures = consultant's price
         b. Plumbing and fixtures   = consultant's price
         c. Heating and chimney     = consultant's price

         9. Site Development—nil

         10. Overhead and Profit
         a. Site overhead   = 5% of cost
         b. Fee and profit  = 10% of cost

         11. Contingencies
         a. Design changes = 3% of cost
         b. Escalation     = 5% of cost
```

Figure 20.1 (cont.)

20.2.3 Element Pricing

Having now prepared the outline specification for all elements in detail, the next step involves a choice to do one of two things first: either develop appropriate unit prices to attach to each of the quantities measured in the elemental analysis process, or do the measurement of the elements. It is generally better to do the measurement, as some minor aspects of cost might emerge for consideration during the measurement process.

In this particular example, the pricing is shown next (Figure 20.2) for convenience, because of its close relationship to the outline specification details listed in Figure 20.1. The measurement was in fact done before the pricing. Furthermore, all the elements are not shown priced in detail, only a few of the most complicated ones, because the primary objective here is to show the principles and techniques involved in pricing elements in general. Students can and should develop unit prices of their own, for ones not explained in detail, for comparison with the complete list of prices shown in the elemental analysis sheets for this particular building in this book.

To assist readers to make their own analyses, the price development is given for the first item (a. Lower floor) shown in the following elemental analysis, and exemplifies the technique to determine an adequate unit price for the first component of the element. All the other components in all the other elements are similarly derived.

It may be said that opinions will differ as to the appropriate amounts to allow for each item in the analysis. In general, it is prudent to allow generous amounts of money for each item under consideration, for several reasons: first, there are always minor items or details in any proposed building design which will be overlooked or omitted; second, normal items which are overvalued can always be trimmed back or even deleted if necessary at a later date; and third, the designer is

Chap. 20 / Pricing Examples—Selected Elements

Price Analysis of Selected Elements

2. *Structure*
One can take a representative area, such as 10 m² (or 100 sf), and work out in detail the required amount of concrete and gravel, or lumber and plywood (as described under contractor's methods), price out these full quantities, then reduce these costs to a unit price per square meter, as indicated.

    a. Lower floor

Concrete:	$13.01/m²
Vapor barrier:	3.99/m²
Skim coat:	5.68/m²
Gravel base:	2.81/m²
Total	$25.49/m²

    b. Upper floor

Joists:	$ 7.03/m²
Hangers:	1.80 each
Plywood:	7.67/m²
Bridging:	1.54/m²
Total	$18.04/m²

3. *Exterior Cladding*
The technique to price this element is the same as described for Structure above.

    c. Walls above ground

Stud frame:	$ 8.82/m²
Dampcourse:	0.04/m²
Sheathing:	6.79/m²
Building paper:	0.91/m²
Insulation:	4.73/m²
Vapor barrier:	0.71/m²
Vinyl siding:	17.76/m²
Total	$39.76/m²

    e. Exterior doors
The simplest and most common way to price elements such as doors and windows is to extract basic prices from current construction cost data books, and then to modify these basic prices to suit the locality, quality, and idiosyncracies of the project in hand.

Prehung door and frame:	$250.00 each
Glass panel (+20%):	× 1.2
Adjusted cost:	$300.00
Add lockset:	67.50 each
Total installed cost	$367.50 door

Door area = 910 × 2030 mm = 1.85 m²

Unit price = 367.50/1.85 = $198.65/m²
(This price could be rounded off @ $199.00/m²)

5. *Vertical Movement*
Generally, stairs are best worked out in some detail, as the price can vary enormously relative to the configuration and complexity of the construction. The more one breaks down the measured components of the stair, the more latitude or lee-

Figure 20.2 Price analysis.

> way one imparts to the various unit prices used to determine the cost of the composite whole.
>
> a. Stairs
>
> | | | | | | |
> |---|---|---|---|---|---|
> | Stringers: | 0.06 m³ | @ $304.00/m³ | = | $ | 18.25 |
> | Risers: | 12 each | @ 10.33 | = | | 123.96 |
> | Treads: | 12 each | @ 16.57 | = | | 198.87 |
> | Finish: | 24 each | @ 1.53 | = | | 36.62 |
> | Handrail: | 3.33 m | @ 16.43 | = | | 54.72 |
> | Install: | 36 hr | @ 17.00 | = | | 612.00 |
> | Total cost, supply and install | | | = | | $1044.42 |
>
> Unit price (1044.42/3.324) = $314.21/meter
>
> **7.** *Fittings and Equipment*
> These are priced in a similar manner to doors and windows, by selecting a basic price from a data book and modifying it to suit the project.
>
> a. Fittings
>
> | | | | |
> |---|---|---|---|
> | Kitchen base: | 2.0 m | @ $339.70 | = $680.00 |
> | | 1.3 m | @ 415.38 | = 540.00 |
> | Countertop: | 3.3 m | @ 75.63 | = 250.00 |
> | Kitchen wall: | 3.0 m | @ 181.93 | = 546.00 |
> | Bath vanity: | 1.0 m | @ 124.77 | = 125.00 |
>
> The unit prices are adjusted to suit the specifics of the proposed design, but the results can be rounded off safely to the nearest dollar. The remainder of the fittings components, such as shelf units, bathroom accessories, and so on, are similarly priced. In this example, such costs would amount to about $300.

Figure 20.2 (cont.)

not bidding to build the job on a competitive basis but is only suggesting a possible budget for the owner's guidance. Ideally, the owner's budget should exceed the contractor's bid.

Finally, it should be noted that the actual dollar values used in this specific example are of little significance and should not be used in a real-life situation; the technique of developing a logical economical argument for the end result is the point to note.

Price Analysis for Concrete Slab[a]

Component	Cost/m²
Material	$ 5.82
Waste (5%)	0.29
Placing	2.00
Screeding	1.40
Finishing	2.50
Curing	1.00
Total cost/m²	$13.01

[a]Using assumed 1985 costs.

Chap. 20 / Pricing Examples—Selected Elements

The unit price of $13.01 is therefore entered into the price analysis for the first item in the first element, in Figure 20.2 above.

20.2.4 Element Summary

The list of elements, the measured quantities of each element, and the unit price developed for each element are now ready for insertion into the Elemental Analysis Summary Sheet. Although there is a general pattern recommended for such summary sheets, each project may require a slightly modified version of the general pattern, to accommodate individual characteristics of the project design. The main features of a typical Summary Sheet are presented in Figure 20.3, with a brief comment on the significance of each item shown in the notes that follow.

ELEMENTAL ANALYSIS – SUMMARY SHEET							
Project Title:				Designer:			
Gross Floor Area:				Estimator:			
Date:		Analysis No.:		Checker:			
Element	Ratio	Quantity	Unit	Price	Amount	Cost/m^2	Total

Figure 20.3 Summary form.

Notes on Figure 20.3

1. The project is identified by a short title.
2. The gross floor area (G.F.A.) is calculated for reference.
3. The names of the designer, estimator, and checker are entered.
4. The date on which the summary is prepared is recorded.
5. The project or job number is entered.
6. The number of the particular analysis is entered.
7. Appropriate titles are selected for column headings.
8. The components can be arranged in many other appropriate ways.

Column titles: The titles and subtitles of the main elements of the building go into the elements column.

The *ratio* column lists the quantity of the particular building element expressed as a simple proportion of the gross floor area. There are advantages to calculating these ratios:

1. They can act as confirmation that the proportion of each element is generally of the right order of magnitude compared to similar previous projects.

2. They can be used to predict costs for elements which have not or cannot be accurately measured for one reason or another, where, for example, drawings have not been sufficiently detailed.
3. They can be used to reduce all costs for each element to a unit price related to the gross floor area of the building, again for comparison with previous work or for prediction of costs of the present work.

The *quantity* is the probable amount or extent of the element, followed by the measurement unit.

The *unit price* of the element comes next, followed by the arithmetical extension of the quantity by the unit price.

The *cost per area* is the amount of the element cost, divided by the gross floor area (G.F.A.). In this way, it is possible to relate prices, one to another, and also to compare prices on one project with another, as all prices are reduced to one constant, the G.F.A.

The *gross rate* is the combined area rate for all components in the element, and the final *total* column is the cumulative cost.

It is worthwhile to experiment with different configurations of the column contents; the one shown is representative of only one way to arrange the page for convenience.

20.3 WORKED EXAMPLE

In Figure 20.4, a Summary Form is shown, modified to suit an Elemental Analysis for the Wood-Framed House under consideration, and complete with all relevant data. Such work would normally be prepared by hand or by electronic spreadsheet; the example shown here has been typed primarily to enhance legibility.

It will be seen that in the worked example, data have been inserted into or calculated for each of the nine columns on pages 1 and 2 of the form. On page 3, the ratio and cost per area columns are left blank. Although it is certainly arithmetically possible to produce data for these two columns on page 3, the usefulness of such results is questionable.

It may also be noted that at the grand total level, the gross rate of $377.25 when multiplied by the gross area of 156.30 m^2 produces a final total of $58,964, not the $58,969 shown. This $5 difference is caused by the necessity to round-off intermediate element rates to the nearest cent, and is unimportant.

What is more important is whether or not the owner and the designer want to pay $59,000 for such a building. Having established this price, it is now possible to consider simpler or less expensive options, by going back through all of the element calculations, by making reductions in quality, by reducing the overall area of the building, by refining the cost information, or by reconsidering a number of allowances which have been included, among other possibilities.

There are always differences of opinion as to how to handle the percentages for overhead and profit, and for the design, escalation, and post-contract contingencies. Some estimators will calculate these percentages as shown in the example, basing each on the basic subtotal cost; other estimators will calculate each percentage separately, based on a cumulative total cost. Such decisions can have a significant effect on the grand total. In this example, the difference in the final total can be as much as $1700, which represents about 3% of the grand total.

Figure 20.5 shows a typical blank Elemental Analysis Form, included for use by the reader in preparing forms to do such analysis on future projects.

Project Title: Metric House Gross Floor Area: 156.30 m² Date: March 1985 Analysis No.: 01					Designer: CMHC Staff Estimator: GMH Checker: LFH				
Measurement					Pricing				
Element	Ratio	Quantity	Unit	Price	Amount	Cost/m²	Gross	Total
1. Substructure a. Normal foundation b. Basement foundation c. Special foundation Total	0.15 1.08 0.06	22.75 170.18 10.08	m² m³ m²	298.14 2.12 27.33	7683 361 275	43.40 2.31 1.76	47.46	7,418
2. Structure a. Lowest floor b. Upper floor c. Roof construction Total	0.49 0.51 0.63	77.25 79.05 98.12	m² m² m²	25.49 18.04 21.48	1969 1426 2108	12.60 9.12 13.49	35.16	5,503
3. Exterior Cladding a. Roof finish b. Walls below ground c. Walls above ground d. Windows and frames e. Doors and screens Total	0.63 0.52 0.59 0.04 0.02	98.13 80.90 91.99 6.47 3.69	m² m² m² m² m²	17.53 4.30 39.76 165.84 199.00	1720 348 3658 1073 734	11.00 2.23 23.40 6.87 4.70	48.20	7,534
Project No: 85.03 Page 1 of 3					Carried forward:		130.82	20,455

Figure 20.4 Elemental analysis: summary sheet.

Measurement				Pricing				
Element	Ratio	Quantity	Unit	Price	Amount	Cost/m²	Gross	Total
				Brought forward:			130.82	20,455
4. Parts and Doors								
a. Permanent	0.65	101.00	m²	5.20	525	3.36		
b. Structural	0.06	9.90	m	6.81	67	0.43		
c. Glazed parts	0.01	1.40	m²	15.10	22	0.13		
d. Doors and locks	0.06	16.65	m²	43.92	731	4.68		
Total							8.60	1,345
5. Vert. Movement								
a. Staircase	0.02	3.32	m	314.21	1043	6.67	6.67	1,043
6. Int. Finishes								
a. Floor: Trowel	0.50	77.25	m²	4.09	316	2.02		
Carpet	0.42	65.14	m²	17.92	1167	7.47		
Vinyl	0.07	11.48	m²	6.81	78	0.50		
Total							9.99	1,561
b. Ceiling: Gypsum	0.49	77.00	m²	6.24	481	3.12		
Spraytex	0.29	45.00	m²	6.20	278	1.80		
Latex	0.21	32.00	m²	4.84	155	1.00		
Total							5.92	914
c. Walls: Gypsum	2.19	342.75	m²	8.50	2914	18.65		
Latex	2.10	327.37	m²	4.84	1585	10.14		
Wood panel	0.06	8.79	m²	40.46	356	2.28		
Ceramic	0.04	6.71	m²	32.50	218	1.40		
Total							32.47	5,073
Project No.: 85.03 Page 2 of 3				Carried forward:			194.47	30,391

Figure 20.4 (cont.)

Measurement				Pricing				
Element	Ratio	Quantity	Unit	Price	Amount	Cost/m²	Gross	Total
				Brought forward:			194.47	30,391
7. Fittings								
a. Cabinets:								
Kitchen 1		2.00	m	339.70	680			
2		1.30	m	415.38	540			
3		3.00	m	181.93	546			
Countertop		3.30	m	75.63	250			
Bath vanity		1.00	m	124.77	125			
b. Remainder		sum	—	300.00	300			
Total							15.62	2,441
8. Services								
a. Electrical		(consultant's estimate)			3000			
b. Plumbing		(consultant's estimate)			4500			
c. Heating		(consultant's estimate)			4000			
Total							73.58	11,500
9. Site Development		(excluded)			Sub-totals		283.67	44,332
10. Overhead and Profit								
a. Site overhead		44332	15%		6650			
b. Office and profit		"	10%		4433			
Total							70.91	11,083
11. Contingencies								
a. Design		44332	2%		886			
b. Escalation		"	4%		1772			
c. Post-contract		"	2%		886			
Total							22.67	3,554
					Grand Totals:		$377.25	$58,969

Project No.: 85.03   Page 3 of 3

Figure 20.4 (cont.)

ELEMENTAL ANALYSIS FORM

PROJECT DATE PAGE 1 OF 2

PROJECT NO. ARCHITECT

GROSS FLOOR AREA ESTIMATOR

ELEMENT	RATIO	QUANTITY	UNIT RATE	AMOUNT	COST PER M^2/SF	TOTAL
1. SUBSTRUCTURE						
a) Normal Foundations						
b) Basement						
c) Special Foundations						
Caissons etc. $_____						
Underpinning $_____						
Total						
2. STRUCTURE						
a) Lowest Floor Construction						
b) Upper Floor Construction						
c) Roof Construction						
Total						
3. EXTERIOR CLADDING						
a) Roof Finish						
b) Walls Below Ground Floor						
c) Walls Above Ground Floor						
d) Windows						
e) Exterior Doors and Screens						
f) Balconies and Projections						
Total						
4. INTERIOR PARTITIONS AND DOORS						
a) Permanent Partitions & Doors						
b) Moveable Partitions & Doors						
c) Glazed Partitions & Doors						
Total						
5. VERTICAL MOVEMENT						
a) Stairs						
b) Elevators and Escalators						
Total						

Figure 20.5 Element form.

QUESTIONS

20.1. Identify two purposes for accurately recording material selection and cost information used in outline specifications to prepare elemental analysis of buildings.

20.2. Explain why it is good practice for a designer to include a small design contingency in a building elemental analysis. Suggest an appropriate percentage and say how you would justify this allowance to the owner/client.

20.3. Why is it preferable to do measurement before pricing in preparing an elemental analysis of any building?

20.4. Outline a practical method for developing a suitable unit price for a staircase in an elemental analysis.

20.5. State the purpose of the "ratio" column in the elemental analysis estimating form, and list two of three advantages to which such ratios can be put.

ELEMENT	RATIO	QUANTITY	UNIT RATE	AMOUNT	COST PER M²/SF	TOTAL
6. INTERIOR FINISHES						
a) Floor Finishes						
b) Ceiling Finishes						
c) Wall Finishes						
(Wall Ratio _____)						
Total						
7. FITTINGS AND EQUIPMENT						
a) Fittings and Fixtures						
b) Equipment						
Total						
8. SERVICES						
a) Electrical						
b) Plumbing and Drains						
c) Heating, Ventilation and Air Conditioning						
Total						
9. SITE DEVELOPMENT						
a) General						
b) Services						
c) Alterations						
d) Demolition						
Total						
10. OVERHEAD AND PROFIT						
a) Site Overhead						
b) Head Office Overhead and Profit						
Total						
11. CONTINGENCIES						
a) Design Contingency						
b) Escalation Contingency						
c) Post-Contract Contingency						
COMMENTS:						

Figure 20.5 (cont.)

20.4 PRACTICE EXAMPLE

On the following page, preliminary architectural sketch plans for a small post & beam bungalow are reproduced. [See Figures 20.6 (a) and (b)]. A set of fully detailed drawings for this house are also included with others at the end of this book.

It is intended that students (and their instructors) use these drawings as a basis for a practice exercise in elemental analysis of costs, using the previous elemental worked example and its various explanatory figures as a model to follow.

This house was designed and built in California a number of years ago, and the drawings are reproduced here with permission. Further reproduction is prohibited.

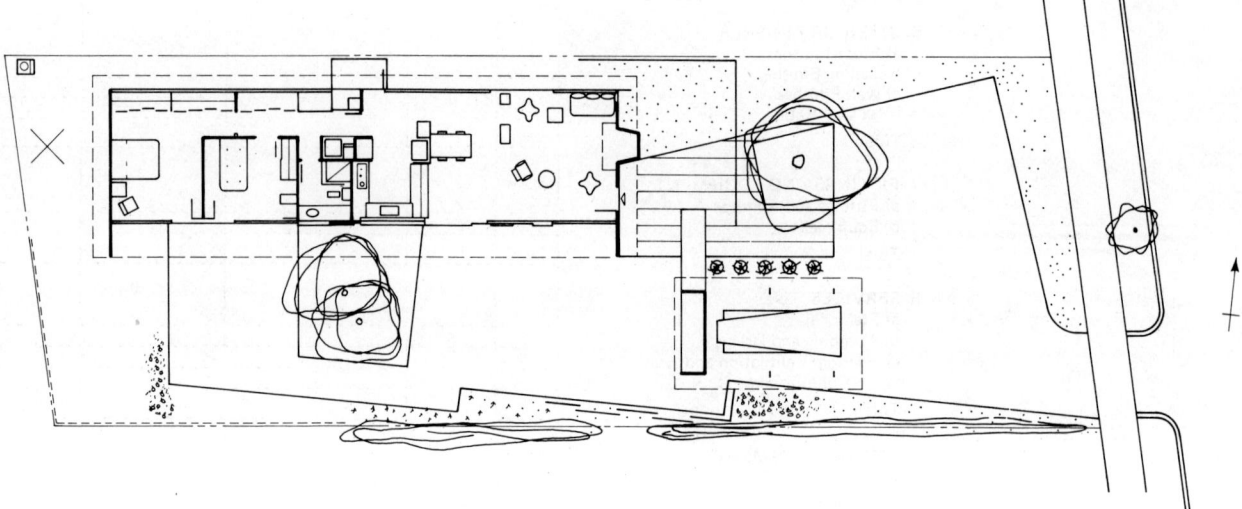

Figure 20.6 (a) Post and Beam House—Plot Plan.

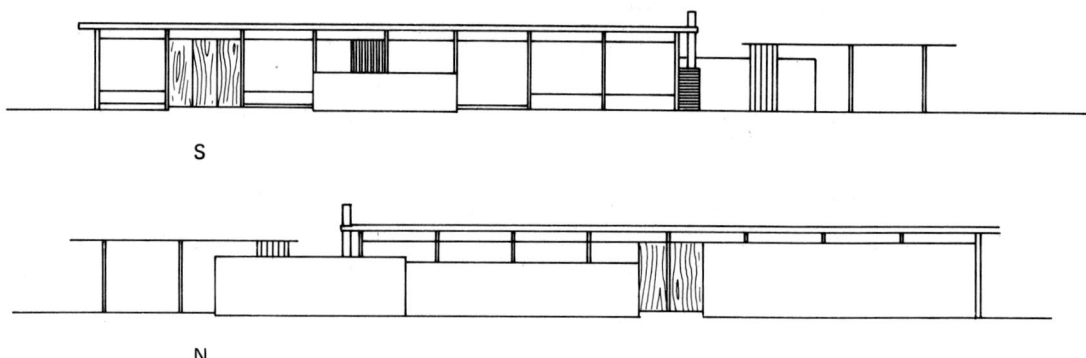

Figure 20.6 (b) Post and Beam House—Elevations.

CHAPTER 21

COMPUTER APPLICATIONS

21.1 INTRODUCTION

Before proceeding into the specific subject matter of this chapter, it is appropriate to state, for the benefit of those who may not already know, that the whole subject of computers abounds with an extraordinarily rich and indeed exotic lexicon of words and phrases to describe every aspect of this fascinating and important societal development. Terms such as "baud," "boot up," "CP/M," "d-base," "goto," "mainframe," "on-line," "RAM," "ROM," "vanilla," and so on are common in the field of computers; in fact, the vocabulary has become so extensive that a new language called "computerese" has evolved to satisfy the special needs of those people who are extensively involved with computers.

However, to understand this chapter, it is not necessary to have extensive knowledge of such terminology. It is really only necessary to know how the computer works *in principle,* and to be able to make a distinction, first between computer *hardware* and *software,* and second between computer *programming* and computer *applications.* Each of these main elements will be explained sufficiently in the following sections to suit the immediate purposes of the author and the reader. In general, this chapter primarily illustrates and concentrates on *applications* of programs specially designed for construction estimating; *programming* is virtually excluded. A minimal amount of necessary terminology will be presented and explained in context. Section 21.2 covers the main elements identified above; Section 21.3 consists of case studies of actual computer applications in construction estimating made available to the author and used with permission.

21.2 MAIN ELEMENTS

21.2.1 Computer Principles

In essence, the computer is a simple device which has a simple life: it consists of a central processing unit (CPU) which can respond to either an electrical current or pulse, or the omission of a pulse. It can also do this task at extremely high speeds, if necessary for long periods of time, virtually without error or fatigue, and with the use of very tiny amounts of power. The basic mode of CPU operation is binary; that is, it counts the number of pulses or omissions in the form of a 1 or a 0, respectively. These counts are called **bits** (from the words "*b*inary dig*its*"), and 7 or 8 of them together (which are called **bytes**) make up the basic units of information processed by the CPU. Numbers and letters might be represented by 8-bit bytes as shown in Table 21.1.

TABLE 21.1
Binary Representations of Numbers and Letters

Numbers	*Letters*	
1 as 00110001	A as 01000001	a as 01100001
2 as 00110010	B as 01000010	b as 01100010
3 as 00110011	C as 01000011	c as 01100011
9 as 00111001	I as 01001001	i as 01101001
and so on	and so on	and so on

The permutations possible with eight digits are enormous. As a result, the power of computers is usually expressed in terms of the number of thousands of bytes (*K*ilobytes) which the CPU can handle in memory, such as 8K, 16K, 32K, 64K 132K, 256K, and so on, each one usually (but not necessarily) double the preceding one. Each byte is arranged to represent any whole number between 0 and 256 or any ASCII (American Standard Code for Information Interchange) alphabetical character or control instruction. The examples given in Table 21.1 show the correct ASCII binary numbers for each of the whole numbers and upper- or lowercase letters listed. When a computer program is written, these bytes are arranged by the programmer into a special sequence to cause the CPU to perform specific functions, such as adding, multiplying, searching, deleting, printing, and so on, in a specific order.

The main advantages of computers are their ability to rapidly accept, file, store, sort, and retrieve or delete large amounts of complex data with minimal error, and in desired and advantageous patterns. Furthermore, most microcomputers are generally easy to operate, and the great majority of programs are "friendly" to the user, to use a computerese word. The main disadvantages are their ability to really foul things up on a mammoth and instantaneous scale (sometimes through no fault of the operator), and the occasional difficulty of learning how to operate particular computer sytems, especially those connected to a mainframe computer.

21.2.2 Hardware and Software

There are actually four terms used in this connection: hardware, software, firmware, and liveware.

Hardware. Computer hardware refers to the actual physical machinery, such as the computer itself and its peripheral or related equipment, such as keyboards, screens, and printers. As stated, the computer itself essentially consists of a device called the central processing unit (CPU), which is an arrangement of silicon chips containing patterns of microscopically sized circuits which permit small electrical currents or pulses to be routed into predetermined or voluntarily selected channels, sequences, or patterns. There are at least four principal types of computers likely to be encountered: *mainframe computers,* which are large machines or a complex of machines requiring their own special air-conditioned rooms, usually owned by large institutions, corporations, or government departments, and frequently able to serve a number of distant or satellite work stations or terminals; *minicomputers,* which are not unlike mainframe computers, but smaller and more movable, although not exactly portable and not usually requiring any special physical accommodation; *microcomputers,* which are usually fully portable and can sit on a ordinary office desk, much like a typewriter; and *pocket computers,* such as programmable calculators, widely used by surveyors, engineers, and accountants.

To operate, the CPU requires auxiliary devices called **peripherals,** the size and complexity of which, together with the capacity of the CPU, determine the type of computer being used. Peripherals serve two distinct purposes: data input and data output. Some peripherals can serve both purposes simultaneously (or more strictly speaking, consecutively).

For **input,** peripherals may take the form of a keyboard or a disk or tape drive which permits ASCII characters to be fed into the CPU. The keyboards look quite similar to ordinary typewriter and calculator keyboards, and are essentially used in the same manner by the computer operator. The disk or tape drives, as the words suggest, are small motorized devices, connected to the computer, which pass the magnetic disk or cassette tape over electronic heads, which read and extract prerecorded ASCII bytes from these sources for input to the CPU. Other input devices involve electronic probes or pens which can be moved about on drawings or other plane surfaces, causing measurable data in the form of coordinates or other quantities to be registered by the CPU.

One such device consists of two highly sensitive linear microphones, set at right angles to each other along two edges of a table or board, and able to pick up and record small sounds emitted at a preset frequency by an electronic probe at specific points in the plane surface. These coordinate data are instantly fed into the CPU for manipulation. Another device is called the **mouse,** which can be moved in any direction on a drawing or chart, causing a ball or wheel inside to rotate; the number of rotations in each direction is registered by the computer. Recent advances in computer technology have resulted in such refinements as the **touch screen,** in which data can be sent to the CPU merely by touching the console screen with a finger or stylus at a specific point, the coordinates of which are registered by the computer. Finally, "voice" and "body" commands are now being developed, by which data are sent to the CPU in reaction to selected spoken words or even bodily movements. Further developments in this field are probably limited only by the human imagination.

For **output,** peripherals may take the form of a small cathode console, like a miniature TV screen, or a printing or plotting device. The console screens are available in a variety of sizes and colors, some having black backgrounds with green characters, some having gray backgrounds with black or white characters, and some being able to handle all colors normally associated with TV screens. The printers act very much like high-speed typewriters, and are available with a variety of character styles, fonts, spacing options, and so on; they are used principally to reproduce numbers and letters in the form of spreadsheets, paragraphs, or small graphical representations, such as charts or tables. The plotters are much the same as the printers but are used primarily to reproduce larger or more complex graphical materials, such as architectural, engineering, or survey drawings. The CPU can also output data for storage to a disk or tape drive. Recent advances in this aspect of the field include printing by inkjet instead of carbon tape or ribbon, using electrostatic devices to print whole pages (instead of individual lines) at one time, and making use of lasers to scan large parts of programs for ultrahigh-speed printing. Another aspect of output involves the transfer of data from one machine to another, either directly or by telephone or other means of communication. A device called a **modem** (from *mo*dulation/*dem*odulation) is used to facilitate such telephone

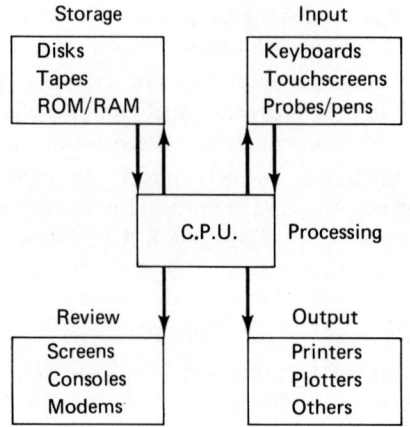

Figure 21.1 Computer peripherals.

transfers. A typical arrangement of common computer hardware is shown symbolically in Figure 21.1.

Software. Computer **software** refers to the disks or tapes on which the programs used to direct the operation of the computer CPU are recorded or stored. These devices consist of plastic magnetic film, either in the form of a small flat thin disk (0.1 mm thick and 11.75 cm or 5.25 in. in diameter) or in the form of a long, flat, thin tape, wound into a cassette, identical to those used in audio tape recorders. Each may be capable of being recorded on one or both sides, depending on machine characteristics. Software usually stores programs which are capable of user modification to some greater or lesser extent. For example, a word-processing program may be purchased and used without change, but it might be to the operator's advantage to modify it to coordinate it with other programs, such as spreadsheets or filing programs, already in use. The term RAM, for **random access memory,** is used in this connection. RAM is usually of a temporary nature. The terms **soft** or **floppy disks** are also used in connection with software programming. The term "software" also embraces the printed documentation in the form of instructional manuals which accompany the disks or tapes, and which explain the programs and their uses.

Firmware. **Firmware** is a term used for programs that have been recorded in a permanent storage device, such as a laser scored digital disk, which cannot be changed by the user. The term ROM, for **read-only memory,** is used in this connection. The term **hard disk** is also often used in this connection. Usually, hard disks are permanently built right into the cabinet that houses the CPU, compared to software disks or tapes, which are invariably interchangeable by insertion and removal into a disk or tape drive. Also, the memory of hard disks is usually measured in terms of megabytes, compared to kilobytes for soft disks or tapes.

Liveware. **Liveware** refers to the human operator. Without question, this component is the most costly, temperamental, uncertain, valuable, and variable of all parts of any computer system. Also, it never comes with printed instructions, and on occasion, breaks down, usually shortly after the computer system breaks down!

21.2.3 Programming

Computer programming involves the development of a series of specific instructions by a person having knowledge of one or more of the languages to which computers respond to cause any particular computer machine to react in a certain manner to prepared or programmed commands. For example, three of these languages are called BASIC, COBOL, and FORTRAN; there are also "high-level" and "low-level" languages available, terms which are used to describe the complexity of the languages understood by various computers. There are also a number of different CPU operating systems available (programs that tell the CPU what to do with data received by it), most of which are incompatible with each other. One of the most common of these is CP/M, which means Control Program for Microcomputers. All of these instructions are usually recorded on a soft plastic disk or magnetic tape. Each computer machine has a small amount of permanent programming built into it, to cause the CPU and the other electronic components to start up and shut down correctly when the main on/off switch of the machine is activated.

The subject of computer programming is totally beyond the scope of this book. There have been literally thousands of books written on the subject in recent years, many of which are of considerable value and others of which are not worth the cost of the paper on which they are printed. Most computer machines are supplied with instructional booklets and even disks with explanatory commands and menus which endeavor to present aspects of programming which are possible to use with the machines as purchased. Many regional schools, community colleges, and universities present courses of varying duration and cost on computer programming, and those readers who are interested in this aspect are directed to such sources for more reliable and practical information on computer programming. It is the opinion of this author, based on experience, that the most proficient, interesting, inexpensive, and enjoyable way to satisfy personal computer needs of this type is to join a local computer club, preferably one focusing directly on the interests of the owners of the specific make and model of computer purchased by the reader.

21.2.4 Applications

Computer applications involves the selection from among a variety of prepared programs of a specific program already capable of doing what the applicant requires. In some cases, minor modification of the selected program is possible and even desirable.

In the general field of computer applications, there are literally tens of thousands of computer programs already in existence. In the particular fields of construction estimating and cost accounting, there are already hundreds of specialized programs, with more being developed every month. All of these exhibit varying degrees of quality and utility, and each has been developed by legitimate or other so-called specialists to serve some real or imaginary specific need of the construction estimator or cost accountant, either or an individual basis or on an industry-wide basis. Programs of most interest to construction estimators probably fall into the categories listed below.

Spreadsheets. A spreadsheet is nothing more than a two-dimensional matrix which consists of vertical columns connecting with horizontal rows, to produce small boxes or "fields" into which data can be inserted. The ordinary quantity take-off paper used by estimators is a good example of a spreadsheet; another common example is a checkbook or bankbook record. The columns are usually identified by letters and the rows by numbers. Most computer spreadsheets can handle a matrix up to 52 columns (each six to eight characters wide) and 256 lines (each one character high), which is more than enough for most purposes. Larger matrices are available. As all columns and lines cannot appear on the computer console screen at once, the program must be able to permit "scrolling" from side to side and forward and backward to permit all parts of the spreadsheet to be viewed and modified. The main advantage of the computer spreadsheet is that the program can command the

CPU to manipulate data inserted into the various field cells; for example, lengths can be multiplied by widths to produce areas, and numbers of areas can be summed or totaled, all automatically, instantly, and correctly. Reflected dimensions can be passed automatically to one or more cells, as required. If a change is made in any cell, the results can also be automatically amended to reflect the effects of the change. End results can be stored on disk or tape, printed out on paper, or abandoned, at the operator's discretion.

A personal observation might be inserted at this point, to the effect that the construction industry as a whole has not yet really responded in any profound way to the tremendous possibilities of using the computer for significant change. Most current spreadsheet programs really only do electronically what the estimators have been doing manually for years. They permit measurement, pricing, and summarizing of results. Granted, they do the work more quickly, possibly more accurately, and certainly more pleasantly, but not really differently. There now exists an opportunity for some real innovations to occur in the fields of construction estimating and economics, as well as design, if some radical thinking could be brought to bear on the topic. The computer itself can assist with some of that "thinking," because most computer programs excel in the "what if" type of game. It may now be possible to tie the financing to the design to the bidding to the costing to the final settlement of accounts in ways not dreamed of heretofore, using electronic technology.

Word processing. Word-processing programs permit the computer equipment to operate much like a very sophisticated typewriter. Words and phrases are typed on the keyboard in the normal fashion and appear on the computer console screen for review. Minor typographical errors are easily amended by deletion and insertion commands. More important, blocks of text can be manipulated by moving them around, deleting them, or repeating them elsewhere. Programs can be purchased that will check for common spelling and grammatical errors; others will automatically prepare tables of contents, indexes, footnotes, search for specific words, insert predetermined strings of words or numbers in predetermined places, and so on. Some programs permit the console screen to be split in two, and thus to allow information or data to be entered into, retrieved from, or transferred between any two files on the same disk or tape. As with spreadsheets, end results can be stored, printed, or abandoned as desired and commanded.

The term "word processing" is somewhat limiting, because such word-processing programs really permit "idea processing," which is a much more valuable commodity. Ideas can be set down (using words or figures) in any order, and then moved around, expanded upon, contracted, or deleted to suit logical argument. The first draft is also the final draft, as words and phrases representing ideas that do not require amendment are not retyped; the remainder are simply refined until a final draft is ready for reproduction by printer or other means.

Specialized Programs. Apart from measurement, pricing, and summarizing, there are of course many programs arranged to suit the needs of estimators involved with cost control, scheduling, purchasing, payroll, and other features involved in the proper management of any construction company. Filing and mailing programs are also very popular and useful for companies with large numbers of customers or clients, or any other extensive data base requiring storage, manipulation, and retrieval, such as labor, material, and equipment output and cost data. In the design field, there are computer programs intended to assist with the graphical aspects of design, in which the data are pictorial or graphical, as distinct from verbal or numerical. Such programs are generally known by the acronym CADD, which stands for "computer-aided design and drafting." CADD programs can also be used to generate data of use to the design estimator, in the form of perimeter lengths of walls, floor areas, building volumes, quantities of specified elements, heating and cooling loads, and the like, which when used with other programs can generate probable costs of proposed solutions to design problems within seconds.

21.2.5 Program Selection

In general, choose software that does exactly what you require or which can be easily modified to do that, and no more. If you buy a program with excess capacity or power that you will not use, you are wasting money. You will also waste time trying to learn aspects of the program that you will not use. In general, choose programs that are menu-driven, as distinct from those which are character-driven. That is, choose a program where a recipe or menu appears when necessary or when commanded on the console screen and shows several options from which the operator can select, compared to programs that require memorization of or reference to a number of instructional commands.

One should always try out a program before purchasing it. Most reputable dealers will not only permit this, but will encourage it. Many computer stores employ people who may know how to sell computer hardware and software, but who are not knowledgeable about (or much interested in) the special needs of the construction estimator. Resist the hard sell; make sure that you have the whole story on the program and its applications. If necessary, retain a reliable construction cost consultant to assist you to identify your specific needs and their satisfaction. If there is a lot of money to be spent on computer equipment and programs, there is also a lot of money to be lost if such acquisitions are poorly handled; a small percentage of such money spent on professional advice is a wise investment.

It is true that computer technology is advancing at a very rapid, indeed exponential rate. For those who hesitate to buy a machine today because of the fear of obsolescence by tomorrow, it should be said that the machine that you buy today will retain all of its existing power and utility for many years to come. Newer or more costly machines may have more advances or features, but these do not detract from the attributes of the machine already purchased. The advantage of buying a machine sooner than later is twofold: first, you have the benefit of using it right away, and second, as you learn how to use it, you also learn about good and bad features, which will assist you to make a better decision when the time comes for upgrading to a more powerful or more sophisticated machine. You will also enjoy the advantage of getting into the mainstream of this astonishing revolution in human affairs, which has been called by some in this twentieth century "Gutenberg 2," in reference and deference to the fact that "Gutenberg 1" saw the introduction of movable type by Johann Gutenberg in Germany in the fifteenth century, which also had an extraordinary outcome, largely unforeseen at the time—I refer to universal literacy.

The proper order of events is the determination of construction estimating needs, the selection of appropriate software, and then the acquisition of the hardware. In general, most microcomputer machines now being manufactured are all of a very high order of quality; there is little to choose between them all. The main considerations therefore revolve around the amount of memory and the nature and quality of the peripherals and other refinements that each manufacturer promotes in conjunction with its own machines. Compatability with other computers is also a significant consideration, both in hardware and software, particularly if the exchange of data between machines is likely to be necessary.

Finally, before presenting some current examples of actual applications in construction estimating, it is deemed appropriate to insert a personal comment at this point to put the reader's mind somewhat at ease. It may be said that although the

Micro computer and dot matrix printer.

manuscript for this book was prepared by the author using word processing and spreadsheet programs on a personal computer and printer, the author knows little of and cares even less about the internal intricacies of computer mechanisms or programming beyond the level of the average layperson. In spite of holding two degrees and speaking two languages, the author is essentially illiterate with respect to understanding computer languages, not unlike the illiterate peasants at Gutenberg when written (or printed) language was being developed. However, just as it is not necessary to know in detail why the internal combustion engine of an automobile works in order to drive a car, neither is it necessary to know in detail why computers and printers work in order to know how to operate them. But just as there are many advantages to knowing how to drive a car, so there are many very great advantages to knowing what a computer and its programs can do for you, and in learning how to make it all work in your favor.

For the record, the equipment used by the author to produce the manuscript for this book consisted of a Kaypro 2 computer, made by Non-Linear Systems, Inc., Solana Beach, California, and a Microline 92 printer, made by the Okidata Corporation, Mount Laurel, New Jersey. The primary programs used were Perfect Writer for word processing and Perfect Calc for spreadsheeting, both produced by Perfect Software, Inc., Berkeley, California. The author set up the equipment and learned how to use the programs without any professional help beyond the instructions supplied with the hardware and software and some timely (and occasionally unasked for) advice from family and friends.

The remainder of this chapter consists of presentation and discussion of some actual case studies of computer applications in use by a variety of construction contracting, design, and consulting firms. A simplified flowchart showing cost information flow in a construction contracting office is shown in Figure 21.2.

21.3 CASE STUDIES

21.3.1 Preamble

The case studies included in this part of this chapter describe actual situations where computer applications are used in daily practice by companies, firms, or organizations of widely different types, sizes, and purposes. These particular case studies were selected to show typical and common applications of computer programs in construction estimating currently in use in the region where the author resides. They

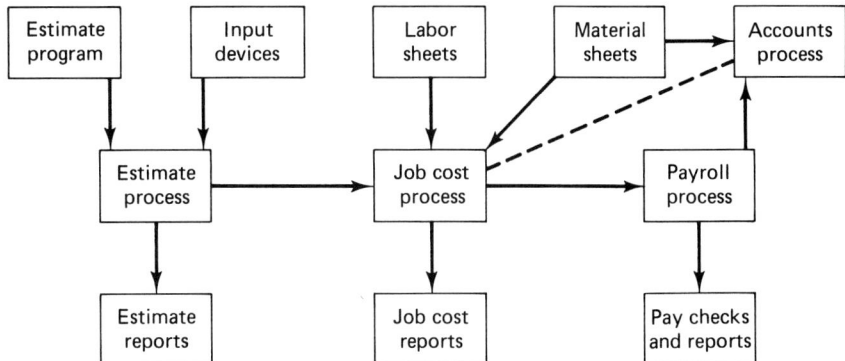

Figure 21.2 Flow of cost information.

are representative of applications to be found throughout North America, although each specific region or locality will no doubt have greater or lesser breadth or depth of hardware and software from which to draw examples for illustrative purposes. All of the case studies presented here are included with the permission of the owners of the companies and programs as identified in each case.

The case studies involve applications developed for use by general contractors, subcontractors, cost consultants, and designers. Notwithstanding the diversity of reasons why each separate application was developed, there are a number of common elements discernible throughout all the cases, not least of which are the elimination of unnecessary paperwork, manual work, and clerical drudgery; the speed of producing useful and reliable results in the form of meaningful data; the significant reduction of opportunities for human error in results through carelessness or forgetfulness; greatly heightened opportunities to manipulate, collate, or correlate data; savings of time, and the generally high quality, consistent, and businesslike appearance of the final hard-copy printout.

It may be of interest to the statistically minded reader to consider the following argument regarding the potential economies to be realized from utilizing computer applications in construction estimating in a small contracting or designing office. Assume one estimator doing take-off, pricing, bidding, and site inspections at $30K (30 thousand dollars) per year, and one technical clerk doing extensions, checking, copying, typing, and so on, at $20K. Assume 45 working weeks per year and 3 weeks per bid, or 15 bids per year, without the use of a computer. The average cost of each bid is therefore one-fifteenth of $50K or $3.33K, allocated $2.00K to the estimator and $1.33K to the clerk.

Studies indicate that the times of the estimator and the clerk using computer applications in estimating are effectively reduced by 35% and 75%, respectively. This translates to a saving of (35% of $2.00K equals $0.70K plus 75% of $1.33K equals $1.00K) $1.70K. $1.70K multiplied by 15 (for 15 bids) equals a saving of about $25K in one year, which is about 50% more than the $16K required for a computer and peripherals with appropriate software suitable for a medium-sized contractor or designer. Not only will the system be paid for in less than one year, but the estimator and the clerk will be much more efficiently utilized, not to mention all the additional advantages that come with the computer and its ability to generate useful reports of every kind.

Costs for construction estimating programs vary widely, depending on their quality, comprehensiveness, popularity, reputation, and so on. At the time of this writing (winter 1985), prices range from less than $500 for the simplest of programs up to $1500 or $2000 for the more sophisticated arrangements. Most programs are either self-explanatory or come with some simple instructions. Short training sessions are also provided by most programmers in connection with their programs, with training times ranging from 4 to 8 hours and training costs ranging from $40 to $60 per hour.

The following titles, used to indicate computer operating systems in the case studies that follow, identify copyrighted and registered trademarks of the companies indicated below:

Trademark	Company or organization
CP/M	Digital Research, Inc.
MS-DOS	Microsoft, Inc.
UNIX	AT&T Technologies, Inc.
IBM	International Business Machines, Inc.

In each case study, only a few programs or reports have been included as being representative of the general types and ranges of applications available. Readers requiring more expanded or detailed information are encouraged to contact the companies identified in the case studies or similar companies in the local region where they reside.

The main benefits to be derived from a close scrutiny of these case studies are probably as follows:

1. Improved *knowledge* of the *subject matter* of the various reports and programs, from a management point of view
2. Better *understanding* of the ways in which such information is *best presented* for use by the estimator, from a computer applications point of view

21.3.2 Case 1: General Contractor

The Company. Information for this case was obtained from a company called CMAC Computer Systems Ltd., Construction Consultants (E. Bachman, President), 135 East 8th Avenue, Vancouver, BC, Canada, V5T 1R8, for their system called Construction III. The company is involved in the construction of all types of residential, commercial, institutional, and industrial buildings, as well as providing a computer consulting service to outside construction companies of any size or configuration.

The System. The Construction III system consists of 30 separate computer programs; these can be loosely structured into three basic areas of interest to construction management personnel:

Design: In this area, there are six programs:

1. Feasibility
2. Proforma
3. Drafting Design
4. Engineering Design
5. Life Cycle Costing
6. Historical Cost Data

Implementation: In this area, there are 10 programs:

1. Mini-Estimator
2. Computender
3. Transmittal Control
4. Critical Path
5. Cost Controller
6. Scheduler
7. Equipment Manager
8. Progress Draw
9. Cost Projection
10. Inventory Controller

Chap. 21 / Computer Applications 271

Management: In this area, there are 14 programs:

1. Directory
3. Cost Flow Projector
5. Time Manager
7. Personal Manager
9. File Manager
11. Acounts Receivable
13. Payroll

2. Cost Projection
4. Cash Requirements
6. Property Manager
8. Vacation Manager
10. Information Manager
12. Accounts Payable
14. General Ledger

It is beyond the scope of this book to describe all of these programs in detail. Two programs of specific interest to estimators involved with measurement and pricing have been selected for further explanation; these are the Mini-Estimator and the Cost Controller programs.

1. *The Mini-Estimator.* This program permits feasibility and conceptual estimates to be prepared from preliminary design drawings or minimal construction program information, to determine approximate costs of proposals. Detailed estimates can also be produced, as more information becomes available in the form of working drawings, leading to final estimates in itemized format, ready for bidding or tendering.

Cost data can be pre-entered and updated as often as necessary. This program interfaces with other programs, such as the Cost Controller program. It also features an analytical facility, permitting many types of "what if" questions to be posed and answered by the estimator. All measurement and pricing data are quickly, simply, and directly entered via a keyboard and screen, with a security control to prevent unauthorized tampering with such data.

Similarly, the Computender program permits preparation of detailed quantity take-offs, materials or cutting lists, even without working drawings being available in many cases, provided that certain design parameters are known. The program will apply pre-entered or updated cost data to quantities or totals, and can compare costs of parts of the current project to previous ones. As with the previous program, this program interfaces with many of the other 30 listed.

2. *The Cost Controller.* This program considers virtually all aspects of project cost control. It permits cost data to be inserted, stored, updated, and retrieved. It permits comparisons to be made between estimated costs and actual costs as they occur. It can also check all invoices against purchase orders or quoted costs. It can keep track of change orders and contract revisions, of budgeted and committed costs, and will warn of cost overruns. It also interfaces with a variety of the accounting programs, as well as with the Progress Draw, Cash Flow Projector, and Scheduler programs, with obvious benefits. The Cost Controller also reports in 30 different formats, suitable for use at various levels of company organization.

Main Features. All programs can be run on computer systems operating under CP/M, MS-DOS, and Unix commands, and are designed to be used either independently or in conjunction with any of the other programs in the system.

The Construction III system can be adapted to meet the client's needs, unlike many other programs in which the client or customer must arrange to make his or her affairs fit into the parameters of the selected program. All programs are menu-driven and are user-friendly, with no previous computer experience necessary to operate them. Help tools and training materials are incorporated into the system. These programs also provide reports suitable for all levels of staff and management and are supported by the parent CMAC company and its staff through a phone-in service which is available to clients.

Notes on Hard-Copy Examples. The hard-copy reproductions that accompany this case study are explained in the corresponding notes listed below.

Figure number	Explanatory notes
21.1A	This shows the Master File (MF) Main Menu, which permits access to the programs listed.
21.1B	This shows Program 01 from the MF Menu, by which codes can be added or deleted.
21.1C	This shows Program 02 from the MF Menu, by which the estimate may be set up. The "M" field under DIMENSIONS allows for the arithmetic functions (such as add or multiply) to be set.
21.1D	This shows Program 04 from the MF Menu, by which unit prices and units of measure for each item can be maintained or modified as necessary.
21.1E	This shows Program 05 from the MF Menu, by which reports by code number may be viewed or printed at any time.
21.1F	This shows the Mini-Estimator (ME) Menu, in which estimating functions are grouped into two categories: maintenance and reporting.
21.1G	This shows Program 05 from the ME Menu, by which subcontractor quotes can be recorded.
21.1H	This shows Program 13 from the ME Menu, which produces an Estimate Summary Report, giving various totals as indicated.
21.1I	This shows the Cost Control (CF) Main Menu, which permits access to the programs listed.
21.1J	This shows Program 02 from the CF Menu, in which pricing data can be recorded and modified.
21.1K	This shows Program 19 from the CF Menu, in which a one-page management report can be produced for each project, with data as noted in the sample.

Reports can be produced for budget evaluations, monthly progress draws, change order tracking, and many other construction management and production requirements.

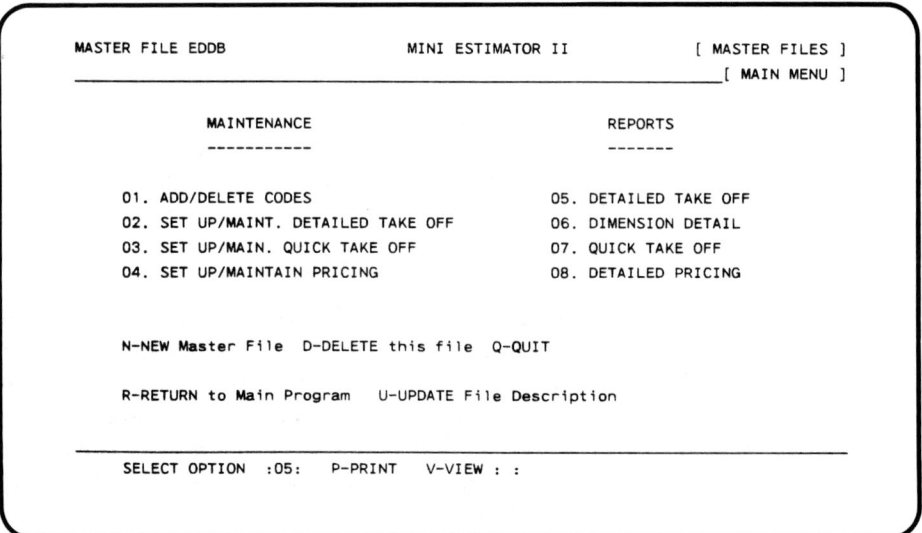

21.1(A) Master file main menu.

Chap. 21 / Computer Applications

```
MASTER FILE EDDB              MINI ESTIMATOR II           [ MASTER FILES ]
                                                          [ 01. DELETE CODES ]

         CODE      CODE                    CATALOGUE
        :NUMBER:   DESCRIPTION  :  ID  :   NUMBER      :  OPTION  :
        +------+----------------+------+----------------+---------+
         01001L    SUPERVISION                                :    :   --
         01005L    MISC LABOR       A
         02002S    EXCAVATION       1     STRIP FOOTINGS
         02002S    EXCAVATION       X     BULK
         03001E    SUB STR.      1FSFTG   20 MPA PUMP MIX
         03001E    SUB STR.      2FSFTG   02X10XRL#4HF
         03001E    SUB STR.      3FSFTG   02X02PEGS
         03001M    SUB STRUCTURE  1 FD W  20 MPA PUMP MIX
         03001M    SUB STRUCTURE  2 FD W  FORM MATERIALS
         06001M    ROUGH CARPENTRY 1EXWAL 02X04X08#2HF
         06001M    ROUGH CARPENTRY 2EXWAL 4X8X3/8SPPLY
         06001M    ROUGH CARPENTRY 3EXWAL 02X04XLF#2HF
         06001S    ROUGH CARPENTRY


  CR -NEXT    P-PREVIOUS    S-SAVE    D-DELETE    G-GO TO    R-RETURN    Q-QUIT
```

21.1(B) Main code file.

```
MASTER FILE EDDB              MINI ESTIMATOR II           [ MASTER FILES ]
                                                [02. SET UP/MAINTAIN DETAILED TAKE OFF]

         CODE      CODE                    CATALOGUE    :    PASSING CO-ORDINATES    :
 :OPT :NUMBER:  DESCRIPTION :  ID  :       NUMBER       : Value 1  Value 2  Value 3 :
 +----+------+--------------+------+--------------------+--------+--------+---------+
 -- : :01001L SUPERVISION       S                         01001LS1 0100LS1  01001LS1
        01005L MISC LABOR       A                         01001L-1    -        -
        02002S EXCAVATION       1     STRIP FOOTINGS         -        -        -
        02002S EXCAVATION       X     BULK                   -        -        -
        03001E SUB STR.      1FSFTG   20 MPA PUMP MIX        -        -        -

    :    D I M E N S I O N S   : EXTENSION :   D E S C R I P T I O N S       :
    : No. 1   M  No. 2  M  No. 3:M  No. 4  : No. 1    No. 2   No. 3 : No. 4  :
    +---------+--------+--------+-----------+--------+--------+--------+---------+
        0.00      0.00     0.00       0.00
        0.00 *  175.00 *   1.25       0.00 MONTHS
        0.00 *    0.00 *   0.00 /    27.00 LENGTH   WIDTH    DEPTH
        0.00 *    0.00 *   0.00 /    27.00 LENGTH   WIDTH    HEIGHT
        0.00 *    0.00 *   0.00 /    27.00 LENGTH   WIDTH    DEPTH

  U-UPDATE   D-DUPLICATE   CR -NEXT  P-PREVIOUS   G-GO TO   R.RETURN   Q-QUIT
```

21.1(C) Estimate set up.

```
MASTER FILE EDDB              MINI ESTIMATOR II              [ MASTER FILES ]
                                      [ 04. SET/MAINT. PRICING] L 1-CATALOGUE ]

                    CATALOGUE      UNIT OF    UNIT
                 :   NUMBER       :MEASURE:   PRICE   :OPTION:
                 +---------------+--------+----------+------+
                    02X04X08#2HF    BD.FT.    0.225:  :  --
                    02X10XRL#4HF    BD.FT.    0.265
                    20 MPA PUMP MIX CU.YDS   52.000
                    4X8X3/8SPPLY    EACH      7.250
                    STRIP FOOTINGS  CU YDS    9.750


U-UPDATE  A-ADD  D-DELETE   S-SAVE   G-GO   CR -NEXT   P-PREV.   R.RETURN  Q-QUIT
```

21.1(D) Units of measurement and pricing.

```
MASTER FILE EDDB              MINI ESTIMATOR II              [ MASTER FILES ]
                                                   [05. DETAILED TAKE OFF REPORT]

   CODE      CODE                   CATALOGUE         D I M E N S I O N S
  NUMBER   DESCRIPTION      ID       NUMBER       No. 1   M  No. 2   M  No. 3
  =====    ============   =======   =============  ======= -======== -========
  01001L   SUPERVISION       S                      0.00      0.00       0.00
  01005L   MISC LABOR        A                      0.00  *  175.00  *   1.25
  01005L   MISC LABOR        A                      0.00  *  175.00  *   1.25
  02002S   EXCAVATION        1      STRIP FOOTINGS  0.00  *    0.00  *   0.00
  02002S   EXCAVATION        X      BULK            0.00  *    0.00  *   0.00
  03001E   SUB STR.         1FSFTG  20 MPA PUMP MIX 0.00  *    0.00  *   0.00
  03001E   SUB STR.         2FSFTG  02X10XRL#4HF    0.00  *    2.00  *   1.66
  03001E   SUB STR.         3FSFTG  02X02PEGS       0.00  /    4.00  *   2.00
  03001M   SUB STRUCTURE    1 FD W  20 MPA PUMB MIX 0.00  *    0.00  *   0.00
  03001M   SUB STRUCTURE    2 FD W  FORM MATERIALS  0.00  *    0.00  *   2.00
  06001M   ROUGH CARPENTRY  1EXWAL  02X04X08#2HF    0.00  /    1.33  *   8.00
  06001M   ROUGH CARPENTRY  2EXWAL  4X8X3/8SPPLY    0.00  /    4.00       0.00
  06001M   ROUGH CARPENTRY  3EXWAL  02X04XLF#2HF    0.00  *    4.00  *   0.66

: :  N-NEXT SCREEN   D-DOWN    G-GO TO CODE  No.   R-RETURN    Q-QUIT
```

21.1(E) Detailed take-off report.

Chap. 21 / Computer Applications

```
   PROJECT NO. 00001           MINI ESTIMATOR II          [ MAIN MENU ]


                   MAINTENANCE                    REPORTS
                   -----------                    -------

        01. ADD/DELETE CODES              08. QUICK ESTIMATE

        02. QUICK ESTIMATE                09. QUANTITY TAKE OFF
                                          10. TAKE OFF ADDITIONS
        03. SET UP/MAINTAIN ESTIMATE
        04. REVISE/UPDATE ESTIMATE        11. DETAILED ESTIMATE
                                          12. CONDENSED ESTIMATE
        05. SET UP/MAINTAIN QUOTES        13. ESTIMATE SUMMARY

        06. UPDATE PRICING and U. of M.   14. QUOTATION SPREAD SHEETS
        07. UPDATE DIMENSION PASSING      15. DETAILED PRICING

        N-NEW Project Selection     D-DELETE this Project     Q-QUIT

   SELECT OPTION ABOVE :  :
```

21.1(F) Mini estimator II menu.

```
   PROJECT NO. 00001        MINI ESTIMATOR II      [05. SET UP/MAINTAIN QUOTES]
                                                   [ CODE NUMBER - 15005S ]

      :   COMPANY INFORMATION      :  INCLUSIONS AND EXCLUSIONS    :   AMOUNTS    :
      ----------------------------+----------------------------------+--------------+
      :Company Name ID :WESTERN:  :            CONTRACT AMOUNT    :    2575.14   :
      :Comp.:WESTERN PLUMBING     :::INC EXTRA 3/4" PIPE           :     247.00   :
      :Name :FRANK SIMMS          :::EXCLUDE   1/2" PIPE           :    -125.89   :
      :Rep.1:MIKE SIMMS           :::                              :       0.00   :
      :phone:(604) 253-6658:      ::                               :       0.00   :
      :Rep.2:                     ::--------------------------------:==============:
      :phone:(   )    -    :      : Phone:Y: Has Plan:Y: Will Bid:Y: :    2696.25   :
      +---------------------------+----------------------------------+--------------+


   BUDGET AMOUNT ............................................. $     0.00

   CONTRACT AWARDED TO ..:              ... CONTRACT AMOUNT $  2696.25
   CONTRACT AWARDED BY ..:STAR CONST.
   COMMENTS ............:ALL WORK INCLUDED

   : :  U-UPDATE   CR -NEXT   P-PREVIOUS   G-GO TO   S-SET UP   A-AWARD   R-RETURN
```

21.1(G) Subcontractor quotation record.

```
PAGE NO. 1        MINI ESTIMATOR II    ESTIMATE SUMMARY REPORT
PROJECT:          HABOUR APARTMENTS              NO: 00001
ADDRESS:          1 HABOUR DRIVE
OWNER:            GALAXY DEVELOPMENTS
REPRESENTATIVE:   TREVOR JONES       PHONE: (606) 254-6896
ARCHITECT:        RAYMOND ASSOCIATES
REPRESENTATIVE:   GENE KENT          PHONE: (000) 000-0000
ESTIMATED BY:     STAR CONSTRUCTION           DATE: 01 MAR 85
REPRESENTATIVE:   JOHN STAR          PHONE: (240) 565-7854
=================================================================
CODE   DESCRIPTION      MATERIAL    LABOR   SUB CONT  EQUIPMENT   TOTAL
01000  PRELIMINARIES        0.00  24196.00      0.00       0.00  24196.00
02000  SITEWORK             0.00      0.00   2902.63       0.00   2902.63
03000  CONCRETE          2507.32      0.00      0.00     348.70   2856.02
04000  MASONRY              0.00      0.00      0.00       0.00       0.00
05000  METALS               0.00      0.00      0.00       0.00       0.00
06000  WOOD & PLASTIC     648.11      0.00      0.00       0.00     648.11
07000  MOISTURE PROT.       0.00      0.00      0.00       0.00       0.00
08000  DOORS & WINDOWS      0.00      0.00      0.00       0.00       0.00
09000  FINISHES             0.00      0.00      0.00       0.00       0.00
10000  SPECIALTIES          0.00      0.00      0.00       0.00       0.00
11000  EQUIPMENT            0.00      0.00      0.00       0.00       0.00
12000  FURNISHINGS          0.00      0.00      0.00       0.00       0.00
13000  SPECIAL CONSTR.      0.00      0.00      0.00       0.00       0.00
14000  CONVEY SYSTEM        0.00      0.00      0.00       0.00       0.00
15000  MECHANICAL           0.00      0.00   1920.00       0.00    1920.00
16000  ELECTRICAL           0.00      0.00      0.00       0.00       0.00
17000  OVERHEAD             0.00      0.00      0.00       0.00       0.00
TOTAL  CONSTR. COSTS     3155.43  24196.00   4822.63     348.70   32522.76

       MARK UP                                                    1853.79
       CONTINGENCY                                                 748.02
18000  DEVELOPMENT          0.00      0.00      0.00       0.00       0.00
TOTAL  PROJECT COSTS     3155.43  24196.00   4822.63     348.70   35124.58
TOTAL  ADJUSTMENTS          0.00  T E N D E R   P R I C E  $    35124.58

TOTAL AREA OF SITE:      22000         BASEMENT PARKING AREA     4300
AREA OF BUILDING:         6090         OTHER BASEMENT AREA        452
BUILDING USABLE AREA:     5604         TOTAL BASEMENT AREA       4752
BUILDING USABLE PERCENT     92         NUMBER OF STALLS           122
NUMBER OF UNITS & SIZE      55, VARIOUS

SUBSTRUCTURE:    $   505438.00  $  106.36 PER AREA  $  4142.93 PER STLL
SUPERSTRUCTURE:  $  3786000.00  $  621.67 PER AREA  $ 68836.36 PER UNIT
CONSTRUCTION:    $    32522.76  $    5.34 PER AREA  $   591.32 PER UNIT
OTHER COSTS:     $     2601.82  $    0.42 PER AREA  $    47.30 PER UNIT
TOTAL PROJECT:   $    35124.58  $    5.76 PER AREA  $   638.62 PER UNIT
```

21.1(H) Estimate summary report.

```
PROJECT NO. 84009            THE COST CONTROLLER                    [MENU]
---------------------------------------------------------------------------
                SET UP/MAINTAIN              REPORTS
                ---------------              -------
           01. CODES/BUDGETS            12. PRICING FILE
           02. PRICING FILE             13. PURCHASE ORDERS
           03. PURCHASE ORDERS          14. INVOICES
           04. SUB-CONTRACTS            15. CHANGE ORDERS
           05. INVOICES                 16. CONTRACT REVISIONS
           06. CHANGE ORDERS            17. LABOR
           07. CONTRACT REVISIONS       18. SOFT COSTS
           08. LABOR                    19. SUMMARY
           09. SOFT COSTS               20. CONSTRUCTION ACTIVITY
           10. BUDGET EVALUATION

           11. UPDATE CURRENT ENTRIES   21. END OF PERIOD PROCESSING

             N-NEW Project    D-DELETE this Project    H-HELP    Q-QUIT

---------------------------------------------------------------------------
           SELECT OPTION :  :
```

21.1(I) Cost control main menu.

```
    PAGE #  1                  P R I C I N G   F I L E            DATE:10/03/1984
                               =========================

    PROJECT:PARK PLACE APARTMENTS              NO.  84009 TYPE:APARTMENTS
    ADDRESS:1234 MAIN STREET, VANCOUVER, B.C.    V6K 3X8        PH:(604)999-1111
    COMPANY:JONES & COMPANY INVESTMENTS                REP:GEORGE JONES
    ADDRESS:1111 GEORGIA STREET, VANCOUVER, B.C. V8W 6M2        PH:(604)111-9999

    COMPANY ID:GENSTAR   COMPANY NAME:GENSTAR LTD.
    REP:RON LANCASTER             APPROVED BY:JOE KAPP
    QUOTE # 1984000001   DATE:09/25/1984  EXPIRY DATE:09/25/1985

           CODE                       CATALOGUE                 UNIT    UNIT    T
           NUMBER   DESCRIPTION       NUMBER         DETAIL     MEASUR  PRICE
           ======   ===============   ===============   ==========   ======  ======== =

           02009M  GRANULAR FILL     BIRDS EYE                  CU.YDS     7.500 T
           02009M  GRANULAR FILL     DRAIN ROCK                 CU.YDS     7.500 T
           02009M  GRANULAR FILL     PIT RUN                    CU.YDS     3.750 T
           02009M  GRANULAR FILL     RIVER SAND                 CU.YDS     4.750 T
           03018M  CONCRETE          20 MPA PUMP MIX            CU.YDS    52.000 T
           03018M  CONCRETE          25 MPA PUMP MIX            CU.YDS    57.000 T
           03018M  CONCRETE          CAL.1/2%                   P 1/2%     1.250 T
           03018M  CONCRETE          HOT WATER                  CU.YDS     0.500 T
           03018M  CONCRETE          OVERTIME                   PER HR    20.000 T
           03018M  CONCRETE          WAITING TIME               PER HR    45.000 T

    COMPANY ID:NORTCOA   COMPANY NAME:NORTH COAST FOREST PRODUCTS LTD.
    REP:BOB LASKO                 APPROVED BY:JOHN PETERSON
    QUOTE # 1984L00002   DATE:09/25/1984  EXPIRY DATE:09/25/1985

           CODE                       CATALOGUE                 UNIT    UNIT    T
           NUMBER   DESCRIPTION       NUMBER         DETAIL     MEASUR  PRICE
           ======   ===============   ===============   ==========   ======  ======== =

           03001M  FORM SUB STR.     4X8X3/4 FORMPLY            EACH      15.000 T
           03001M  FORM SUB STR.     L02X04X16#2FIR             BD.FT.     0.265 T
           06001M  ROUGH CARPENTRY   02X04XLF#2FIR              BD.FT.     0.235 T
           06001M  ROUGH CARPENTRY   02X04XLF#2HF               BD.FT.     0.222 T
           06001M  ROUGH CARPENTRY   4X8X3/8 FIR PLY            EACH       9.700 T
           06001M  ROUGH CARPENTRY   4X8X3/8SHEETING            EACH       9.250 T
           06001M  ROUGH CARPENTRY   4X8X5/8 FIR PLY            EACH      12.500 T
           06001M  ROUGH CARPENTRY   4X8X5/8 SPR.PLY            EACH      11.200 T
           06001M  ROUGH CARPENTRY   L02X04LF#2HF               BD.FT.     0.222 T
           06001M  ROUGH CARPENTRY   L02X04RL#2HF               BD.FT.     0.225 T
           06001M  ROUGH CARPENTRY   L02X04X08#2HF              BD.FT.     0.225 T
           06001M  ROUGH CARPENTRY   L02X04X10#2                BD.FT.     0.258 T
```

21.1(J) Price data record.

```
 PAGE    1              SUMMARY  PROGRESS  REPORT                    DATE 10/03/1984
                        ===============================
PROJECT NAME: PARK PLACE APARTMENTS           NUMBER: 84009    TYPE: APARTMENTS
     ADDRESS: 1234 MAIN STREET, VANCOUVER, B.C.  V6K 3X8       PHONE: (604)999-1111
COMPANY NAME: JONES & COMPANY INVESTMENTS          REPRESENTATIVE: GEORGE JONES
     ADDRESS: 1111 GEORGIA STREET, VANCOUVER, B.C. V8W 6M2     PHONE: (604)111-9999

                              --- BY CODE SERIES ---
                              ==========================

 CODE                ORIGINAL    REVISED    CURRENT     AMOUNT    COST THIS    COST     HOLD BACK   %    COST TO    ACTUAL
 SERIES DESCRIPTION   BUDGET     BUDGET     BUDGET    COMMITTED    PERIOD    TO DATE    TO DATE   COM   COMPLETE   VARIANCE
 ====== =========== ========== ========== ========== ========== ========== ========== ========== ===  ========== ==========
  01   PRELIMINARIES  173375.00  173375.00  175475.00       0.00       0.00       0.00       0.00   0   175475.00   -2100.00
  02   SITE WORK       99820.00   99820.00  100220.00   45000.00   10125.00   10125.00    1125.00  10    90095.00    -400.00
  03   CONCRETE       315302.50  315302.50  315630.00      93.50    1133.50    1133.50       0.00   0   314496.50    -327.50
  04   MASONRY          8520.00    8520.00    8520.00       0.00       0.00       0.00       0.00   0     8520.00       0.00
  05   METALS           6070.00    6070.00    6070.00       0.00       0.00       0.00       0.00   0     6070.00       0.00
  06   WOOD & PLASTIC 308100.00  318100.00  308100.00  119142.35   22492.34   22492.34    3675.00   7   285607.66   10000.00
  07   MOISTURE PROT.  79665.00   79665.00   79900.00       0.00       0.00       0.00       0.00   0    79900.00    -235.00
  08   DOORS & WINDOWS 43659.00   43659.00   43659.00       0.00       0.00       0.00       0.00   0    43659.00       0.00
  09   FINISHES       246654.00  246654.00  248654.00       0.00       0.00       0.00       0.00   0   248654.00   -2000.00
  10   SPECIALTIES      7000.00    7000.00    7000.00       0.00       0.00       0.00       0.00   0     7000.00       0.00
  11   EQUIPMENT       60000.00   60000.00   60000.00       0.00       0.00       0.00       0.00   0    60000.00       0.00
  12   FURNISHINGS     15000.00   15000.00   15000.00       0.00       0.00       0.00       0.00   0    15000.00       0.00
  14   CONVEY SYSTEM   36000.00   36000.00   36000.00       0.00       0.00       0.00       0.00   0    36000.00       0.00
  15   MECHANICAL     201000.00  201000.00  204000.00   25000.00    6600.00    6600.00     900.00   3   197400.00   -3000.00
  16   ELECTRICAL     125000.00  125000.00  125000.00       0.00       0.00       0.00       0.00   0   125000.00       0.00
 ====== =========== ========== ========== ========== ========== ========== ========== ========== ===  ========== ==========
 ****** GRAND TOTALS 1725165.50 1735165.50 1733228.00  189235.85   40350.84   40350.84    5700.00   2  1692877.16    1937.50
 ====== =========== ========== ========== ========== ========== ========== ========== ========== ===  ========== ==========


                              --- BY CODE TYPE ---
                              ==========================

 CODE                ORIGINAL    REVISED    CURRENT     AMOUNT    COST THIS    COST     HOLD BACK   %    COST TO    ACTUAL
 TYPE   DESCRIPTION   BUDGET     BUDGET     BUDGET    COMMITTED    PERIOD    TO DATE    TO DATE   COM   COMPLETE   VARIANCE
 ====== =========== ========== ========== ========== ========== ========== ========== ========== ===  ========== ==========
  L    LABOR          131425.00  131425.00  132825.00       0.00       0.00       0.00       0.00   0   132825.00   -1400.00
  S    SUB-CONTRACTS 1061120.50 1066120.50 1066248.00  188000.00   37550.00   37550.00    5700.00   4  1028698.00    -127.50
  M    MATERIALS      460330.00  465330.00  461465.00    1235.85    2800.84    2800.84       0.00   1   458664.16    3865.00
  E    EQUIPMENT       72290.00   72290.00   72690.00       0.00       0.00       0.00       0.00   0    72690.00    -400.00
 ====== =========== ========== ========== ========== ========== ========== ========== ========== ===  ========== ==========
 ****** GRAND TOTALS 1725165.50 1735165.50 1733228.00  189235.85   40350.84   40350.84    5700.00   2  1692877.16    1937.50
 ====== =========== ========== ========== ========== ========== ========== ========== ========== ===  ========== ==========
```

21.1(K)   Management budget report.

21.3.3 Case 2: Cost Consultant

The Company. Information for this case was obtained from a company called E. B. Foxon Ltd., Construction Consultants (E. B. Foxon, President), 14901 95A Avenue, Surrey, BC, Canada, V3R 7T6, for their system called Estimax. The company provides computerized construction cost services, as well as economic and procedural advice to construction and design firms on a fee basis.

The System. The Estimax system is structured into three basic levels of application:

1. *Small subcontracting firm:* a program with eight files or modules:

   1. Data entry   2. Pricing
   3. Transactions   4. Unit estimating
   5. Reports   6. Trade summary
   7. General summary   8. Estimating guide

2. *Small to medium-sized general construction companies:* a program with 10 files or modules:

   1-8. Same as for level 1
     9. Subtrade file
    10. Bid analysis

3. *Medium-sized to large construction organizations:* a program with 12 files or modules:

1-7. Same as for level 1
8. Cost library
9, 10. Same as for level 2
11. Measurement methods
12. Bid depository

All three levels are compatible with an overall cost control program, which can also be used as a stand-alone site-reporting facility. The program will produce the following reports:

1. Daily time sheet and labor distribution
2. Daily materials report
3. Daily construction report
4. Estimated/actual cost comparisons
5. Cost control report

Main Features. In the words of its designers, the Estimax system is a total take-off and estimating system with cost control and site-reporting facilities. It is also a training vehicle that promotes the disciplines of sound quantity take-off and estimating practices. It is powerful, flexible, and cost-effective.

The Uniform Construction Index (Masterformat) code number system is used throughout Estimax, making the system applicable anywhere in North America. The CIQS Standard Method of Measurement is incorporated, thus encouraging the user to be consistent and accurate in the measurement of construction work. Other estimating guides are also included. The Scope of Work for each trade as determined by the British Columbia Construction Association is automatically presented to the user at every trade level.

Separate or alternative prices are automatically accomplished within the system and through to the final hard-copy printout. Equipment expenditures, cash allowances, and subcontractor records are separately recorded and printed out, together with virtually any other desired aspect of quantity take-off, pricing, and site-reporting functions for any contract under way, at any point in time.

There are programs within the system that permit calculation and recording of single item and unit prices. Estimates prepared for projects are automatically available from within the pricing library, together with transaction programs that permit updating and revision. Mechanical and electrical contractors can cost out the prices for pipework, conduits, hangers, connectors, and other fittings.

Historical data are readily available, concerning personnel, project history, and statistics of various types and in various modes. The data entry program utilizes multiline worksheet applications with automatic metric conversion built in; data may be entered in metric or imperial dimensions. Headings and descriptions repeat automatically until ordered to be changed. There is a useful cross-referencing capability between data entry and a number of other files or modules. The pricing module shows extensions and unit measurements with labor and material pricing. One useful feature permits checking for missing or forgotten items, and all items are automatically extended, checked, and recorded.

All software is IBM compatible and will run on ICL and most other popular computers having PC.DOS, MS.DOS, and CPM/80 operating systems. To operate properly, the programs also require a dBase run-time program.

Notes on Hard-Copy Examples. The hard-copy reproductions that accompany this case study are explained in the corresponding notes listed below.

Figure number	Explanatory notes
21.2A	This report shows specific information about the concrete take-off; similar separate reports can be produced for all other trade work.
21.2B	This report shows detailed information about all trades measured for the project, with each shown in context and in order.
21.2C	This report shows labor costs expressed in dollars allocated to various portions of the project.
21.2D	This report shows materials costs in dollars as allocated to various portions of the job.
21.2E	This report shows equipment costs in dollars as allocated to various portions of the job.
21.2F	This report shows an estimate summary of the general contractor's own work.
21.2G	This report shows the general estimate summary for the entire project.

Other reports can be produced for separate or alternative prices, changes, taxes, and virtually any other specific item of interest.

```
PROJECT NAME          TEST
ESTIMATOR

85/01/09

PAGE NO. 00001    Concrete Report
85/01/09

                         QUANTITIES REPORT

 CODE/ITEM          TRADE         NO.      LENGTH      WIDTH      HEIGHT    WEIGHT      DIA.       WASTE

 3.30/  196 C.I.P.CONCRETE FOUNDATIONS    0.00    121.92000    0.45000    0.20000   0.00000    0.00000    1.05

 3.30/  197 C.I.P.CONCRETE FOUNDATIONS    0.00    121.92000    0.30000    2.13000   0.00000    0.00000    1.10

 3.30/  198 C.I.P.CONCRETE ELEV.SLABS     0.00      6.14680    6.14680    6.14680   0.00000    0.00000    1.10

 3.30/  199 C.I.P.CONCRETE COLUMNS       23.00      0.20000    0.20000    3.65000   0.00000    0.00000    1.05

 3.30/  200 C.I.P.CONCRETE COLUMNS       12.00      3.00000    3.00000    3.00000   0.00000    0.00000    2.00

 ** TOTAL **
                                         35.00    253.18680   10.09680   15.12680   0.00000    0.00000    6.30
```

21.2(A) Sample trade take-off summary.

```
PROJECT NAME          TEST
ESTIMATOR

85/01/09

PAGE NO. 00001        TRADE Report
85/01/09


CODE/ITEM          TRADE                  NO.      LENGTH      WIDTH      HEIGHT     WEIGHT      DIA.       WASTE

1.00/   238  GENERAL REQUIREMENTS       25.00      1.21000    2.43000    0.00000    0.00000    0.00000     0.00

3.00/   193  FORMWORK FOUNDATIONS        0.00      9.99998    0.00000    0.00000    1.00000    0.00000     1.05

3.10/   194  REINFORCING FOUNDATIONS     0.00      0.00000    0.00000    0.00000  680.40000    0.01000     1.10

3.20/   195  REINFORCING FOUNDATIONS     0.00      0.00000    0.00000    0.30000   90.72000    0.01000     1.10

3.30/   196  C.I.P.CONCRETE FOUNDATIONS  0.00    121.92000    0.45000    0.20000    0.00000    0.00000     1.05

3.30/   197  C.I.P.CONCRETE FOUNDATIONS  0.00     21.92000    0.30000    2.13000    0.00000    0.00000     1.10

3.30/   198  C.I.P.CONCRETE ELEV.SLABS   0.00      6.14680    6.14680    6.14680    0.00000    0.00000     1.10

3/30/   199                              23.00                            3.65000    0.00000    0.00000
                                                                                                0.00000
```

21.2(B) Summary of all trade measurements.

```
PROJECT NAME          TEST
ESTIMATOR

85/01/09

PAGE NO. 00001        Labour Summary
85/01/09

                              Labour Summary

Trade                         Function            Totals         percentage

ARCHITECTURAL EQUIPMENT       COOKING              363.60           0.40

ARCHITECTURAL EQUIPMENT       STORAGE              360.00           0.40

BOXES                         PULL BOX             320.00           0.35

BRICK MASONRY                 INTERN WALLS        3673.60           4.10

BRICK MASONRY                 CLEAN & POINT        125.37           0.14

BRICK MASONRY                 BACKING/FACINGS     1176.00           1.31

C.I.P.CONCRETE COLUMNS        8X8 INTERNAL          77.55           0.08

C.I.P.CONCRETE COLUMNS                            7270.56           8.12

C.I.P.CONCRETE ELEV. SLABS
```

21.2(C) Summary of labor costs.

```
PROJECT NAME          TEST
ESTIMATOR

85/01/09

PAGE NO. 00001     Materials Summary
85/01/09

                        Material Summary

Trade                    Material              totals      Percentage

ARCHITECTURAL EQUIPMENT  30" RANGE             5252.00        3.08

ARCHITECTURAL EQUIPMENT  REFRIGERATOR          6400.00        3.76

BOXES                    HINGED CABINETS       1000.00        0.58

BRICK MASONRY            12X4X6 JUMBO          5510.40        3.23

BRICK MASONRY            12X4X6 JUMBO           100.30        0.05

BRICK MASONRY            STANDARD COMMON       1470.00        0.86

C.I.P.CONCRETE COLUMNS   35001B                  88.12        0.05

C.I.P.CONCRETE COLUMS                                         4.85
```

21.2(D) Summary of materials costs.

```
PROJECT NAME          TEST
ESTIMATOR

85/01/09

PAGE NO. 00001     Equipment Summary
85/01/09

                        Equipment Summary

Trade                    Function              totals      Percentage

ARCHITECTURAL EQUIPMENT  COOKING                  0.00        0.00

ARCHITECTURAL EQUIPMENT  STORAGE                  0.00        0.00

BOXES                    PULL BOX                 0.00        0.00

BRICK MASONRY            INTERN WALLS           120.00        0.40

BRICK MASONRY            CLEAN & POINT          120.00        0.40

BRICK MASONRY            BACKING/FACING         150.00        0.50

C.I.P.CONCRETE COLUMNS                                        0.00

C.I.P.CONCRETE COLUMNS
```

21.2(E) Summary of equipment costs.

```
3
PROJECT NAME          TEST
ESTIMATOR

85/01/09

PAGE NO. 00001     Indexed estimate summary
85/01/09


CODE/ITEM  TRADE                          Function    Labour   Labour    Material  Equipment   Grand Total
                                                      Unit     Time      Unit

1.00/   238 GENERAL REQUIREMENTS          BARRIER     10.00    0.0000    20.00     0.0000      2205.21

3.00/   193 FORMWORK FOUNDATIONS          FOOTINGS    10.00    0.0000    20.00     20.0000     15456.95

3.10/   194 REINFORCING FOUNDATIONS       FOOTINGS    0.00     250.0000  22.00     0.0000      447.56

3.20/   195 REINFORCING FOUNDATIONS       FOOTINGS    0.00     200.0000  2.00      4.0000      55.62

3.30/   196 C.I.P.CONCRETE FOUNDATIONS    FOOTINGS    10.00    0.0000    30.00     20.0000     664.80

3.30/   197 C.I.P.CONCRETE FOUNDATIONS                22.00    0.0000    25.00     0.0000      4447.12

3/30/   198 C.I.P.CONCRETE FOUNDATIONS                                             0.0000
```

21.2(F) Summary of contractor's own work.

```
PROJECT NAME          TEST
ESTIMATOR

85/01/09

PAGE NO. 00001     General Summary
85/01/09

                              General Summary

UCI        Trade              Labour Total    Material Total   Equipment Total   Grand Total


* Subtotal For GENERAL REQUIREMENTS
  1.00 GENERAL REQUIREMENTS     735.07          1470.14           0.00             2205.21

** SUBTOTAL **
                                735.07          1470.14           0.00             2205.21


* Subtotal For FORMWORK FOUNDATIONS
  3.00 FORMWORK FOUNDATIONS     3091.39         6182.78           6182.78          15456.95
** SUBTOTAL **
                                3091.39         6182.78           6182.78          15456.95


* Subtotal For REINFORCING FOUNDATIONS
  3.10 REINFORCING FOUNDATIONS  250.00          197.56            0.00             447.56
  3.20 REINFORCING FOUNDATIONS  54.00           0.54              1.08             55.62
** SUBTOTAL **
                                304.00          198.10            1.08             503.18
```

21.2(G) Summary of the general estimate.

21.3.4 Case 3: Estimating Service

The Company. Information for this case was obtained from a company called Automation Bid Center (Winfred Liem, President), 825 McBride Boulevard, New Westminster, BC, V3L 5B5, for their system called Autobid. The company primarily offers estimating, management, and accounting software, as well as computer training to contractors and subcontractors. It also sells hardware and support for a complete turnkey package.

The System. The Autobid system consists of several fairly large "supermicro" computers having 10- and 20-megabyte hard disk storage, to which are connected a network of several satellite computers and hardcopy high-speed printers. There are approximately 40 linked or interfaced programs, structured to provide information to the estimator either in summarized form or in as much detail as required, and either on console screen or printed in hard-paper copy. The titles of most of these programs are similar to those listed in Case 2, and include estimating, job costing, payroll, and several accounting functions, as well as word processing. The amount of data held by the system, in the form of numbers of job items, standards, cost data, and so on, is limited primarily by the power of the computer used and the amount of disk space purchased. For example, an estimator can set up a standard checklist of items relative to his or her particular trade interests, to ensure that no significant item of work is forgotten during the estimating process; such lists can be as comprehensive as necessary and amended when appropriate as the length of the list is essentially limited only by the amount of memory available on the computer disk.

The program is created by the Autobid staff. The data base or catalog for each program is created by the estimator, using labor, material, equipment, and cost standards for each industry application known to the estimator. Taxes, fringe benefits, contingencies, profit margins, and similar factors can be determined and included in the system for automatic application at any desired point in the estimating process. The data base can be easily, quickly, and indeed automatically updated as often as necessary.

Main Features. Measurements may be entered from drawings into an estimate in one of three ways: through the keyboard, through a wheel-type measuring probe, or through a sonic digitizer as described in Section 21.2. Areas and volumes measured with the probe and the digitizer are automatically calculated, thus eliminating that task for the estimator. The sonic digitizer can also replace the keyboard, as it permits the estimator to make all keyboard selections using the digitizing pen to probe a menu designed by the estimator and located on the digitizer work surface. Automatic conversion is also available; thus units may be measured in imperial or metric dimensions, or units may be measured in feet or pounds and totals produced in yards or tons, as desired. Measured items can also be priced at the time of entry, or later, when a total is accumulated for any item assembly.

According to its designers, to operate the system requires no previous computer knowledge. Each program is menu-driven, with additional help in the form of instructional manuals and trained personnel available at all times. The system permits each estimator to approach the measurement and pricing process in a way that is familiar and comfortable; there is no need to adopt artificial or cumbersome or new techniques. Judgment factors can be applied by the estimator to the measurement, pricing, and summarizing processes, directly at the keyboard. Groups of items can be assembled by the estimator to suit the interests of the work to be done, such as concrete with formwork, wall framing with insulation, drywall with paint, conduit with integral wiring, and so on. One-off or unusual items can be inserted or deleted at any point in the take-off or pricing process. Lists to order materials

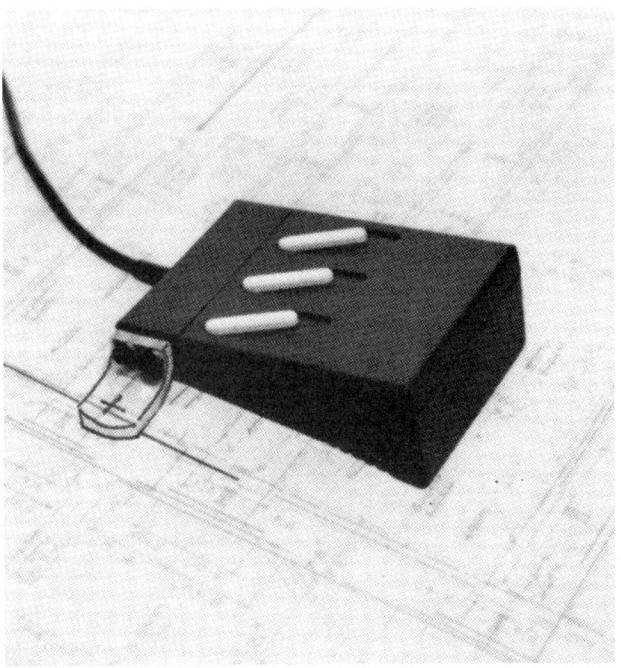

Digitizer probe.

or predict labor distributions can quickly be produced from measured work items. If any item in the estimate has to be changed, the system permits automatic recalculation of the entire estimate, viewable on the console screen, before printing or reprinting.

All information or data provided by each estimator for use in the system is confidential to that estimator; there is no access to such input by unauthorized persons, such as competitors. As can be imagined, security is a considerable problem when dealing with masses of relatively sensitive cost and company information. After doing work on the system, all information is backed up onto a spare disk, which is always in the possession of the estimator, before the original information is deleted from the system.

Notes on Hard-Copy Examples. The hard-copy reproductions that accompany this case study are explained in the corresponding notes listed below. The numbered footnotes for each explanation correspond to the legend numbers shown on each figure.

Figure number	*Explanatory notes*
21.3A	The Parts Code accepts and reports basic materials and labor standards on file.
01	12-digit alphanumeric code used to access material, labor, and equipment standards
02	20-digit alphanumeric code to define the "part"
03	Unit of measure for take-off
04	Probe code; identifies the units of measure required when using the probe
05	Unit of measure for pricing
06	Conversion factor, to convert quantities into a second unit of measure
07	Unit material cost
08	Time required to install one unit of this part
09	Number of parts that can be installed in 1 hour

(continued)

Figure number	Explanatory notes
21.3B	The Price Category reports all parts grouped by price and permits global updating of coded prices.

        01   Price category code
        02   Material markup code
        03   Date of last revision of prices

21.3C   The Assembly Report shows fixed relationships between parts, as well as total cost and labor hours for the assembly.

        01   12-digit alphanumeric code
        02   21-digit alphanumeric description
        03   Base value of the assembly
        04   Probe code, as before
        05   Part code from the catalog
        06   Quantity per base value
        07   Unit of measure for take-off
        08   Probe code for the part code
        09   Unit of measure for pricing
        10   Conversion factor, as before
        11   Unit material cost
        12   Hours to install one unit of this part
        13   Unit material cost
        14   Hours to install one unit of assembly

21.3D   The Labor Code reports cost information on individuals or work crews

        01   2-digit alphanumeric code
        02   20-digit alphanumeric code
        03   Base rate for the labor crew
        04   Percent fringe benefit burden
        05   Markup applied to the base rate
        06   Date of most recent revision

21.3E   The Material Mark-up Code reports markup standards applied to each code.

        01   2-digit alphanumeric code
        02   20-digit alphanumeric code
        03   Percent discount applied to unit material cost
        04   First percent tax markup
        05   Second percent tax markup
        06   Percent markup applied to unit material cost
        07   Date of most recent revision

21.3F   The Estimate Part Code reports total quantities and labor and material costs for each part.

        01   Part description
        02   Estimated quantity for each part
        03   Unit of measure for pricing
        04   Loaded unit material cost
        05   Labor rate used in the estimate
        06   Labor difficulty factor applied
        07   Standard rate from the catalog
        08   Material cost for each part
        09   Labor cost for each part

21.3G   The Estimate Summary reports total labor hours and costs, materials costs, and equipment costs for each part; subtotals may be selected by Drawing Reference and Estimate Section.

        01   Section number selected
        02   Drawing reference; totals are given for each part for this drawing reference
        03   Loaded unit material costs
        04   Unit labor hours from the catalog

Chap. 21 / Computer Applications

Figure number	Explanatory notes
21.3H	The Subtrade Report lists all subtrade quotations entered for a given estimate number.

           01   Estimate number
           02   Number of the subtrade type assigned
           03   Description of the subtrade type
           04   Actual name of the subcontractor
           05   Quoted material cost
           06   Quoted labor cost
           07   Any relevant information or notes

21.3I   The Final Bid Report lists total direct costs and labor hours, as well as totals for selected subtrade prices. Indirect costs can be calculated and added to arrive at the final bid price.

           01   Project description
           02   Indirect costs (up to 10 can be added)
           03   Direct costs summation
           04   "What if" calculations, for productivity, cost ratios, efficiency factors, etc.
           05   Subtrade material and labor totals
           06   Final bid price and unit cost

21.3J   Example of an actual take-off work sheet, prepared by a company using this system.

Other more detailed reports are of course available from the system for a variety of management and procedural functions.

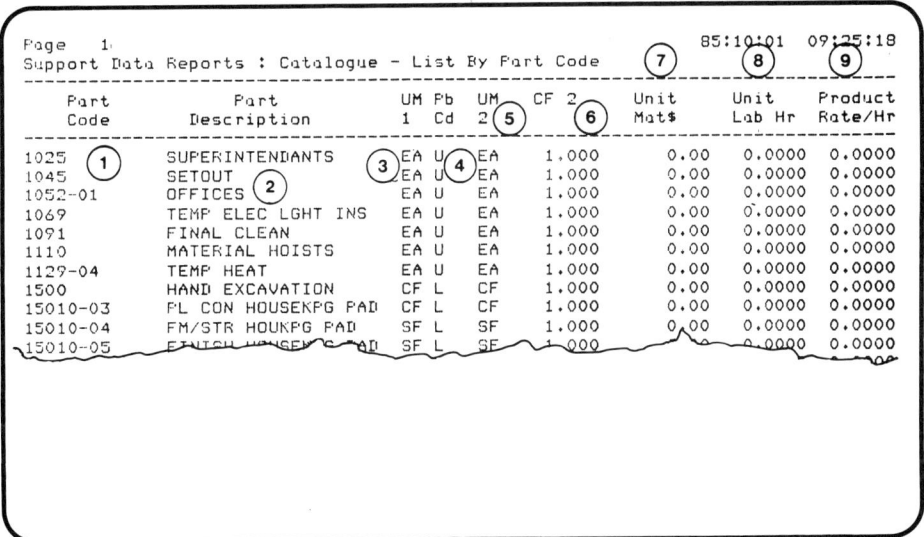

21.3(A)   The parts code.

"The broken lines across the bottom of Figures 21.3 (A) and (B) indicate that these screens can be scrolled forwards or backwards as required to disclose more information."

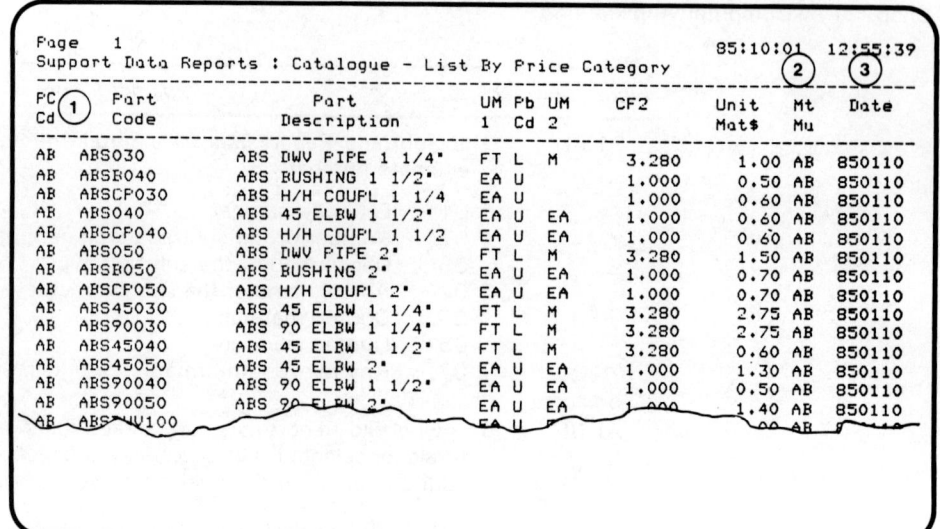

21.3(B) The price catalog.

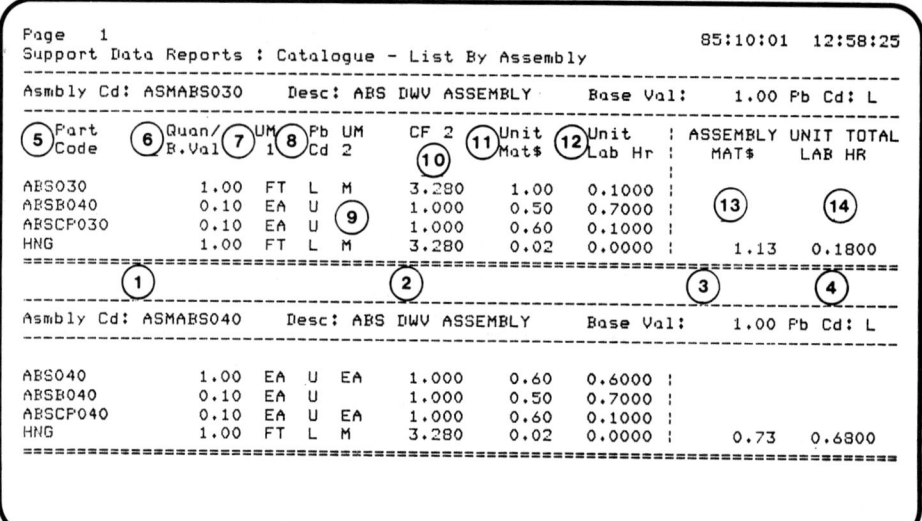

21.3(C) The assembly report.

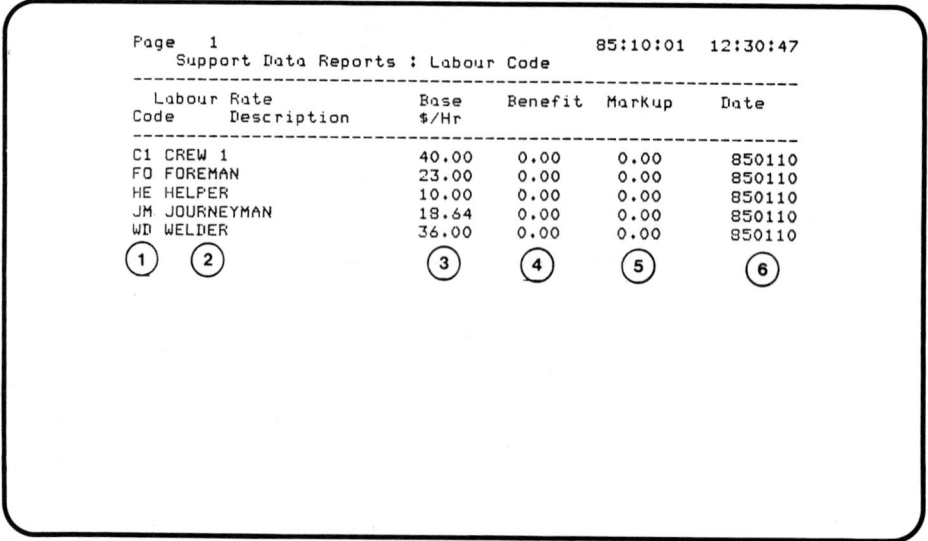

21.3(D) The labor code.

```
Page    1                                    85:10:01  12:30:13
      Support Data Reports : Materials Mark-Up
----------------------------------------------------------------
      Markup              Disc    Tax1    Tax2   Markup    Date
   Code  Description       (3)    (4)     (5)    (6)      (7)

   AB  ABS DWV PIPE        10.00   0.00    0.00   0.00   850110
   AC  ABS DWV SOIL PIPE   15.00   0.00    0.00   0.00   850110
   CH  CHECK VALVE         15.00   0.00    0.00   0.00   850110
   CU  COPPER PIPE         35.00   0.00    0.00   0.00   850110
   NT  NET WELDING         25.00   0.00    0.00   0.00   850110
   (1)    (2)
```

21.3(E) The material mark-up code.

```
Page    2                                              85:10:01  09:36:48
Estimate Reports : ESTIMATE DETAIL by Est-P.Code
-------------------------------------------------------------------------
Est No: 85001   (2)    (3)   (4)    (5)   (6)    (7)      (8)     (9)

     Description      Qty   UM    Unit    Lab    LDF    Product   Total   Total
 (1)                         2    Mat $   Rate          Rate/hr   Mat $   Lab $

POLY U/S SOG         1541   SF    0.27    0.05   0.00    1.00      416      77
SCREED SOG           1541   SF    0.00    0.15   0.00    1.00        0     231
SCREEDS STRUCT SLBS   162   SF    0.00    0.25   0.00    1.00        0      41
CURING SLBS          1703   SF    0.30    0.06   0.00    1.00      511     102
FORM DR OPENINGS       20   FT    0.75    7.94   0.00    1.00       15     160
CONC PUMP               8   HR   75.00    0.00   0.00    1.00      600       0
LUMBER 2 X 6            3   FT    1.84    6.62   0.00    1.00        6      21
LUMBER 2 X 2           32   FT    0.98    3.30   0.00    1.00       32     106
LUMBER 2 X 4           93   FT    1.23    3.30   0.00    1.00      115     308
PLYWOOD 3/4"           49   SF    9.53    5.29   0.00    1.00      467     259
=========================================================================
                                                Estimate Total   60756   29120
```

21.3(F) The estimate part code.

```
Page    2                                              85:10:01  10:05:08
Estimate Reports : ESTIMATE SUMMARY by Est-Sec-D.Ref-P.Code
-------------------------------------------------------------------------
Est No: 85001          Sec: (1)         D.Ref: (2)

   Description      Qty   UM   Unit   Unit    Total   Total   Total   Total
                          2    Mat$   Lab Hr  Mat $   Lab Hr  Lab $   Equip $
                               (3)    (4)

FINISH STRUCT SLBS    162  SF   2.69  0.0000    437     0       0       0
POLY U/S SOG         1541  SF   0.27  0.0000    416     0      77       0
SCREED SOG           1541  SF   0.00  0.0000      0     0     231       0
SCREEDS STRUCT SLBS   162  SF   0.00  0.0000      0     0      41       0
CURING SLBS          1703  SF   0.30  0.0000    511     0     102       0
FORM DR OPENINGS       20  FT   0.75  0.0000     15     0     160       0
CONC PUMP               8  HR  75.00  0.0000    600     0       0       0
=========================================================================
                                      DWG Ref.  60136   0    28426      0
```

21.3(G) The estimate summary.

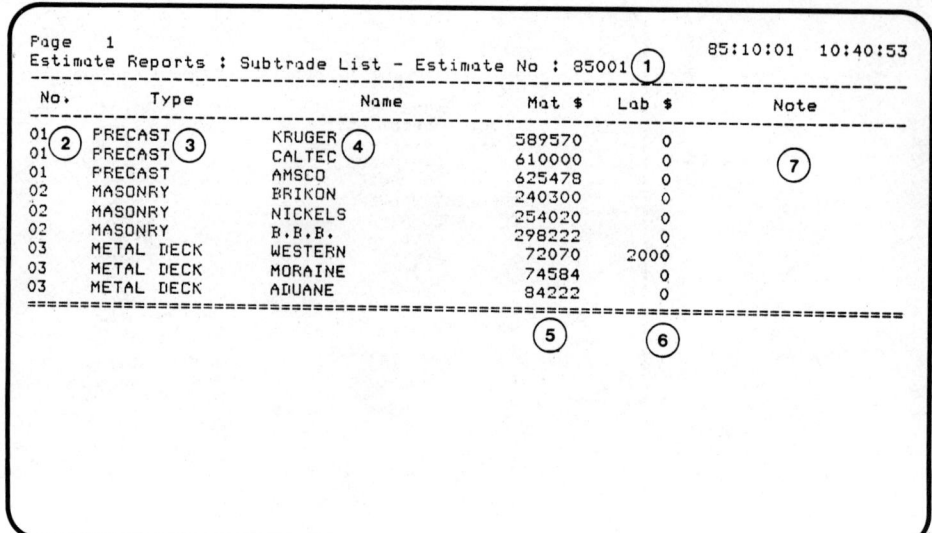

21.3(H) The subtrade report.

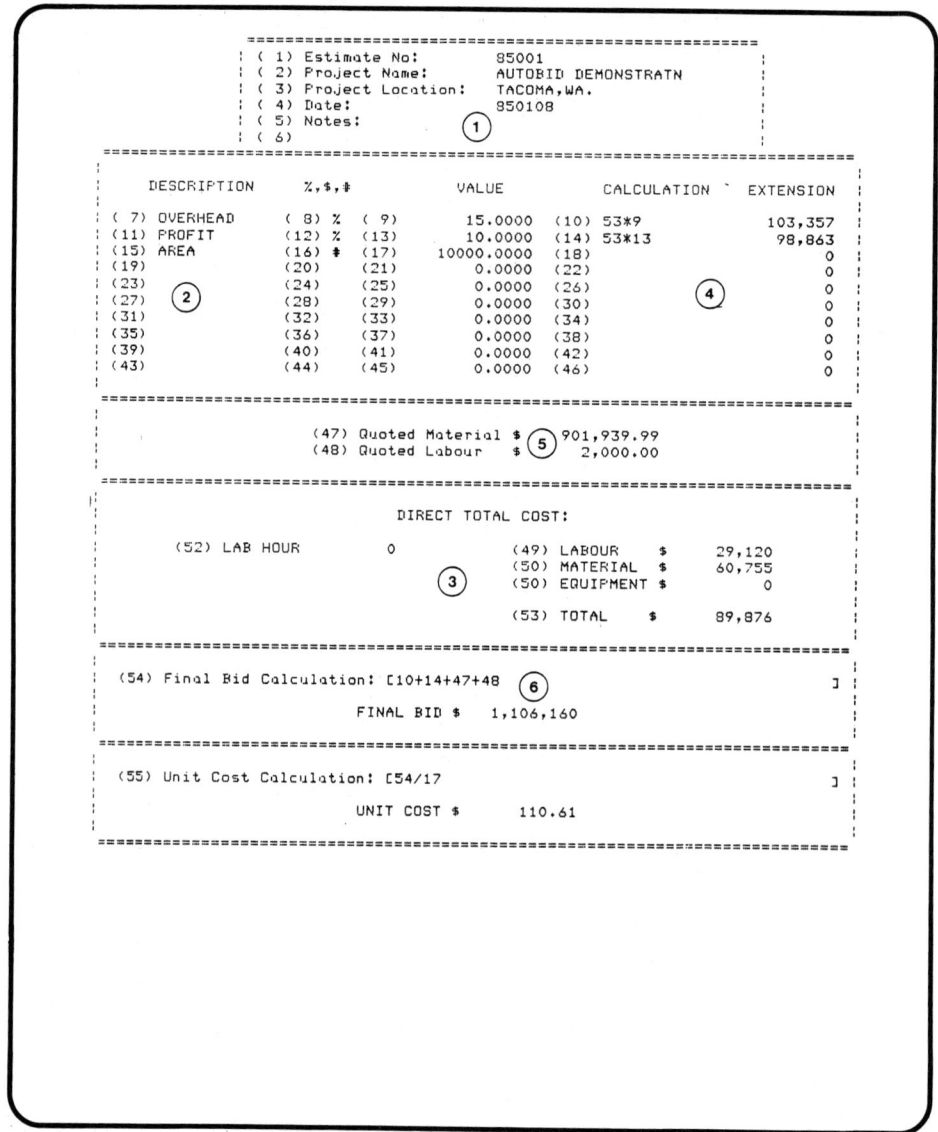

21.3(I) The final bid report.

```
3
Report: Take-Off Worksheet                                              Date-Time 840703 08:51
Selection:                                                              Page:    1
-----------------------------------------------------------------------------------------------
:Estimate No.: 7849         Project Name :              Location:                             :
:Take-Off ID : 30401A -30405C       Drawing Ref: First  -Last                                 :
-----------------------------------------------------------------------------------------------

-----------------------------------------------------------------------------------------------
:Take-Off ID : 30401A      Take-Off ID Description:CONT STRIP FTGS     UM: M                  :
-----------------------------------------------------------------------------------------------
:Drawing    : Mlt : Length : Width : Height : Perimeter :  Area   : SSA 1   : SSA 2   : Volume :         :
:Reference  :     :        :       :        :           :         : 2(1xh)  : (perimxh):        : 2 x L  :
-----------------------------------------------------------------------------------------------
: FTG STEP  :  1  :   0.50 :  0.70 :  0.55  :    2.40   :   0.35  :   0.55  :   1.32  :   0.19 :   1.00:
: HNDICAP   :  1: :   3.20 :  0.92 :  0.20  :    8.24   :   2.94  :   1.28  :   1.65  :   0.59 :   6.40:
: HNDICAP   :  1: :   1.00 :  0.92 :  0.20  :    3.84   :   0.92  :   0.40  :   0.77  :   0.18 :   2.00:
: HNDICAP   :  1: :   2.50 :  0.92 :  0.20  :    6.84   :   2.30  :   1.00  :   1.37  :   0.46 :   5.00:
: HNDICP R  :  1: :   3.50 :  0.92 :  0.20  :    8.84   :   3.22  :   1.40  :   1.77  :   0.64 :   7.00:
: SERV RAM  :  2: :  10.00 :  0.92 :  0.20  :   43.68   :  18.40  :   8.00  :   8.74  :   3.68 :  40.00:
: TYPE F    :  1: :  97.80 :  0.50 :  0.30  :  196.60   :  48.90  :  58.68  :  58.98  :  14.67 : 195.60:
: TYPE G    :  1: :   9.42 :  0.70 :  0.30  :   20.24   :   6.59  :   5.65  :   6.07  :   1.98 :  18.84:
: TYPE M    :  1: : 183.91 :  0.80 :  0.30  :  369.42   : 147.13  : 110.35  : 110.83  :  44.14 : 367.82:
: TYPE N    :  1: :  96.55 :  0.90 :  0.25  :  194.90   :  86.90  :  48.28  :  48.73  :  21.72 : 193.10:
: WHSE RAM  :  1: :  13.00 :  0.92 :  0.20  :   27.84   :  11.96  :   5.20  :   5.57  :   2.39 :  26.00:
===============================================================================================
: Total     : 12 : 421.38 :  9.12 :  2.90  :  882.84   : 329.61  : 240.78  : 245.78  :  90.65 : 862.76:
-----------------------------------------------------------------------------------------------
```

21.3(J) Actual take-off example.

21.3.5 Case 4: Design Consultant

The Company. Information for this case was obtained from a company called Microlight Computer Systems Ltd. (J. W. Whalen, Director), 4438 Valencia Avenue, North Vancouver, BC, Canada, V7N 4B1, for their system called Technique CADD. The company provides computer graphics services to architects, engineers, and contractors in both the public and private sectors.

The System. The Technique system consists of a number of disks, one containing the basic CADD (computer-aided design and drafting) program, and others containing drawing files and library files for use with the program. There is also a tutorial manual for rapid instruction in the system. The system requires use of a computer having 256K storage capacity, a color screen monitor, two 360K double-sided disk drives, a dot-matrix printer for verbal and numerical reports, and a size-D plotter for drawings. All input is by keyboard entry.

The CADD program is intended primarily to produce graphics or drawings. It therefore has functions or modes to permit lines of various types (such as straight, circled, solid, broken, thick, and thin) to be produced by the computer on a screen. These lines can be moved around, modified, or erased as desired. Line ends and midpoints can easily be established. Areas enclosed by the lines can be expanded, contracted, or otherwise changed in shape or size at will, using modes similar to panning and zooming in photography. Grids of any size can be placed over or removed from the drawings if and as necessary. Dimensioning and labeling of lengths or areas is achieved simply and rapidly using the keyboard. Once the graphics on the screen are in a form acceptable to the designer, they can be printed directly by a dot-matrix printer, or they can be written to a disk, ready for fine-line printing on a high-speed plotter. The library program features full-size storage of prepared drawings, rotation of drawings about a point, repeated duplication of portions of drawings, mirroring of images, and automatic scaling of drawings in either imperial or metric dimensions.

Main Features. According to its developers, no previous computer experience is necessary to be able to operate this system within minutes of its being set up. The development of drawings can proceed in layers, with up to 40 layers being possible, using a multicolored coded reference system. It is simple to "flip" between layers, to compare one with another. An electronic version of the manual "cut and paste" technique to assemble satisfactory design solutions is also incorporated into this program, but with some distinct advantages over the old manual technique, in terms of scale, speed, and flexibility.

Of special interest to estimators is the capacity of the system to rapidly and accurately generate lengths, areas, and volumes of all or parts of the building being drawn. The results of such calculations can be used for all sorts of analysis, involving structural design, thermal calculations, economic considerations, and so on. The data thus generated can also be easily manipulated by the computer, for example, to show cumulative totals, totals in ascending or descending orders of magnitude, areas relative to perimeters, and so on, with the results being instantly printed out on paper if required. Furthermore, cost data can be input, to show the cost effects of any proposed design solution or amendment immediately on the screen, on a "what if" or tentative basis, to discover the greatest cost benefit to the designer or owner of the proposed building under review.

Notes on Hard-Copy Examples. The hard-copy reproductions that accompany this case study are explained in the corresponding notes listed below.

Figure number	Explanatory notes
21.4A	Portion of a typical residential floor plan produced by CADD equipment. Imperial or metric dimensions can be overlaid and printed, on command.
21.4B	Area calculations automatically produced by CADD for designated portions of the plan above.
21.4C	Printer graphic screen dump, showing the living room of the building highlighted and identified with its area noted in square meters.
21.4D	Output from the Datafile, showing three manipulations: 1. Areas of specified rooms in ascending order 2. Areas of specified rooms in alphabetical order 3. Volumes of specified areas in ascending order Price data could now be applied to measurements.

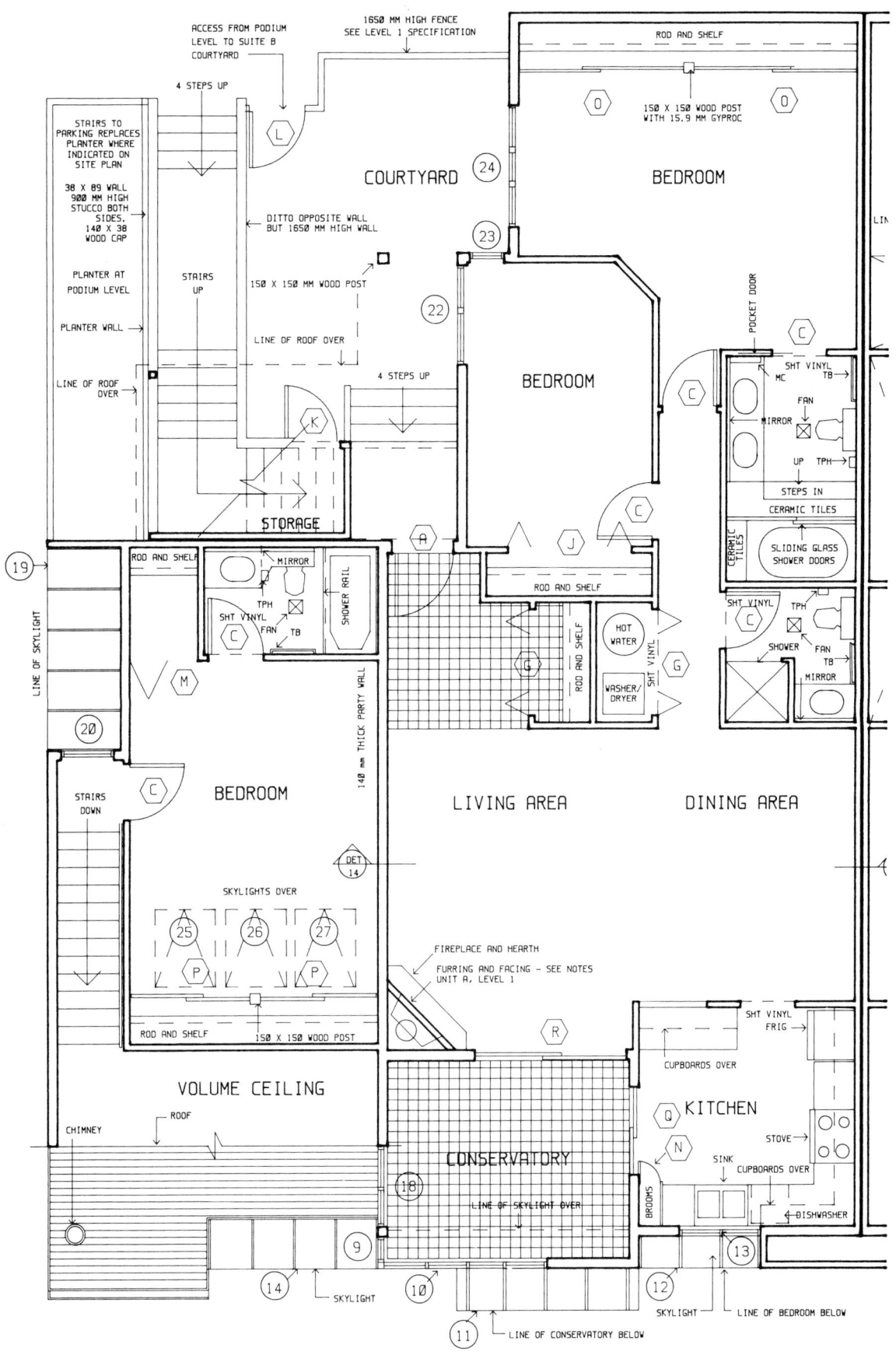

21.4(A) Portion of a typical floor plan.

294 Part IV / Applications

```
           L I S T   O F   A R E A   T A G S
           1    NAME:BEDROOM 1        21.18 SQ.M. TOTAL
           2    NAME:BEDROOM 2        10.73 SQ.M. TOTAL
           3    NAME:BATHROOM 1        6.12 SQ.M. TOTAL
           4    NAME:BATHROOM 2        3.67 SQ.M. TOTAL
           5    NAME:CLOSET            4.25 SQ.M. TOTAL
           6    NAME:HALLWAY           8.78 SQ.M. TOTAL
           7    NAME:LIVING RM        16.31 SQ.M. TOTAL
           8    NAME:DINING RM        12.64 SQ.M. TOTAL
           9    NAME:CONSERV'Y         9.89 SQ.M. TOTAL
          10    NAME:KITCHEN           9.87 SQ.M. TOTAL
          11    NAME:LEVEL1 BED       19.58 SQ.M. TOTAL
          12    NAME:LEV1.BATH         3.83 SQ.M. TOTAL
          13    NAME:LEV1.CLOS.        1.61 SQ.M. TOTAL

           Press any key....
```

21.4(B) Automatic area calculations.

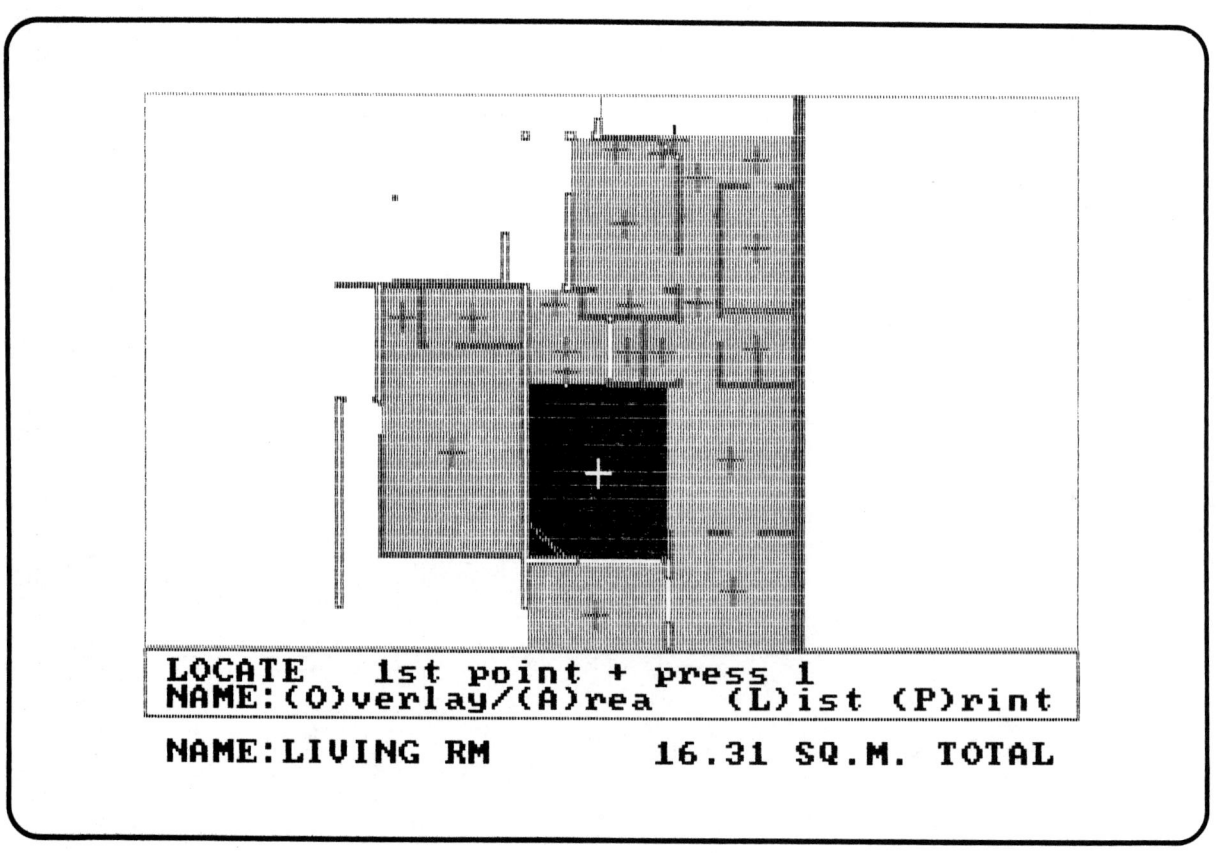

21.4(C) Photo of Console Screen.

```
1. FILENAME      LAYER NAME    AREA NAME    TOTAL SQ.M.    DATE       TIME
   -----------   ----------    ---------    -----------    -------    -------
   NEWPORT.1D    LEVEL 2       LEV1.CLOS.        1.61      6-22-85    3:54p
   NEWPORT.1D    LEVEL 2       BATHROOM 2        3.67      6-22-85    3:54p
   NEWPORT.1D    LEVEL 2       LEV1.BATH         3.83      6-22-85    3:54p
   NEWPORT.1D    LEVEL 2          CLOSET         4.25      6-22-85    3:54p
   NEWPORT.1D    LEVEL 2       BATHROOM 1        6.12      6-22-85    3:54p
   NEWPORT.1D    LEVEL 2         HALLWAY         8.78      6-22-85    3:54p
   NEWPORT.1D    LEVEL 2         KITCHEN         9.87      6-22-85    3:54p
   NEWPORT.1D    LEVEL 2        CONSERV'Y        9.89      6-22-85    3:54p
   NEWPORT.1D    LEVEL 2       BEDROOM 2        10.73      6-22-85    3:54p
   NEWPORT.1D    LEVEL 2       DINING RM        12.64      6-22-85    3:54p
   NEWPORT.1D    LEVEL 2       LIVING RM        16.31      6-22-85    3:54p
   NEWPORT.1D    LEVEL 2       LEVEL1 BED       19.58      6-22-85    3:54p
   NEWPORT.1D    LEVEL 2       BEDROOM 1        21.18      6-22-85    3:54p
                                              --------
                                                128.46

2. FILENAME      LAYER NAME    AREA NAME    TOTAL SQ.M.    ROOM HT.   VOLUME CU.M
   -----------   ----------    ---------    -----------    --------   -----------
   NEWPORT.1D    LEVEL 2       BATHROOM 1        6.12        2.20       13.45
   NEWPORT.1D    LEVEL 2       BATHROOM 2        3.67        2.20        8.08
   NEWPORT.1D    LEVEL 2       BEDROOM 1        21.18        2.40       50.82
   NEWPORT.1D    LEVEL 2       BEDROOM 2        10.73        2.40       25.75
   NEWPORT.1D    LEVEL 2          CLOSET         4.25        2.40       10.21
   NEWPORT.1D    LEVEL 2        CONSERV'Y        9.89        2.10       20.78
   NEWPORT.1D    LEVEL 2       DINING RM        12.64        2.40       30.33
   NEWPORT.1D    LEVEL 2         HALLWAY         8.78        2.40       21.06
   NEWPORT.1D    LEVEL 2         KITCHEN         9.87        2.40       23.70
   NEWPORT.1D    LEVEL 2       LEV1.BATH         3.83        2.40        9.19
   NEWPORT.1D    LEVEL 2       LEV1.CLOS.        1.61        2.40        3.85
   NEWPORT.1D    LEVEL 2       LEVEL1 BED       19.58        2.40       46.99
   NEWPORT.1D    LEVEL 2       LIVING RM        16.31        3.20       52.20
                                              --------               ---------
                                                128.46                  316.42

3. FILENAME      LAYER NAME    AREA NAME    TOTAL SQ.M.    ROOM HT.   VOLUME CU.M
   -----------   ----------    ---------    -----------    --------   -----------
   NEWPORT.1D    LEVEL 2       LEV1.CLOS.        1.61        2.40        3.85
   NEWPORT.1D    LEVEL 2       BATHROOM 2        3.67        2.20        8.08
   NEWPORT.1D    LEVEL 2       LEV1.BATH         3.83        2.40        9.19
   NEWPORT.1D    LEVEL 2          CLOSET         4.25        2.40       10.21
   NEWPORT.1D    LEVEL 2       BATHROOM 1        6.12        2.20       13.45
   NEWPORT.1D    LEVEL 2        CONSERV'Y        9.89        2.10       20.78
   NEWPORT.1D    LEVEL 2         HALLWAY         8.78        2.40       21.06
   NEWPORT.1D    LEVEL 2         KITCHEN         9.87        2.40       23.70
   NEWPORT.1D    LEVEL 2       BEDROOM 2        10.73        2.40       25.75
   NEWPORT.1D    LEVEL 2       DINING RM        12.64        2.40       30.33
   NEWPORT.1D    LEVEL 2       LEVEL1 BED       19.58        2.40       46.99
   NEWPORT.1D    LEVEL 2       BEDROOM 1        21.18        2.40       50.82
   NEWPORT.1D    LEVEL 2       LIVING RM        16.31        3.20       52.20

   Newport Terrace Housing                       128.46                  316.42
   Summary of areas and volumes                --------               ---------
                                              Total area             Total volume
```

21.4(D)   3 manipulations of data.

21.3.6 Case 5: Specialty Contractor

The Company. Information for this case was obtained from a company called Management Data Base Systems Corporation (T. D'Sena, President), 3857 Sunset Street, Burnaby, BC, Canada, V5G 1T4, for their system called Bid-Rite. The company provides specialty computer estimating and cost-accounting services for plumbing, electrical, mechanical, and indeed most other types of contractors, on a fee basis. It also installs its system in such offices on a contract basis.

The System. The Bid-Rite system first requires that the contractor enter all of the commonly used data pertinent to his or her operation, such as plumbing, electrical, or other material parts descriptions, current prices, and standard labor rates, into a file on a computer disk to form the primary data base, popularly called the "price book." The system allows modifications to be made to the data-base information as and when necessary. Assistance is available to the contractor by the computer company during this stage of preparation. Pricing information can be extracted from the construction company records or from annually issued price data books, such as *Pricemaster,* published by Pricemaster, PO Box 307, Islington, Ontario, Canada, M9A 4X3, *Building Construction Cost Data,* published by R. S. Means Company, Inc., 100 Construction Plaza, Kingston, Mass. 02364, or similar sources, for insertion into the system.

The take-off process involves use of a device called the Mandat Micro-Measure, which consists of a digitizer probe having a scale which can be set by the estimator to suit the drawings from which the measurement is being done, as well as three small buttons which the estimator presses to record things measured lineally, such as start and finish of runs of pipe, and things measured numerically, such as fittings, bends, branches, and so on. These distances and numbers are automatically fed into the computer directly from and by the probe, and are extended into labor hours and labor and material costs for each item identified by the estimator. Irregular areas and volumes can also be measured and recorded with the device, if necessary.

The estimator selects an item from the prepared pricebook file, identifies its relation to the particular estimate being prepared, sets the scale for the drawing, uses the Micro-Measure to measure the lengths or count the number of instances of the item occurring on the drawings, and the computer does the rest.

The system requires a minimum computer capacity of 256K RAM and MS-DOS operating system, with two 5¼-in. disk drives and 400K memory for programs. It is, however, recommended that a 10-megabyte hard disk be used to store the programs and data base, because of the added convenience, flexibility, and security which such capacity creates.

Main Features. The system has a unique search feature which permits rapid selection of parts without knowledge of the part code number in the system. The system also sports a take-off worksheet component, which lets the estimator focus on specific aspects of the work of one particular trade or specialty at one time, exactly as the estimator would do in the traditional manual mode of measurement.

The system accommodates virtually simultaneous electronic take-off and paper printout of measurements on the item-by-item or real-time basis, with automatic extension of dimensions, which is a convenience for the estimator trying to keep track of the take-off process. It also permits virtually instant and error-free job setup from the estimating system to the job cost system for those projects for which a bid is converted into a contract.

The system is arranged to handle last-minute adjustments, such as new or revised quotations received right up to bid day. It permits entry of specification items in advance of bid-day deadlines; furthermore, it will not permit duplications and will report any specification items that were overlooked during the measurement process. As with most similar systems, Bid-Rite incorporates automatic conversion from any scale to any other, and from metric to imperial dimensions, and vice versa.

Notes on Hard-Copy Examples. The hard-copy reproductions that accompany this case study are explained in the corresponding notes listed below.

Figure number	Explanatory notes
21.5A	This shows part of the Main Index File, used to locate general groups of items contained in the Sub Index File.
21.5B	This file is used to locate specific groups of items of contained in the Parts File.

Chap. 21 / Computer Applications

Figure number	Explanatory notes
21.5C	This file contains individual parts with their respective material costs and labor hours.
21.5D	This is an example of one specific assembly of parts, as presented on the screen to the estimator.
21.5E	This file permits global adjustments to markup factors on material costs on file.
21.5F	This file shows current labor rates and fringe benefits on file, permitting global adjustment to such entries at any time.
21.5G	This shows a small portion of a very extensive checklist file which reminds the estimator of items to consider during estimate preparation.
21.5H	This shows the take-off worksheet, into which actual dimensions are entered by the estimator from the drawings, using the computer keyboard and the Micro-Measure device.
21.5I	This shows one example of many types of reports able to be extracted instantaneously from the system.
21.5J	This shows an Estimate Summary, with identification detail, labor and material summaries, and all markup clearly shown.

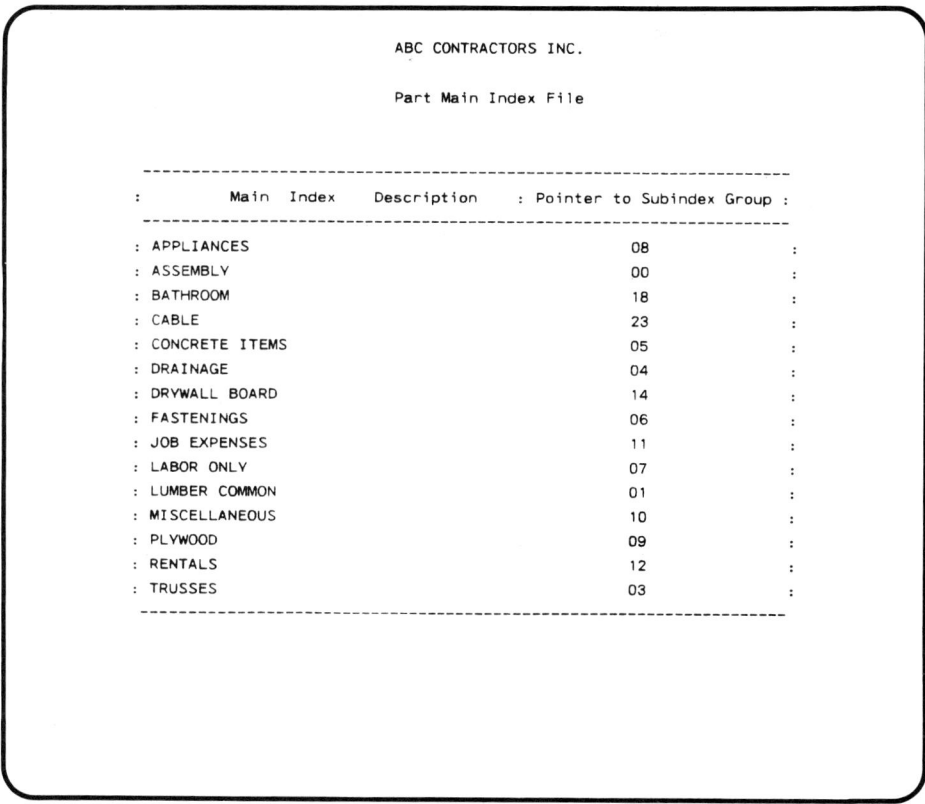

21.5(A) Main index file.

```
                              ABC CONTRACTORS INC.

                              Part Sub Index File


     ------------------------------------------------------------------
     :              Index    Description      : Pointer to Part Group  :
     ------------------------------------------------------------------
     : CONCRETE FORM, COLUMN                     00001                 :
     : CONCRETE FORM, PILE CAP                   00002                 :
     : CONCRETE FORM, WALL FOOTING               00004                 :
     : WALL ASSEMBLY, 88 5/8" 16" OC             00003                 :
     : FIR                                       01001                 :
     : STUDS                                     01003                 :
     : UTILITY                                   01005                 :
     : BELGIAN                                   03001                 :
     : HOWE                                      03002                 :
     : DOWNSPOUT, ELBOW                          04003                 :
     : DOWNSPOUT, PIPE                           04002                 :
     : DRAIN FILE                                04004                 :
     : EAVE TROUGH                               04001                 :
     : ANCHOR BOLTS                              05001                 :
     : CONCRETE                                  05002                 :
     : PORTLAND CEMENT                           05003                 :
     : PRECAST CONCRETE ROOF SLABS               05005                 :
     : REINFORCEMENT BA                                                :
     ------------------------------------------------------------------
```

21.5(B) Sub index file.

```
Section   : Report                 ABC CONTRACTORS INC.                      Date: 120484
Selection : Price Book Report      Listing by Part Description               Page:  1
-----------------------------------------------------------------------------------------------------
: Pointer  Part Description     : Part        : UM1: Unit        :  Cross      :Prb.: PC: Matl.:Lab: JC  Date :
:                                : Code        :    : Mat.$ Lab. Hrs. Reference: :Code:    :Mkp:Dct:Cd.:Cat: mmddyy:
-----------------------------------------------------------------------------------------------------
:01001 LUMBR COM FIR 1 X 4, 10'    LUMCF010410    FT   0.19   0.000              L   LU   BO  NA  NA  MA  091784 :
:01001 LUMBR COM FIR 1 X 4, 12'    LUMCF010412    FT   0.19   0.000              L   LU   BO  NA  NA  MA  091784 :
:01001 LUMBR COM FIR 1 X 4, 14'    LUMCF010414    FT   0.19   0.000              L   LU   BO  NA  NA  MA  091784 :
:01001 LUMBR COM FIR 1 X 4, RANDOM LUMCF0104 R    FT   0.19   0.000              L   LU   BO  NA  NA  MA  010180 :
:01001 LUMBR COM FIR 2 X 4, 10'    LUMCF020410    FT   0.25   0.000              L   LU   BO  NA  NA  MA  091784 :
:01001 LUMBR COM FIR 2 X 4, 12'    LUMCF020412    FT   0.25   0.000              L   LU   BO  NA  NA  MA  091784 :
:01001 LUMBR COM FIR 2 X 4, 14'    LUMCF020414    FT   0.25   0.000              L   LU   BO  NA  NA  MA  091784 :
:01001 LUMBR COM FIR 2 X 4, RANDOM LUMCF0204 R    FT   0.25   0.000              L   LU   BO  NA  NA  MA  010180 :
:01003 STUDS 88 5/8"               LUMST885       EA   0.85   0.000              U   LU   BO  NA  NA  MA  062184 :
:01003 STUDS 90 5/8"               LUMST905       EA   0.91   0.000              U   LU   BO  NA  NA  MA  062184 :
:01003 STUDS 92"                   LUMST92        EA   0.95   0.000              U   LU   BO  NA  NA  MA  062184 :
:01003 WOOD                        THNW           FT   4.00   1.000              L   NA   NA  NA  NA  MA  103184 :
:01005 LUMBR COM UTILITY 2 X 6, 12' LUMCU020612   FT   0.24   0.000              L   LU   BO  NA  NA  MA  091784 :
:01005 LUMBR COM UTILITY 2 X 6, 14' LUMCU020614   FT   0.24   0.000              L   LU   BO  NA  NA  MA  091784 :
:01005 LUMBR COM UTILITY 2 X 8, 8'  LUMCU020808   FT   0.24   0.000              L   LU   BO  NA  NA  MA  091784 :
:01005 LUMBR COM UTILITY 2 X 8, 10' LUMCU020810   FT   0.24   0.000              L   LU   BO  NA  NA  MA  091784 :
:01005 LUMBR COM UTILITY 2 X 8, 12'                FT   0.24   0.000                        BO  NA  NA  MA  091784 :
:01005 LUMBR COM UTILITY 2 X 10, 10'               FT   0.35   0.000                                              :
:      LUMBR COM UTILITY                                0.35

       UM = Unit of Measurement    CF = Conversion Factor    Prb. = Probe    Mat. = Material    JC = Job Cost
             Tmp. = Temporary    Lab. = Labor    Hrs = Hours    PC = Price Category Code    Cat. = Category
```

21.5(C) Parts file.

```
3
Section   : Report                   ABC CONTRACTORS INC.                    Date: 120484
Selection : Price Book Report       Listing by Assembly Code                  Page:   1
--------------------------------------------------------------------------------
:         Assembly        Code    : CFFORM         Date Revised : 103184  Probe Code = A      :
: Group: 00 Class:001 Description : COLUMN FOOTING FORM 5'X5'X1'6"  UM = SF  Ref. Base Value :  100.00:
--------------------------------------------------------------------------------
: Assembly    :Qty. Per:     Part  Description   :UM1:    Unit      :  Cross       :Prb.:PC:Mat.:Lab.: JC : Date  :
: Sub Part    :Base Val:                         :   : Mat.$  Lab. Hrs. Reference :Code: :Code:Code:Cat.:mmddyy:
--------------------------------------------------------------------------------
:LACA             5.00 LABOR CARPENTER            EA   0.00   1.000             U    NA  NA   CA   LA  091784:
:LAHE             3.50 LABOR HELPER               EA   0.00   1.000             U    NA  NA   HE   LA  091784:
:LUMCU020410    150.00 LUMBR COMMON UTILITY 2 X 4, 10'  FT  0.20  0.000          L    LU  BO   NA   MA  091884:
:LUMCU021210    100.00 LUMBR COM UTILITY 2 X12, 10'    FT   0.51   0.000         L    LU  BO   NA   MA  062184:
:NAC3             2.00 NAILS COMMON 3"            LB   0.77   0.000             U    NA  BO   NA   MA  090784:
:     ASSEMBLY CODE: [CFFORM       ]       Non Labor Cost / 001:   82.54    Labor Hrs. / 001 :    8.50      :
--------------------------------------------------------------------------------
```

21.5(D) Assembly costs file.

```
                        ABC CONTRACTORS INC.
                        Material Markup File

          --------------------------------------------------
          : Material    :   Description    : Matl Mkup : Date Last :
          : Markup Code :                  :    %      : Modified  :
          --------------------------------------------------
          :    BO       BOUGHT OUT MATERIALS    0.00      060884   :
          :    ER       EQUIPMENT RENTALS      10.00      060884   :
          :    JE       JOB EXPENSES           10.00      060884   :
          :    MA       MATERIALS A            12.00      060884   :
          :    MB       MATERIALS B            13.00      060884   :
          :    MC       MATERIALS C            14.00      060884   :
          :    ST       SUB TRADES             10.00      060884   :
          --------------------------------------------------
```

21.5(E) Material mark-up file.

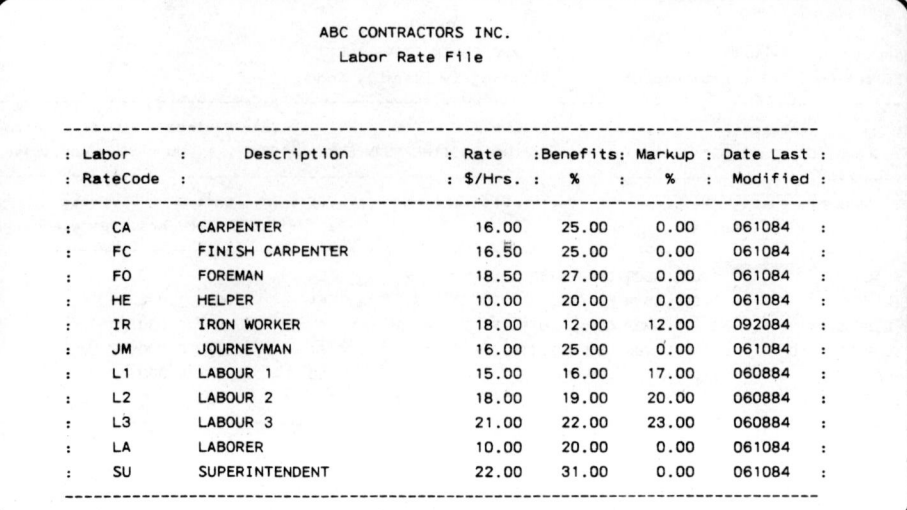

ABC CONTRACTORS INC.
Labor Rate File

Labor RateCode	Description	Rate $/Hrs.	Benefits %	Markup %	Date Last Modified
CA	CARPENTER	16.00	25.00	0.00	061084
FC	FINISH CARPENTER	16.50	25.00	0.00	061084
FO	FOREMAN	18.50	27.00	0.00	061084
HE	HELPER	10.00	20.00	0.00	061084
IR	IRON WORKER	18.00	12.00	12.00	092084
JM	JOURNEYMAN	16.00	25.00	0.00	061084
L1	LABOUR 1	15.00	16.00	17.00	060884
L2	LABOUR 2	18.00	19.00	20.00	060884
L3	LABOUR 3	21.00	22.00	23.00	060884
LA	LABORER	10.00	20.00	0.00	061084
SU	SUPERINTENDENT	22.00	31.00	0.00	061084

21.5(F) Labor rates and fringe benefits.

```
Section   : Report                      ABC CONTRACTORS INC.                        Date: 120484
Selection: Check List Report                 Check List                             Page: 2
---------------------------------------------------------------------------------------------
: Check :  Job Division              :     Job Item              :U.M.:  Part   :   Part Description   :
:List # : No.    Description         : No.  Description         Qty  :    :  Code   :                      :
---------------------------------------------------------------------------------------------
:   1      6  CARPENTRY                673  TRUSS JOIST           1.00  EA  0    CHECKLIST SETUP ITEM   :
:   1      6  CARPENTRY                675  MILLWORK              1.00  EA  0    CHECKLIST SETUP ITEM   :
:   1      6  CARPENTRY                676  FINISH CARPENTRY      1.00  EA  0    CHECKLIST SETUP ITEM   :
:   1      7  PROTECTION               701  ROOFING-SHINGLE-TILE  1.00  EA  0    CHECKLIST SETUP ITEM   :
:   1      7  PROTECTION               702  WATER & DAMPROOFING   1.00  EA  0    CHECKLIST SETUP ITEM   :
:   1      7  PROTECTION               703  MASONARY INSULATION   1.00  EA  0    CHECKLIST SETUP ITEM   :
:   1      7  PROTECTION               704  FOUNDATION INSULAT    1.00  EA  0    CHECKLIST SETUP ITEM   :
:   1      7  PROTECTION               706  SOUND-ACOUST INSULAT  1.00  EA  0    CHECKLIST SETUP ITEM   :
:   1      7  PROTECTION               708  GENERAL SHEET METAL   1.00  EA  0    CHECKLIST SETUP ITEM   :
:   1      8  DOORS & HARDWARE         801  HM DOORS & FRAMES     1.00  EA  0    CHECKLIST SETUP ITEM   :
:   1      8  DOORS & HARDWARE         802  WOOD DOORS & FRAMES   1.00  EA  0    CHECKLIST SETUP ITEM   :
:   1      8  DOORS & HARDWARE         803  O.H. & ROLLING DRS    1.00  EA  0    CHECKLIST SETUP ITEM   :
:   1      8  DOORS & HARDWARE         805  FINISH HARDWARE       1.00  EA  0    CHECKLIST SETUP ITEM   :
:   1      8  DOORS & HARDWARE         806  GLASS & GLAZING       1.00  EA  0    CHECKLIST SETUP ITEM   :
:   1      9  FINISHES                 901  LATH-PLASTER-STUCO    1.00  EA  0    CHECKLIST SETUP ITEM   :
:   1      9  FINISHES                 902  SPEC COATINGS-SPRAY   1.00  EA  0    CHECKLIST SETUP ITEM   :
:   1      9  FINISHES                 903  DRYWALL & STUDS       1.00  EA  0    CHECKLIST SETUP ITEM   :
:   1      9  FINISHES                 904  ACOUST TILE & SUSP    1.00  EA  0    CHECKLIST SETUP ITEM   :
:   1      9  FINISHES                 905  INTEGRATED CEILING    1.00  EA  0    CHECKLIST SETUP ITEM   :
:   1      9  FINISHES                                            1.00  EA       CHECKLIST SETUP ITEM   :
"   1      9  FINISHES                                            1.00            CHECKLIST SETUP ITEM   :
```

21.5(G) Portion of check list file.

```
Section   : Report                    ABC CONTRACTORS INC.                              Date: 1204 84
Selection: Take-off Detail               Takeoff Worksheet                               Page: 1
--------------------------------------------------------------
:Estimate No.   : 1100      Project Name :PACIFIC INDUSTRIAL PARK   :
:Date of Estimate: 020584   Customer Name:WESTPARK DEVELOPMENT INC.:
-----------------------------------------------------------------------------------------------------
:Computer:   Computer ID   :   Drawing   : Takeoff:Probe:Length : Width : Height : Count :Perimeter: Area    : Volume:
:  ID    :   Description   :   Reference :  Unit  :Code :        :       :        :       :         :         :       :
-----------------------------------------------------------------------------------------------------
:  1       CONCRETE SLAB     DWG A1          FT      V    125.00   30.00    1.50    1.00    310.00   3750.00  5625.00:
:  1       CONCRETE SLAB     DWG A2          FT      V     10.00   15.00    1.00    1.00     50.00    150.00   150.00:
:  1       CONCRETE SLAB     DWG B1          FT      V    102.00   11.00    1.10    1.00    226.00   1122.00  1234.20:
: Total for Computer ID [1   ]    586.00 = Perimeter     5022.00 = Area      763.60 = S.S.A.     7009.20 = Volume:
=====================================================================================================
:  2       CEILINGS          ROOM 1          FT      A     12.50   11.00            1.00              47.00    137.50 :
:  2       CEILINGS          ROOM 2          FT      A     12.00   15.00            1.00              54.00    180.00 :
:  2       CEILINGS          ROOM 3          FT      A     13.00   12.00            1.00              50.00    156.00 :
:  2       CEILINGS          ROOM 4          FT      A     12.50   11.00            1.00              47.00    137.50 :
: Total for Computer ID [2   ]    198.00 = Perimeter      611.00 = Area                                               :
=====================================================================================================
:  3       OUTSIDE WATERLINE DWG A1          FT      L    125.00                    1.00                              :
:  3       OUTSIDE WATERLINE DWG A1          FT      L    133.00                    1.00                              :
:  3       OUTSIDE WATERLINE DWG A2          FT      L    101.00                    1.00                              :
: Total for Computer ID [3   ]    359.00=Length                                                                       :
=====================================================================================================
```

21.5(H) Take-off worksheet.

```
Section   : Report                    ABC CONTRACTORS INC.                              Date: 120484
Selection: Est. Detail Summary        Estimate Detail Summary                           Page:    2
-----------------------------------------------------------------------------------------------------
: Division No. :   3         Description   : CONCRETE                          :
-----------------------------------------------------------------------------------------------------
: Job Item No. : 351         Description   : FOOTINGS       Job Item Qty.:  1.00    Unit of Measure : EA  :
-----------------------------------------------------------------------------------------------------
:   Part    : Assembly Sub Part :    Description     :UM: Quantity: U.Mat : U. Lab.:  Total  :  Total    :
:   Code    : Part Code  : Qty  :                    : 1: Required:  $ *  : Hrs. *:  Matl. $ :Lab. Hrs.  :
-----------------------------------------------------------------------------------------------------
:  LAFO                         :LABOR FOREMAN         EA    12.00    0.00    1.00     0.00     12.00 :
:  LAHE                         :LABOR HELPER          EA    24.00    0.00    1.00     0.00     24.00 :
:  C02000                       :CONCRETE - 2000 PSI   CF  1320.00    2.95    0.00  3894.00      0.00 :
-----------------------------------------------------------------------------------------------------
                                                      : JOB ITEM :  351    TOTAL   7818.39    451.58 :
                                                      -----------------------------------------------

      No. = Number    UM = Unit of Measurement    Matrl. = Material   Lab. = Labor   Hrs. = Hours
      NOTE THAT UNIT MATRL. $ & UNIT LABOUR HRS. HAVE THE DIFFICULTY & PRICE FACTORS INCLUDED ALREADY
```

21.5(I) Job item summary.

```
------------------------------------------------------------------
:Estimate No. : 1100       Project Name  : PACIFIC INDUSTRIAL PARK :
:Date Estimate: 020584     Customer Name : WESTPARK DEVELOPMENT INC. :
:                  Prepared By :  MR. SUMMERS                     :
------------------------------------------------------------------

------------------------------------------------------------------
:Project Location   : PRINCE RUPERT    : Customer Address : 2233 PAPIER ST   :
:Customer Quote No.: 34                :                    BURNABY, B.C.    :
:Due Date           : 032184           :                    V5G LT9          :
:Contact Person     : S. SULLIVA       : Customer Tel. No. 604-433-4583      :
------------------------------------------------------------------
:                                      :                                    :
:              LABOUR                  :              MATERIAL              :
:                                      :                                    :
:  1. Hours              541           :  6. Cost $             22054       :
:  2. Cost $            9062           :  7. Markup (Approx. % * )  0.80    :
:  3. Markup (Approx. % *)  0.61       :  8. Markup $             176       :
:  4. Markup $            56           :                                    :
:                        ------        :                          ------    :
:  5. Sub Total         9117           :  9. Sub Total           22231      :
:                       ======         :                         ======     :
------------------------------------------------------------------

        ----------------------------------------------------------
        :                                                        :
        : 10. Labor + Material Total ( 5 + 9 )        31348      :
        : 11. Overhead ( 15.00 % of 2 )                1359      :
        :                                             -----      :
        : 12. Sub Total ( 10 + 11 )                   32707      :
        : 13. General & Admin. Expense ( 12.00 % of 12 ) 3925    :
        :                                             -----      :
        : 14. Sub Total ( 12 + 13 )                   36632      :
        : 15. Profits    ( 10.00% of 14 )              3663      :
        :                                             -----      :
        : 16. Bid Price ( 14 + 15 )                   40296      :
        : 17. Bond ( 0.02% of 16 )                        8      :
        : 18. Insurance ( 0.03% of 16 )                  12      :
        :                                             -----      :
        : 19. Final Total ( 16 + 17 + 18 )            40304      :
        :                                             =====      :
        :                                                        :
        ----------------------------------------------------------

 * Approximate % because of labour & Material Markup mix. MARKUP $ IS CORRECT.
```

21.5(J)   General estimate summary.

21.4 CONCLUSION

21.4.1 Inferences

It is true to say that every company or organization will have its own specific and peculiar configuration of financial and human resources. It will therefore be obvious that many more case studies could have been included in this chapter. However, the object was not to cover every possible base, but rather to suggest some general features and refinements of interest to most people involved in construction estimating and budgeting studies. From the foregoing examples, it should not be difficult to develop the main parameters and criteria for computer applications for any given situation. In fact, this activity may be undertaken as an assignment by the more advanced student of estimating, in conjunction with other field studies in project management.

21.4.2 A Final Word

Having come this far in this book, the reader will (or should) have a clearer view of the estimating process as a whole, its several components, and its relationship to other aspects of building technology. It is to be hoped that the book has given the reader a systematic base on which to build further knowledge and skill, and has also removed much of the confusion and doubt that unnecessarily surrounds the economic side of the design and construction processes. In life, the elimination of ignorance reduces doubts and fears and replaces them with knowledge and confidence. However, we all face the paradoxical problem that as more and more knowledge becomes available, we are in danger of becoming more and more ignorant, because of the difficulty of keeping abreast of it all. The object of this book was to impart knowledge and thus reduce ignorance, at least in the area of construction estimating techniques.

QUESTIONS

21.1. Distinguish between the two phrases "programming for applications" and "applications of programming" as they apply to construction estimating.

21.2. Differentiate between the terms "bit" and "byte" as they are used in computer terminology. Explain the meaning of the word "binary" in this context.

21.3. Identify four main advantages to the construction estimator to be gained from the use of a computer system.

21.4. Identify and briefly describe three common types of peripherals used for *both* input and output to the computer CPU (six in all).

21.5. Draw and label the parts of a single spreadsheet matrix such as might be used by a construction estimator.

21.6. Explain why the decision to purchase a particular brand of computer should follow decisions regarding the acquisition of software.

PART IV
RESEARCH ASSIGNMENTS

The research assignments that follow are intended to give readers a channel through which to acquire knowledge of current local issues and aspects relative to the contents of the chapters contained in this part of the book. The basic objective in all of these assignments is to allow readers to compare theory with practice, by studying this and other similar books, and by encountering opinions and activities, other than those of themselves and this author, on the stated topics. A guide to content, procedure, and evaluation for these assignments has been included in Section 1.4. A review of these guidelines is recommended before starting work on the assignment projects.

CHAPTER 16: INTRODUCTION TO APPLICATION

Investigate the supply and installation of the work of any one of the five trades discussed in this chapter. Show by photographic illustration and by neatly drawn sketches what the work comprises and how it is done. Report on materials and equipment used, crew sizes established, productivity and waste, working conditions, and costs.

CHAPTER 17: MEASUREMENT—CONTRACTORS

Using the checklist and the first page of the measurement shown in Figure 17.10 as a guide, complete the rough carpentry work for the Wood-Framed House shown on the drawings in Part VI numbered MET-3. Compare your measured quantities of lumber, plywood, and sheathing paper with the totals listed in the accompanying notes.

CHAPTER 18: MEASUREMENT—DESIGNERS

Investigate estimating methods, other than elemental analysis, currently used by designers to determine project construction budgets in the area where you reside. If elemental analysis is used, compare or contrast it to the techniques suggested in this book. Specifically report on measurement methods, sources of cost data, and reliability of results. Comment on the academic or other training and experiential background of personnel doing this type of work.

CHAPTER 19: PRICING EXAMPLES—SELECTED TRADES

Investigate labor productivity to place concrete in slabs, beams, and columns on a small to medium-sized reinforced-concrete structure. Specifically report on crew sizes and configurations, ancillary equipment, and output. Develop a table of factors to make allowance for all likely attributes, such as size or complexity of the project, weather and time of year, quality of the labor force, and so on. Introduce cost factors wherever pertinent.

CHAPTER 20: PRICING EXAMPLES—SELECTED ELEMENTS

Using the metric model in the book as a guide, develop a specification outline, suitable for use in elemental analysis, for either the Post-and-Beam House *or* the Retail Store, for which drawings are included in Part VI. Develop quantities and unit prices for any five elements in that building *other than* those selected for demonstration in the book in connection with the Wood-Frame House. Show intermediate details of element quantity and price development, and identify sources of cost data. Prepare an appropriate Summary Sheet; calculate ratios for the five elements that you have developed in detail.

CHAPTER 21: COMPUTER APPLICATIONS

Identify the types of computers and titles of software programs used by (1) architectural designers and (2) construction estimators for the development of budgets and estimates for building projects in the region where you live. Specifically report on hard and soft costs, arising out of acquisition and training for use, respectively.

PART V APPENDICES

APPENDIX A
ABBREVIATIONS

A.1 INTRODUCTION

In Section 8.2.9 the subject of abbreviation of words and symbols was introduced in connection with writing short descriptions of items of work in measurement. In this appendix, some of the more common abbreviations encountered in estimating are listed under separate headings. The symbols used to designate metric and imperial quantities are listed in Section 9.2.3.

A.2 DRAWINGS

bgdg	bridging	flrg	flooring
bldg	building	flshg	flashing
blkwk	blockwork	fndn	foundation
brkwk	brickwork	ftg	footing
brr	bearer	grd	grade
b/w	both ways	grnd	ground
chmny	chimney	o.c.	on centers
clg	ceiling	opng	opening
col	column	ptn	partition
cont	continuous	pvg	paving
dpfg	dampproofing	reinf	reinforced
dtl	detail	wndw	window
e.f.	each face	wpf	waterproof
elev	elevation	xbdrg	cross-bridging
e/w	each way		

A.3 DESCRIPTIONS

a/b	as before	lab	labor
avg	average	l&m	labor and material
bbl	barrel	lngth	length
bdth	breadth	mat	material
circ	circular	max	maximum
ct	coat	meas	measure
ctg	coating	min	minimum
dbl	double	MP	mean perimeter
ddt	deduct	n.e.	not exceeding
dim	dimension	pcs	pieces
ditto	as above, repeat	proj	projection
excav	excavation	qty	quantity
exstg	existing	RC&W	raking cutting and waste
fin	finished	rnd	round
flt	float	str	straight
fmwk	formwork	temp	temperature
hlw	hollow	temp	temporary
ht	height	trwld	troweled
intl	internal	wi	with
intr	interior	wl	wall

A.4 MATERIALS

ag	agricultural tile	ins	insulation
a&g	asphalt and gravel	lam	laminated
blk	block	mldg	molding
brk	brick	ms	mild steel
cmt	cement	mtr	mortar
conc	concrete	rebar	reinforcing steel
galv	galvanized	shthg	sheathing
hdwd	hardwood	t&g	tongue and groove

A.5 MEASUREMENT

A	area	o/a	overall
bdth	breadth	rad	radius
dpth	depth, deep	SFCA	square feet contact area
fbm	foot board measure	SQ	one square (100 sf)
lngth	length	vol	volume
M	thousand	wdth	width
MP	mean perimeter	wt	weight
n.w.	narrow widths		

A.6 SPECIFICATIONS

AD	air dried	no.	number
DIV	division	S1S	surfaced one side
G1S	good one side	S2S	surfaced two sides
G2S	good two sides	SEC	section
ga	gage	specs	specifications
KD	kiln dried	WF	wide flange

A.7 PUNCTUATION

@	at	÷	divide
#0	number zero	&	and
0#	zero pounds	/	times (also divide)
:	shall be	0'	zero feet
°	degree	0"	zero inches
$	dollar	%	percent
−	minus or deduct	>	greater than
+	add	<	smaller than
×	multiply		

APPENDIX B
MASTERFORMAT

B.1 INTRODUCTION

About 20 years ago, the Construction Specifications Institute (CSI) in the United States and its counterpart in Canada, Construction Specifications Canada (CSC), jointly produced a reference system in an attempt to improve organization of the production, distribution, filing, and retrieval of construction literature of every sort and type. The first edition of the system was contained in a document titled "The Uniform Construction Index," which had three parts: first, information of interest to specification writers in the form of guidelines for the organization of construction contract documents; second, information to assist manufacturers of construction materials, products, or systems to produce standardized product literature in a form acceptable to the specification writers; and third, a numbered code system intended for use by contractors and others to keep track of the various parts of each construction project.

The numbered code system was also used to integrate the three parts of the Uniform System, along with an alphabetical index. The system consisted of 16 major Divisions, with each Division containing a number of Sections, each of which was further subdivided into Broadscope and Narrowscope titles. After some years of use, a revised edition was published simultaneously in the United States and Canada, with the new title "Masterformat" and one additional Division (00) added, to cover all the front-end documentation normally included in contracts. The short excerpt from Masterformat shown on the following pages includes all 17 Divisions and shows the assigned numbers and titles of all Broadscope Sections within each Division. It is reproduced here with permission.

The worked examples in measurement and pricing of various parts of the projects which have been included in this book are identified, wherever necessary, with their appropriate Masterformat number and title. The complete list is included here to assist students to organize and identify further practice exercises according to this international standard of reference. For further information on this topic, readers are encouraged to contact either the Construction Specifications Institute or Construction Specifications Canada, both of whose postal addresses are given in Appendix D.

B.2 EXCERPT FROM MASTERFORMAT

DIVISION 0 — BIDDING AND CONTRACT REQUIREMENTS	
00010	PRE-BID INFORMATION
00100	INSTRUCTIONS TO BIDDERS
00200	INFORMATION AVAILABLE TO BIDDERS
00300	BID/TENDER FORMS
00400	SUPPLEMENTS TO BID/TENDER FORMS
00500	AGREEMENT FORMS
00600	BONDS AND CERTIFICATES
00700	GENERAL CONDITIONS OF THE CONTRACT
00800	SUPPLEMENTARY CONDITIONS
00950	DRAWINGS INDEX
00900	ADDENDA AND MODIFICATIONS
DIVISION 1 — GENERAL REQUIREMENTS	
01010	SUMMARY OF WORK
01020	ALLOWANCES
01030	SPECIAL PROJECT PROCEDURES
01040	COORDINATION
01050	FIELD ENGINEERING
01060	REGULATORY REQUIREMENTS
01070	ABBREVIATIONS AND SYMBOLS
01080	IDENTIFICATION SYSTEMS
01100	ALTERNATES/ALTERNATIVES
01150	MEASUREMENT AND PAYMENT
01200	PROJECT MEETINGS
01300	SUBMITTALS
01400	QUALITY CONTROL
01500	CONSTRUCTION FACILITIES AND TEMPORARY CONTROLS
01600	MATERIAL AND EQUIPMENT
01650	STARTING OF SYSTEMS
01660	TESTING, ADJUSTING, AND BALANCING OF SYSTEMS
01700	CONTRACT CLOSEOUT
DIVISION 2 — SITEWORK	
02010	SUBSURFACE INVESTIGATION
02050	DEMOLITION
02100	SITE PREPARATION
02150	UNDERPINNING
02200	EARTHWORK
02300	TUNNELLING
02350	PILES, CAISSONS AND COFFERDAMS
02400	DRAINAGE
02440	SITE IMPROVEMENTS
02480	LANDSCAPING
02500	PAVING AND SURFACING
02590	PONDS AND RESERVOIRS
02600	PIPED UTILITY MATERIALS AND METHODS
02700	PIPED UTILITIES
02800	POWER AND COMMUNICATION UTILITIES
02850	RAILROAD WORK
02880	MARINE WORK
DIVISION 3 — CONCRETE	
03050	CONCRETING PROCEDURES
03100	CONCRETE FORMWORK
03150	FORMS
03180	FORM TIES AND ACCESSORIES
03200	CONCRETE REINFORCEMENT
03250	CONCRETE ACCESSORIES
03300	CAST-IN-PLACE CONCRETE
03350	SPECIAL CONCRETE FINISHES
03360	SPECIALLY PLACED CONCRETE
03370	CONCRETE CURING
03400	PRECAST CONCRETE
03500	CEMENTITIOUS DECKS

Figure B.1 Masterformat.

03600 GROUT
03700 CONCRETE RESTORATION AND CLEANING

DIVISION 4—MASONRY

04050 MASONRY PROCEDURES
04100 MORTAR
04150 MASONRY ACCESSORIES
04200 UNIT MASONRY
04400 STONE
04500 MASONRY RESTORATION AND CLEANING
04550 REFRACTORIES
04600 CORROSION RESISTANT MASONRY

DIVISION 5—METALS

05010 METAL MATERIALS AND METHODS
05050 METAL FASTENING
05100 STRUCTURAL METAL FRAMING
05200 METAL JOISTS
05300 METAL DECKING
05400 COLD-FORMED METAL FRAMING
05500 METAL FABRICATIONS
05700 ORNAMENTAL METAL
05800 EXPANSION CONTROL
05900 METAL FINISHES

DIVISION 6—WOOD AND PLASTICS

06050 FASTENERS AND SUPPORTS
06100 ROUGH CARPENTRY
06130 HEAVY TIMBER CONSTRUCTION
06150 WOOD-METAL SYSTEMS
06170 PREFABRICATED STRUCTURAL WOOD
06200 FINISH CARPENTRY
06300 WOOD TREATMENT
06400 ARCHITECTURAL WOODWORK
06500 PREFABRICATED STRUCTURAL PLASTICS
06600 PLASTIC FABRICATIONS

DIVISION 7—THERMAL AND MOISTURE PROTECTION

07100 WATERPROOFING
07150 DAMPROOFING
07200 INSULATION
07250 FIREPROOFING
07300 SHINGLES AND ROOFING TILES
07400 PREFORMED ROOFING AND SIDING
07500 MEMBRANE ROOFING
07570 TRAFFIC TOPPING
07600 FLASHING AND SHEET METAL
07800 ROOF ACCESSORIES
07900 JOINT SEALANTS

DIVISION 8—DOORS AND WINDOWS

08100 METAL DOORS AND FRAMES
08200 WOOD AND PLASTIC DOORS
08250 DOOR OPENING ASSEMBLIES
08300 SPECIAL DOORS
08400 ENTRANCES AND STOREFRONTS
08500 METAL WINDOWS
08600 WOOD AND PLASTIC WINDOWS
08650 SPECIAL WINDOWS
08700 HARDWARE
08800 GLAZING
08900 GLAZED CURTAIN WALLS

DIVISION 9—FINISHES

09100 METAL SUPPORT SYSTEMS
09200 LATH AND PLASTER
09230 AGGREGATE COATINGS
09250 GYPSUM WALLBOARD
09300 TILE
09400 TERRAZZO
09500 ACOUSTICAL TREATMENT
09550 WOOD FLOORING
09600 STONE AND BRICK FLOORING
09680 CARPETING
09700 SPECIAL FLOORING
09760 FLOOR TREATMENT
09800 SPECIAL COATING
09900 PAINTING
09950 WALL COVERING

DIVISION 10—SPECIALTIES

10100 CHALKBOARDS AND TACKBOARDS
10150 COMPARTMENTS AND CUBICLES
10200 LOUVERS AND VENTS

Figure B.1 (cont.)

10240	GRILLES AND SCREENS	11170	WASTE HANDLING EQUIPMENT
10250	SERVICE WALL SYSTEMS	11190	DETENTION EQUIPMENT
10260	WALL AND CORNER GUARDS	11200	WATER SUPPLY AND TREATMENT EQUIPMENT
10270	ACCESS FLOORING	11300	FLUID WASTE DISPOSAL AND TREATMENT EQUIPMENT
10280	SPECIALTY MODULES		
10290	PEST CONTROL		
10300	FIREPLACES AND STOVES	11400	FOOD SERVICE EQUIPMENT
10340	PREFABRICATED STEEPLES, SPIRES, AND CUPOLAS	11450	RESIDENTIAL EQUIPMENT
		11460	UNIT KITCHENS
10350	FLAGPOLES	11470	DARKROOM EQUIPMENT
10400	IDENTIFYING DEVICES	11480	ATHLETIC, RECREATIONAL, AND THERAPEUTIC EQUIPMENT
10450	PEDESTRIAN CONTROL DEVICES		
10500	LOCKERS		
10520	FIRE EXTINGUISHERS, CABINETS, AND ACCESSORIES	11500	INDUSTRIAL AND PROCESS EQUIPMENT
		11600	LABORATORY EQUIPMENT
10530	PROTECTIVE COVERS	11650	PLANETARIUM AND OBSERVATORY EQUIPMENT
10550	POSTAL SPECIALTIES		
10600	PARTITIONS	11700	MEDICAL EQUIPMENT
10650	SCALES	11780	MORTUARY EQUIPMENT
10670	STORAGE SHELVING	11800	TELECOMMUNICATION EQUIPMENT
10700	EXTERIOR SUN CONTROL DEVICES		
		11850	NAVIGATION EQUIPMENT
10750	TELEPHONE ENCLOSURES		
10800	TOILET AND BATH ACCESSORIES		
10900	WARDROBE SPECIALTIES		

DIVISION 12—FURNISHINGS

12100	ARTWORK
12300	MANUFACTURED CABINETS AND CASEWORK
12500	WINDOW TREATMENT
12550	FABRICS
12600	FURNITURE AND ACCESSORIES
12670	RUGS AND MATS
12700	MULTIPLE SEATING
12800	INTERIOR PLANTS AND PLANTINGS

DIVISION 11—EQUIPMENT

11010	MAINTENANCE EQUIPMENT
11020	SECURITY AND VAULT EQUIPMENT
11030	CHECKROOM EQUIPMENT
11040	ECCLESIASTICAL EQUIPMENT
11050	LIBRARY EQUIPMENT
11060	THEATER AND STAGE EQUIPMENT
11070	MUSICAL EQUIPMENT
11080	REGISTRATION EQUIPMENT
11100	MERCANTILE EQUIPMENT
11110	COMMERCIAL LAUNDRY AND DRY CLEANING EQUIPMENT
11120	VENDING EQUIPMENT
11130	AUDIO-VISUAL EQUIPMENT
11140	SERVICE STATION EQUIPMENT
11150	PARKING EQUIPMENT
11160	LOADING DOCK EQUIPMENT

DIVISION 13—SPECIAL CONSTRUCTION

13010	AIR SUPPORTED STRUCTURES
13020	INTEGRATED ASSEMBLIES
13030	AUDIOMETRIC ROOMS
13040	CLEAN ROOMS
13050	HYPERBARIC ROOMS
13060	INSULATED ROOMS
13070	INTEGRATED CEILINGS
13080	SOUND, VIBRATION, AND SEISMIC CONTROL
13090	RADIATION PROTECTION
13100	NUCLEAR REACTORS
13110	OBSERVATORIES

Figure B.1 (cont.)

13120	PRE-ENGINEERED STRUCTURES	14300	HOISTS AND CRANES
13130	SPECIAL PURPOSE ROOMS AND BUILDINGS	14400	LIFTS
		14500	MATERIAL HANDLING SYSTEMS
13140	VAULTS	14600	TURNTABLES
13150	POOLS	14700	MOVING STAIRS AND WALKS
13160	ICE RINKS		
13170	KENNELS AND ANIMAL SHELTERS	14800	POWERED SCAFFOLDING
		14900	TRANSPORTATION SYSTEMS
13200	SEISMOGRAPHIC INSTRUMENTATION		

DIVISION 15—MECHANICAL

13210	STRESS RECORDING INSTRUMENTATION
13220	SOLAR AND WIND INSTRUMENTATION
13410	LIQUID AND GAS STORAGE TANKS
13510	RESTORATION OF UNDERGROUND PIPELINES
13520	FILTER UNDERDRAINS AND MEDIA
13530	DIGESTION TANK COVERS AND APPURTENANCES
13440	OXYGENATION SYSTEMS
13540	THERMAL SLUDGE CONDITIONING SYSTEMS
13560	SITE CONSTRUCTED INCINERATORS
13600	UTILITY CONTROL SYSTEMS
13700	INDUSTRIAL AND PROCESS CONTROL SYSTEMS
13800	OIL AND GAS REFINING INSTALLATIONS AND CONTROL SYSTEMS
13900	TRANSPORTATION INSTRUMENTATION
13940	BUILDING AUTOMATION SYSTEMS
13970	FIRE SUPPRESSION AND SUPERVISORY SYSTEMS
13980	SOLAR ENERGY SYSTEMS
13990	WIND ENERGY SYSTEMS

15050	BASIC MATERIALS AND METHODS
15200	NOISE, VIBRATION, AND SEISMIC CONTROL
15250	INSULATION
15300	SPECIAL PIPING SYSTEMS
15400	PLUMBING SYSTEMS
15450	PLUMBING FIXTURES AND TRIM
15500	FIRE PROTECTION
15600	POWER OR HEAT GENERATION
15650	REFRIGERATION
15700	LIQUID HEAT TRANSFER
15800	AIR DISTRIBUTION
15900	CONTROLS AND INSTRUMENTATION

DIVISION 16—ELECTRICAL

16050	BASIC MATERIALS AND METHODS
16200	POWER GENERATION
16300	POWER TRANSMISSION
16400	SERVICE AND DISTRIBUTION
16500	LIGHTING
16600	SPECIAL SYSTEMS
16700	COMMUNICATIONS
16850	HEATING AND COOLING
16900	CONTROLS AND INSTRUMENTATION

DIVISION 14—CONVEYING SYSTEMS

14100	DUMBWAITERS
14200	ELEVATORS

Figure B.1 *(cont.)*

APPENDIX C
SELECTED BIBLIOGRAPHY

C.1 INTRODUCTION

Literature pertaining to construction technology, technique, and procedure is very extensive. The titles in the following lists have been selected as a guide to reliable American, British, and Canadian books to which reference may be made for further study and information on the topics and issues included in this book. The exclusion of other titles from the list does not imply their unreliability, although it is the experience of the author that the range of quality is about as great as the range of quantity of books published in this field. Five titles in each of 10 categories have been alphabetically listed.

C.2 ON CONTRACTS

1. Collier, Keith F., *Construction Contracts.* Reston, Va.: Reston Publishing Co., Inc., 1979.
2. Goldsmith, Immanual, *Canadian Building Contracts.* Agincourt, Ontario: Carswell Company Ltd., 1976.
3. Hudson, A., *Building and Engineering Contracts.* London: Sweet & Maxwell Ltd., 1970.
4. Jabine, William, *Case Histories in Construction Law.* Boston: CBI Publishing Co., Inc., 1970.
5. Walker, N., Walker, E., and Rohdenburg, T., *Legal Pitfalls in Architecture, Engineering, and Building Contracts.* New York: McGraw-Hill Book Company, 1979.

C.3 ON CONTRACTING

1. Clough, Richard, *Construction Contracting.* New York: John Wiley & Sons, Inc., 1975.
2. Deatherage, George E., *Company Organization and Management.* New York: McGraw-Hill Book Company, 1964.

3. Godel, Jules B., and Seymour, Berger, *Estimating and Project Management for Small Construction Firms.* New York: Construction Publishing Company, Inc., 1976.
4. Wass, Alonzo, *Construction Management and Contracting.* New York: McGraw-Hill Book Company, 1975.
5. Zehner, J. R., *Builder's Guide to Contracting.* New York: McGraw-Hill Book Company, 1975.

C.4 ON CONSTRUCTION

1. Cannon, Kenneth F., and Hatley, Frederick G., *Building Construction Technology.* Scarborough, Ontario: McGraw-Hill Ryerson Ltd., 1982.
2. Ching, Francis D. K., *Building Construction Illustrated.* New York: Van Nostrand Reinhold Company, Inc., 1975.
3. Hornbostel, Caleb, and Hornung, William J., *Materials and Methods for Contemporary Construction.* Englewood Cliffs, N.J.: Prentice-Hall, Inc., 1982.
4. Huntington, Whitney C., and Mickadeit, Robert, *Building Construction: Materials and Types of Construction.* New York: John Wiley & Sons, Inc., 1975.
5. Smith, Ronald R., *Principles and Practices of Light Construction.* Englewood Cliffs, N.J.: Prentice-Hall, Inc., 1970.

C.5 ON COST CONTROL

1. Neil, James M., *Construction Cost Estimating for Project Control.* Englewood Cliffs, N.J.: Prentice-Hall, Inc., 1982.
2. Gooch, Kenneth O., and Caroline, John, *Construction for Profit.* Reston, Va.: Reston Publishing Co., Inc., 1980.
3. Kharbanda, Om Prakash, et al., *Project Cost Control in Action.* Englewood Cliffs, N.J.: Prentice-Hall, Inc., 1981.
4. Walker, Frank R., *Practical Accounting for Contractors.* Chicago: Frank R. Walker Publishing Co., Inc., 1980.
5. Ward, Sol, and Litchfield, Thorndike, *Cost Control in Design and Construction.* New York: McGraw-Hill Book Company, 1980.

C.6 ON COST DATA

1. Hanscomb, Roy, & Associates, *Yardsticks for Costing.* Toronto: Southam Business Publications, Ltd. 1983.
2. Landsdowne, David K., *Construction Cost Handbook.* Scarborough, Ontario: McGraw-Hill Ryerson Ltd., 1983.
3. Means, Robert S., *Building Construction Cost Data.* Duxbury, Mass.: Robert Snow Means Company, Inc., 1983.
4. Richardsons, *General Construction Estimating Standards.* San Francisco, Calif.: Richardsons Engineering Services, Inc., 1983.
5. Sarriel, Edward, *The National Construction Estimator.* Carlsbad, Calif.: Craftsman Book Company, 1983.

C.7 ON ECONOMICS

1. Bathurst, Peter E., and Butler, David A., *Building Cost Control: Techniques and Economics.* London: William Heinemann Ltd., 1980.
2. Ferry, Douglas J., and Brandon, Peter S., *Cost Planning of Buildings.* St. Albans, Herts., England: Granada Publishing Ltd., 1980.

3. Helyar, Frank W., *Construction Estimating and Costing.* Scarborough, Ontario: McGraw-Hill Ryerson Ltd., 1978.
4. Hunt, William D., *Creative Control of Building Costs.* New York: McGraw-Hill Book Company, 1967.
5. Seeley, Ivor H., *Building Economics.* London: The Macmillan Press Ltd., 1978.

C.8 ON ESTIMATING

1. Adrian, James J., *Construction Estimating.* Reston, Va.: Reston Publishing Co., Inc., 1982.
2. Collier, Keith F., *Fundamentals of Construction Estimating and Cost Accounting.* Englewood Cliffs, N.J.: Prentice-Hall, Inc., 1974.
3. Peurifoy, Robert L., *Estimating Construction Costs.* New York: McGraw-Hill Book Company, 1975.
4. Vance, Mary A., *Selected List of Books on Building Cost Estimating.* Monticelli, Ill.: Vance Bibliographies, 1979.
5. Walker, Frank R., *The Building Estimator's Reference Book.* Chicago: Frank R. Walker Publishing Co., Inc., 1980.

C.9 ON MANAGEMENT

1. Adrian, James J., *CM: The Construction Management Process.* Reston, Va.: Reston Publishing Co., Inc., 1981.
2. Barrie, Donald S., and Paulson, Boyd C., *Professional Construction Management.* New York: McGraw-Hill Book Company, 1978.
3. McNulty, Alfred P., *Management of Small Construction Projects.* New York: McGraw-Hill Book Company, 1982.
4. O'Brien, James J., and Zilly, Robert G., *Contractor's Management Handbook.* New York: McGraw-Hill Book Company, 1971.
5. Reiner, Laurence E., *A Handbook of Construction Management.* Englewood Cliffs, N.J.: Prentice-Hall, Inc., 1972.

C.10 ON MATERIALS

1. Dagostino, Frank R., *Materials of Construction.* Reston, Va.: Reston Publishing Co., Inc., 1982.
2. Godel, Jules B., *Sources of Construction Information.* Metuchen, N.J.: Scarecrow Press, Inc., 1978.
3. Hornbostel, Caleb, *Construction Materials: Types, Uses, and Applications.* New York: John Wiley & Sons, Inc., 1978.
4. Maguire, Byron W., *Construction Materials.* Reston, Va.: Reston Publishing Co., Inc., 1981.
5. Watson, Donald A., *Construction Materials and Processes.* New York: McGraw-Hill Book Company, 1978.

C.11 ON SPECIFICATIONS

1. Hardie, Glenn M., *Construction Contracts and Specifications.* Reston, Va.: Reston Publishing Co., Inc., 1981.
2. Lewis, Jack, *Construction Specifications.* Englewood Cliffs, N.J.: Prentice-Hall, Inc., 1975.

3. Meier, Hans W., *Construction Specifications Handbook*. Englewood Cliffs, N.J.: Prentice-Hall, Inc., 1975.
4. Rosen, Harold J., *Construction Specifications Writing*. New York: John Wiley & Sons, Inc., 1974.
5. Watson, Donald A., *Specifications Writing for Architects and Engineers*. New York: McGraw-Hill Book Company, 1964.

APPENDIX D
CONSTRUCTION ASSOCIATIONS

D.1 INTRODUCTION

One of the characteristics of the construction industry is its tremendous diversification and fragmentation. Official, semiofficial, and unofficial associations, societies, institutes, clubs, and the like abound at every national, regional, and local level. All of the major construction professions involved, such as architecture, engineering, management, economics, labor, and so on, are represented. Virtually all technical interests, such as supervision, specifications, and purchasing, are provided for. Every trade, from aluminum work to welding work, has organizations of one kind or another to look after its interests. Although there is some weakness in this proliferation of interests, stemming from enormous problems in communication, standardization, and polarity, there are also strengths, in that attention can be sharply focused on the problems and issues that confront the various sectors of the construction industry, both within their own spheres of interest and in their relationships with other sectors of the industry.

Some of the benefits for people belonging to an appropriate construction organization are the opportunity to share and compare technical news and opinions with like-minded colleagues, to keep up to date in professional work and daily activities with new developments in the field, to speak to government and other authorities with a more powerful voice on matters of importance, to have a say in the affairs of the particular sector of the industry, and to enjoy the prestige that is usually associated with such membership.

In this appendix, a short list of selected organizations of particular interest to people involved with estimating and construction economics has been prepared. The list identifies only *national associations* in both Canada and the United States. Most of these national bodies have regional or local chapters or offices, to attend to the needs of their members. Most such local chapters or branches hold monthly meetings throughout the fall, winter, and spring seasons, at which topics of interest to the membership are presented by a variety of local or national authorities. More specific information about these organizations and their locals can be obtained by writing to the addresses given in the lists. The addresses are listed alphabetically, and were correct and current at the date of publication of this book.

D.2 CANADA

1. Associate Committee for Building Codes
   National Research Council
   Ottawa, Ontario K1A 0S9

2. Canadian Construction Association
   85 Albert Street
   Ottawa, Ontario K1P 6A4

3. Canadian Construction Documents Committee
   85 Albert Street
   Ottawa, Ontario K1P 6A4

4. Canadian Construction Women
   Construction House
   2675 Oak Street
   Vancouver, British Columbia V6H 2K3

5. Canadian Government Standards Board
   Government Publishing Center
   Supply and Services Canada
   Ottawa, Ontario K1A 0S9

6. Canadian Institute of Quantity Surveyors
   P.O. Box 124, Station 124
   Toronto, Ontario M4G 3Z3

7. Canadian Labor Congress
   2841 Riverside Drive
   Ottawa, Ontario K1V 8X7

8. Canadian Standards Association
   178 Rexdale Boulevard
   Rexdale, Ontario M9W 1R3

9. Construction Management Institute
   5799 Yonge Street, Suite 901
   Willowdale, Ontario M2M 3V3

10. Construction Specifications Canada
    1 St. Claire Avenue West, Suite 1206
    Toronto, Ontario M4V 1K6

11. Royal Architectural Institute of Canada
    328 Somerset Street West
    Ottawa, Ontario K2P 0J9

D.3 UNITED STATES

1. American Association of Cost Engineers
   308 Monongahela Building
   Morgantown, WV 26505

2. American Federation of Labor (AFL-CIO)
   Building & Construction Trades Department
   815 16 Street NW, Suite 603
   Washington, DC 20006

3. American Institute of Architects
   1735 New York Avenue NW
   Washington, DC 20006

4. American National Standards Institute
   1430 Broadway
   New York, NY 10018
5. American Society for Testing and Materials
   P.O. Box 390
   Glen Ellyn, IL 60137
6. American Society of Professional Estimators
   441 South 48th Street
   Tempe, AZ 85281
7. American Subcontractors Association
   8401 Corporate Drive
   Landover, MD 20785
8. Associated General Contractors of America
   1957 E Street NW
   Washington, DC 20006
9. Association of Women in Construction
   327 South Adams Street
   Fort Worth, TX 76104
10. Construction Specifications Institute
    601 Madison Avenue
    Alexandria, VA 22314
11. National Conference of States on Building Codes
    481 Carlisle Drive
    Herndon, VA 22070
12. National Estimating Society
    904 Bob Wallace Avenue, Suite 213
    Hunstville, AL 35801
13. Professional Construction Estimators Association
    P.O. Box 1107
    Cornelius, NC 28031

APPENDIX E
CHECKLISTS

E.1 INTRODUCTION

The use of checklists by estimators is widespread and common. This practice comes about by the nature of the estimator's work, in which it is necessary to make sure that nothing of any significance is overlooked or alternatively included when it should have been excluded. To try to remember the thousand and one possibilities which may or may not arise would indeed be a herculean task. Thus most estimators prepare or acquire fairly extensive checklists of the more general aspects of their work, together with detail about particular aspects of their special interests, to which they can refer when preparing an estimate.

In this appendix, some of the more common items that might appear on most estimator's lists are included for reference. The items have been arranged according to the recommended Masterformat Division numbers, and within the Divisions, in alphabetical order.[1]

E.2 DIVISION 00 CONTRACT

Addenda
Bid forms
Bonds and insurance
Contract forms
Drawings
Instructions to bidders
Occupancy
Permits
Specifications
Surveys
Time and timing

---

[1] See also "Construction Contracts and Specifications," Chapter 17, Recommended Section Content, for further information on this topic (Reference 1 in Section C.11).

E.3 DIVISION 01 ADMINISTRATION

Access to site	Storage space
Accommodation	Submissions
Allowances	Superintendent
Alternatives	Temporary facilities
First aid	Temporary services
Hoisting	Testing
Parking	Traffic control
Scaffolding	

E.4 DIVISION 02 SITE WORKS

Clearing	Paving
Demolition	Piles
Drainage	Protection
Excavation	Soils reports
Filling	

E.5 DIVISION 03 CONCRETE

Accessories	Finishes
Formwork	Precast concrete
Cast-in-place concrete	Reinforcement
Curing	Toppings

E.6 DIVISION 04 MASONRY

Clay brickwork	Pointing
Cleaning	Reinforcement
Concrete blockwork	Stonework
Mortar	

E.7 DIVISION 05 METALS

Decking	Framing
Expansion joints	Joists
Fabricated work	Ornamental work
Fastenings	Welding
Finishes	

E.8 DIVISION 06 WOOD AND PLASTICS

Architectural woodwork	Ornamental plastic
Fasteners	Rough carpentry
Finish carpentry	Structural plastic
Heavy timber work	Wood treatments

App. E / Checklists

E.9 DIVISION 07   THERMAL AND MOISTURE

Accessories	Preformed roofing
Built-up roofs	Sealants
Calking	Shingles
Dampproofing	Synthetic roofing
Fireproofing	Tiles
Flashings	Waterproofing
Insulation	

E.10 DIVISION 08   DOORS AND WINDOWS

Curtain walls	Glazing
Doors	Hardware
Entrances	Storefronts
Frames	Windows
Glass	

E.11 DIVISION 09   FINISHES

Acoustic systems	Specialty finishes
Carpeting	Stucco work
Coatings	Terrazzo
Flooring	Tile work
Painting	Wall boards
Papering	Wall coverings
Plastering	Wood floors

E.12 DIVISION 10   SPECIALTIES

Accessories	Louvers
Cabinets	Partitions
Chalkboards	Postal items
Corner guards	Screens
Cubicles	Shelving
Flagpoles	Sun control
Grilles	Tackboards
Identification	Vents
Lockers	Wardrobes

E.13 DIVISIONS 11–16

The work of these remaining divisions is usually handled by specialty subtrades on behalf of the main contractor; refer to the Masterformat for more detail.

APPENDIXES
RESEARCH ASSIGNMENTS

The research assignments that follow are intended to give readers a channel through which to acquire knowledge of current local issues and aspects relative to the contents of the chapters contained in this part of the book. The basic objective in all of these assignments is to allow readers to compare theory with practice, by studying this and other similar books, and by encountering opinions and activities, other than those of themselves and this author, on the stated topics. A guide to content, procedure, and evaluation for these assignments has been included in Section 1.4. A review of these guidelines is recommended before starting work on the assignment projects.

CONTRACTOR'S PRACTICE

The work of certain trades, selected from the working drawings included in this book, have been measured and priced in detail in previous chapters. Using these examples as models, measure and price the work of *all remaining trades* in all projects. Prepare a Final Summary for each project, ready for bidding to the owner.

DESIGNER'S PRACTICE

Certain elements, based on the drawings for the Wood-Frame House, have been selected for demonstration of elemental measurement and price analysis in this book. Using these examples as models, measure and price *the elements* in the Post-and-Beam House and the Retail Store Project. Prepare a Final Summary for each project, ready for presentation to the owner.

PART VI

THE PROJECT DRAWINGS

INTRODUCTION

Working drawings for three partial projects and three complete projects are included in this section of the book. The worked examples on measurement and pricing shown throughout the rest of the book are based on or related to these project drawings. These six projects have been selected because of the simplicity of their construction, the typicality of their specific detailing, and their aptness to the particular objectives of this book. In addition, the three completed projects were chosen for the excellence of their general design. The six projects are identified for reference in Table PD.1. It may be noted that two of these projects are drawn in metric dimensions, two in imperial, and two in both metric and imperial, and the project identification numbers reflect this feature.

TABLE PD.1

Project Identification

Project title	Metric	Imperial
1. Residential Foundation	MET-1	IMP-1
2. Retaining Wall Project	MET-2	IMP-2
3. Fireplace Construction		IMP-3
4. Wood-Framed House	MET-3	
5. Post-and-Beam House		IMP-4
6. Retail Store Building	MET-4	

Projects 1 to 3

The drawings for the first three partial projects were specially developed by the author for this book. No permission is required for their reproduction in this book; further reproduction is prohibited without written permission from the publisher.

Project 4

This is a wood-framed residence, with a full concrete basement and a pitched roof, drawn to metric standards applicable in Canada. The building was designed by the Canada Mortgage and Housing Corporation, which is an agency of the federal government of Canada, similar to the Federal Housing Authority in the United States. The design illustration first appeared in the CMHC publication "House Designs for Classroom Use" in June 1980, and it is reproduced with permission, with the following three specific qualifications to which the attention of readers is drawn:

1. The material is provided by way of general information only. CMHC expresses no guarantee or warranty, express or implied, for the efficacy of any advice contained herein for any particular application.
2. Any illustrations, sketches, and diagrams included in this material have been prepared by way of general illustration only.
3. They are not intended to be plans for the construction or creation of any of the subject systems or devices, and CMHC assumes no responsibility for any use of such illustration as building plans.

Project 5

This is a simple post-and-beam residence, having a concrete slab-on-grade floor system and a flat roof, drawn to imperial standards applicable in the United States. The building was designed by H. Douglas Byles and Eugene Weston, Architects. The design illustration first appeared in *California Arts & Architecture* magazine in May 1950, and it is reproduced with permission; further reproduction or use beyond this book is prohibited.

Project 6

The final project is a masonry and wood-framed commercial building, having a concrete slab-on-grade floor and long-span trussed joists supporting a flat roof. The building was designed by Bemben and Kuzych, Architects. The design illustration first appeared in "Construction Contracts and Specifications" in January 1981, and it is reproduced with permission. Further reproduction or use beyond this book is prohibited.

WORKING DRAWINGS

In each case, the original working drawings have been redrawn especially for this book, to improve legibility of lines, letters, and numbers, to eliminate minor discrepancies or ambiguities, to clarify specific details appropriate to the subject matter, and to show acceptable drafting standards and practices by example.

SPECIFICATIONS

The amount of specification detail has been deliberately kept to a minimum, sufficient to give necessary information about particular materials, products, or systems about which the student estimator has to be told to understand the worked examples shown in the book. Such information is not intended to limit the latitude for students and instructors to develop other specific information relative to these drawings, if they wish to modify the given projects to a fairly large degree, for educational purposes.

The first three projects require little more explanation or specification beyond that which is already shown on or inferable from the drawings. The remaining three buildings do require some more information to be provided before comprehensive estimating can occur. The following notes will provide a guide to such information.

In Figure 20.1 a detailed Element Outline is shown for the Wood-Framed House. This element outline can also be used as an outline specification if detailed measurement and pricing is proposed to be done for trades in this building other than the ones exemplified in the book. The outline can also be used as a model to assist students to prepare a similar outline specification for the Post-and-Beam House and the Retail Store Building, prior to doing detailed or elemental estimating on either or both of these two projects.

Before preparing such outlines, reference should be made to current local building codes appropriate to the region in which the reader resides. Most of the basic safety, health, and welfare aspects of construction are governed by such laws. However, there are a few residual specification features or options in each of these buildings about which personal choices and decisions can be easily made. For example, one could choose a particular type of facing brick for a fireplace, or special exotic hardwood paneling to decorate a living room, or choose wood windows instead of metal, and so on. Specification notes can also be made regarding the possible wishes of the hypothetical owners for whom these buildings are being designed and built. For example, the owner may wish to do his or her own landscaping, or painting, or installation of fixtures, and so on.

Furthermore, although one house shows drywall and another shows lath and plaster, it would not be difficult to exchange these two finishes in each of the buildings, or to replace either with yet another (such as plywood paneling), to create more opportunities to use these drawings for additional practice exercises in estimating.

WORKED EXAMPLES

In the first three partial projects, all trades have been measured and priced in detail elsewhere in the book. In the first two of these projects, one has been measured in metric dimensions and the other in imperial dimensions, to show both systems of measurement.

In the remaining three complete projects, the specific estimating uses to which these drawings have been put in this book are indicated in Table PD.2. No work has been measured or priced in the Retail Store Building; this project is intended exclusively for student practice work. Elemental analysis was made of selected elements of the Wood-Framed House.

TABLE PD.2
Estimating Titles

Measurement section	Wood-Framed House	Post-and-Beam House
1. Excavation	Yes	No
2. Formwork	Yes	Yes
3. Concrete	Yes	Yes
4. Carpentry	Yes	Yes
5. Drywall	No	Yes

It will be seen from Table PD.2 that only a few of the possible trades have been measured or priced in the book. The intention is that students should put into practice the precepts and principles of estimating enunciated in this book by measuring and pricing the remaining trades, using these working drawings and the worked examples as models for format and content.

As a side note, it has been mentioned that the drawings could also be used to develop exercises in specification preparation, prior to the estimating exercises. It is also obviously intended that students (and their instructors or mentors) should apply the principles and practices shown in connection with these included projects to sets of drawings and specifications for other projects to which they may have access for educational or practical purposes.

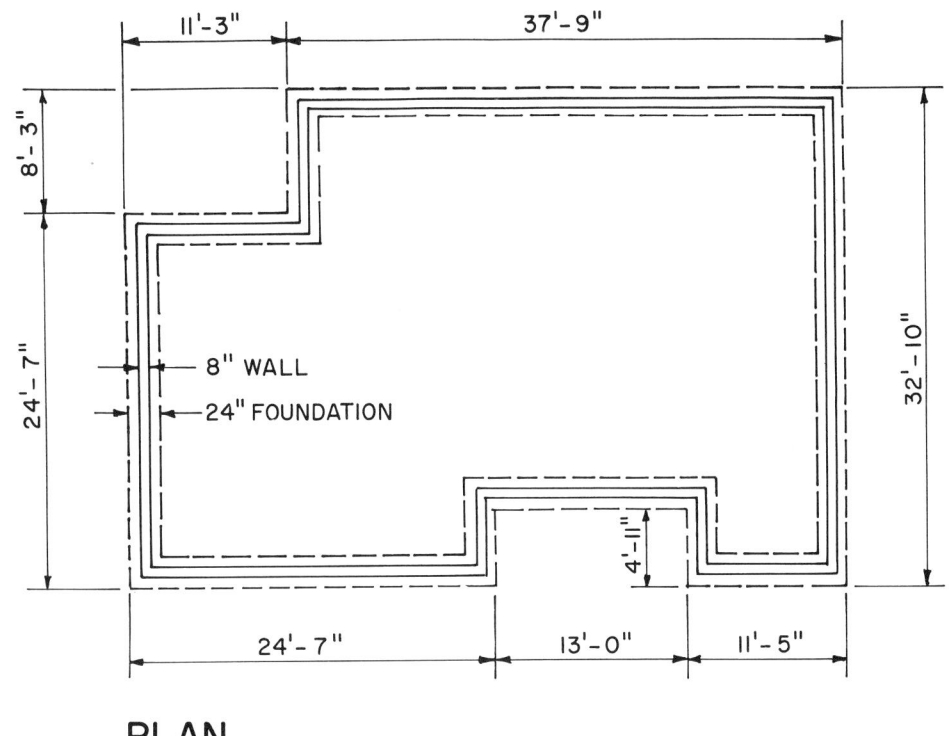

PLAN

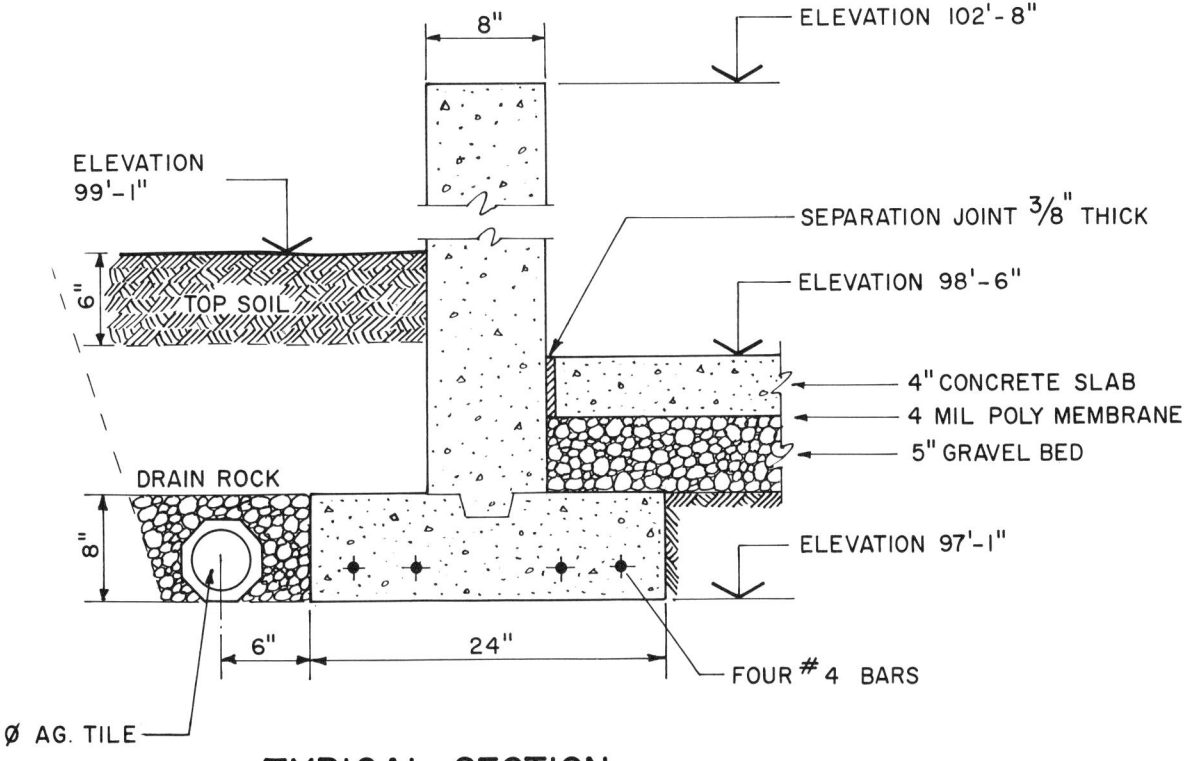

TYPICAL SECTION
RESIDENTIAL FOUNDATION   (Page 1 of 1)   IMP.-1

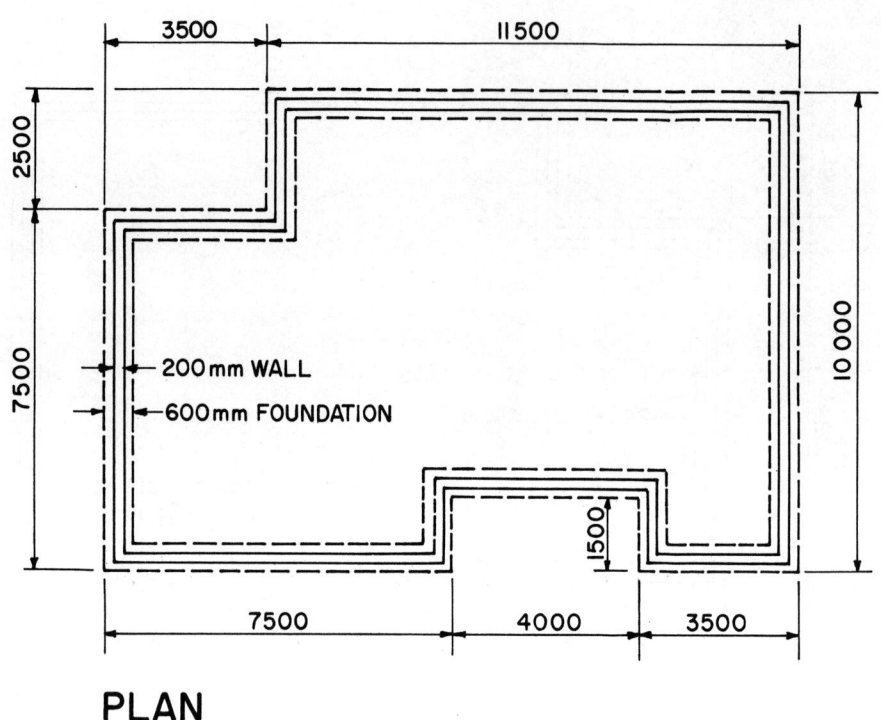

PLAN

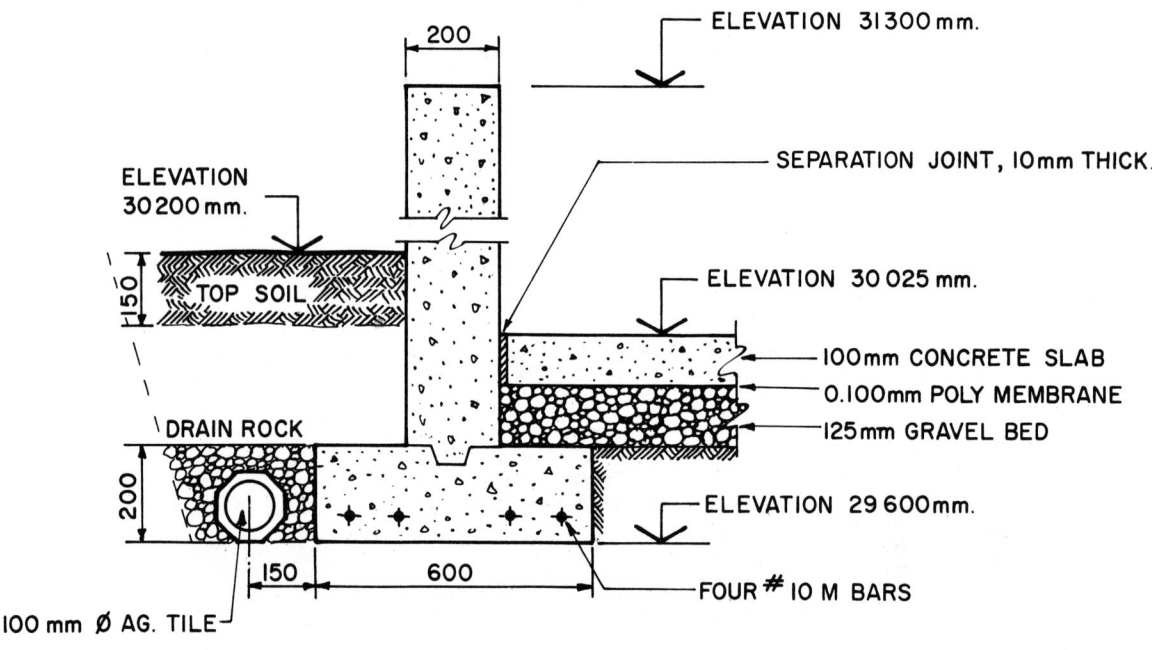

TYPICAL SECTION

RESIDENTIAL FOUNDATION (Page 1 of 1) MET.-1

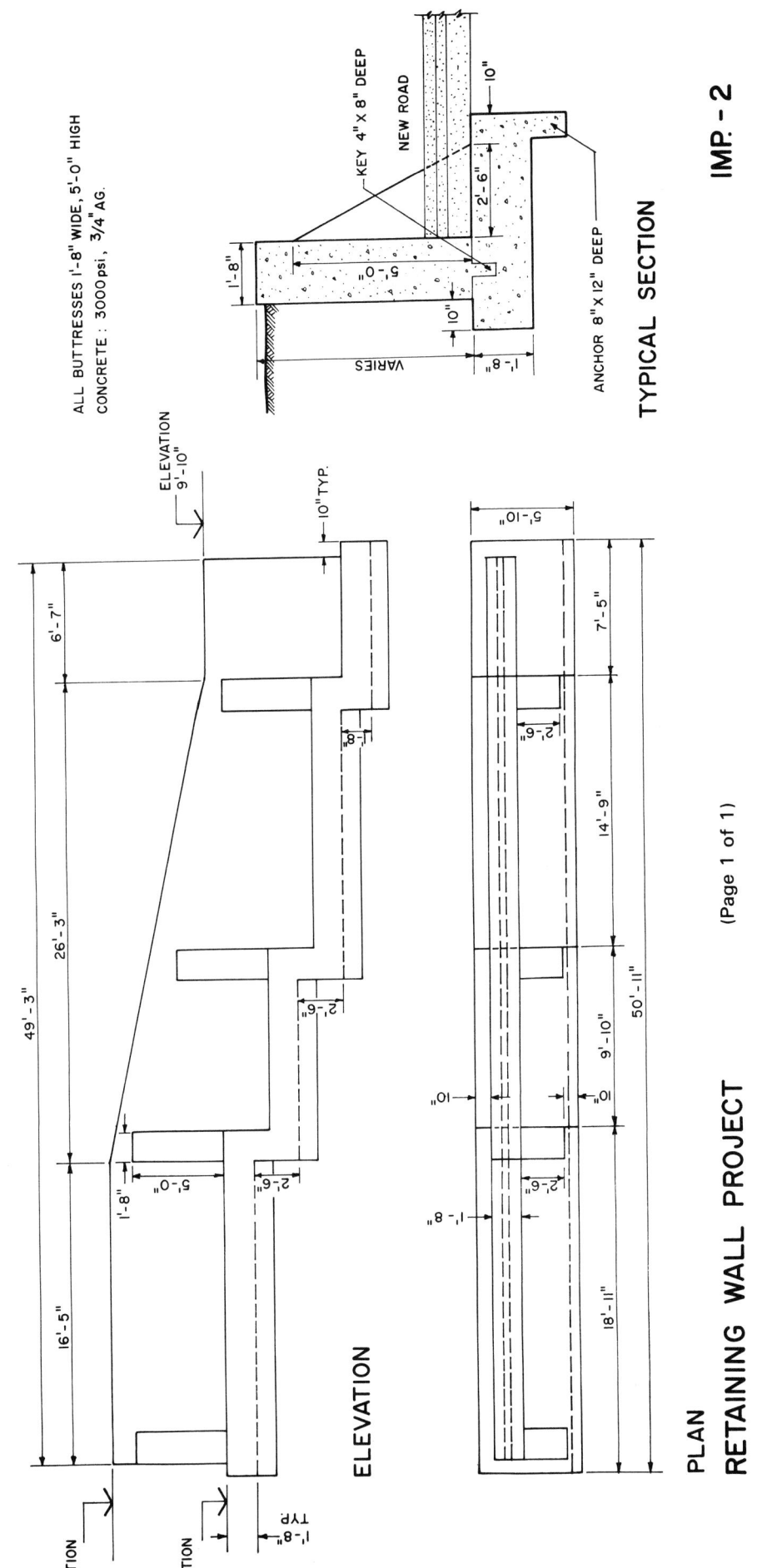

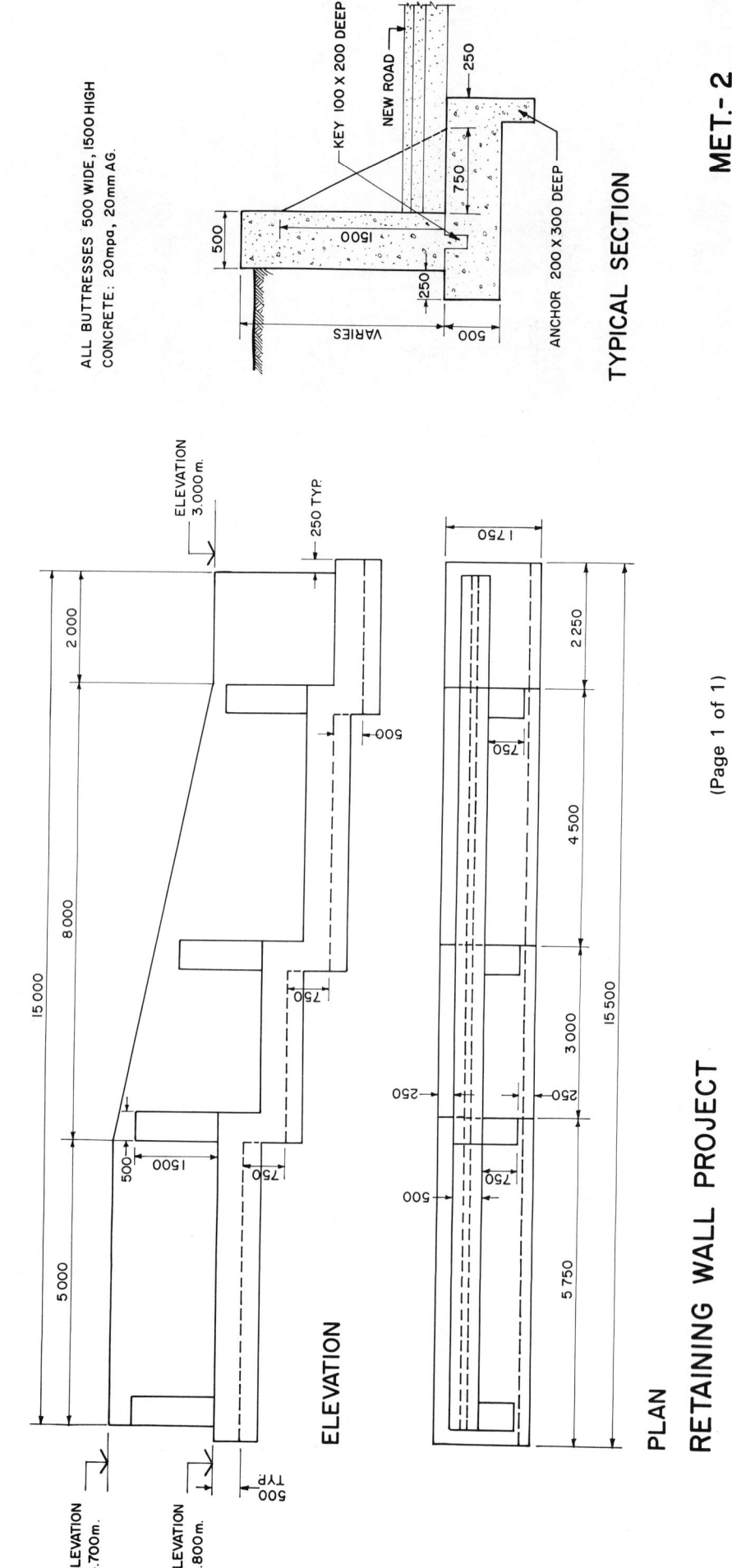

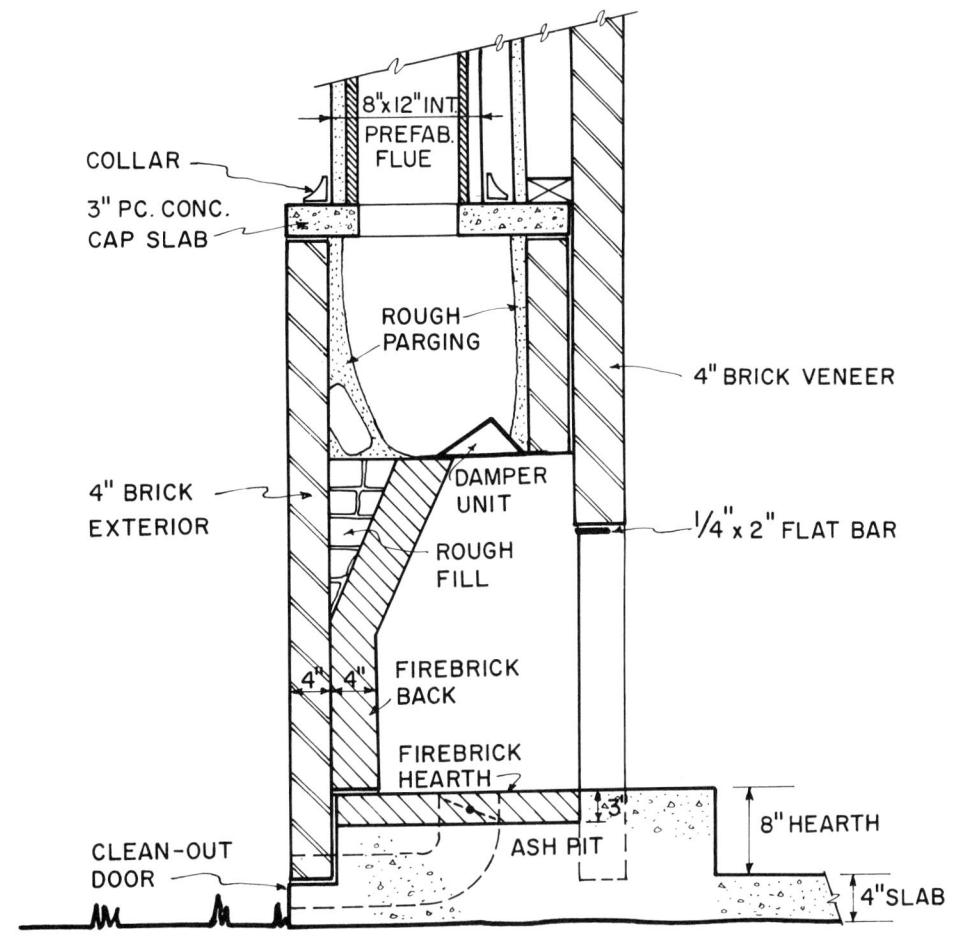

FIREPLACE CONSTRUCTION (Page 1 of 1) IMP. 3

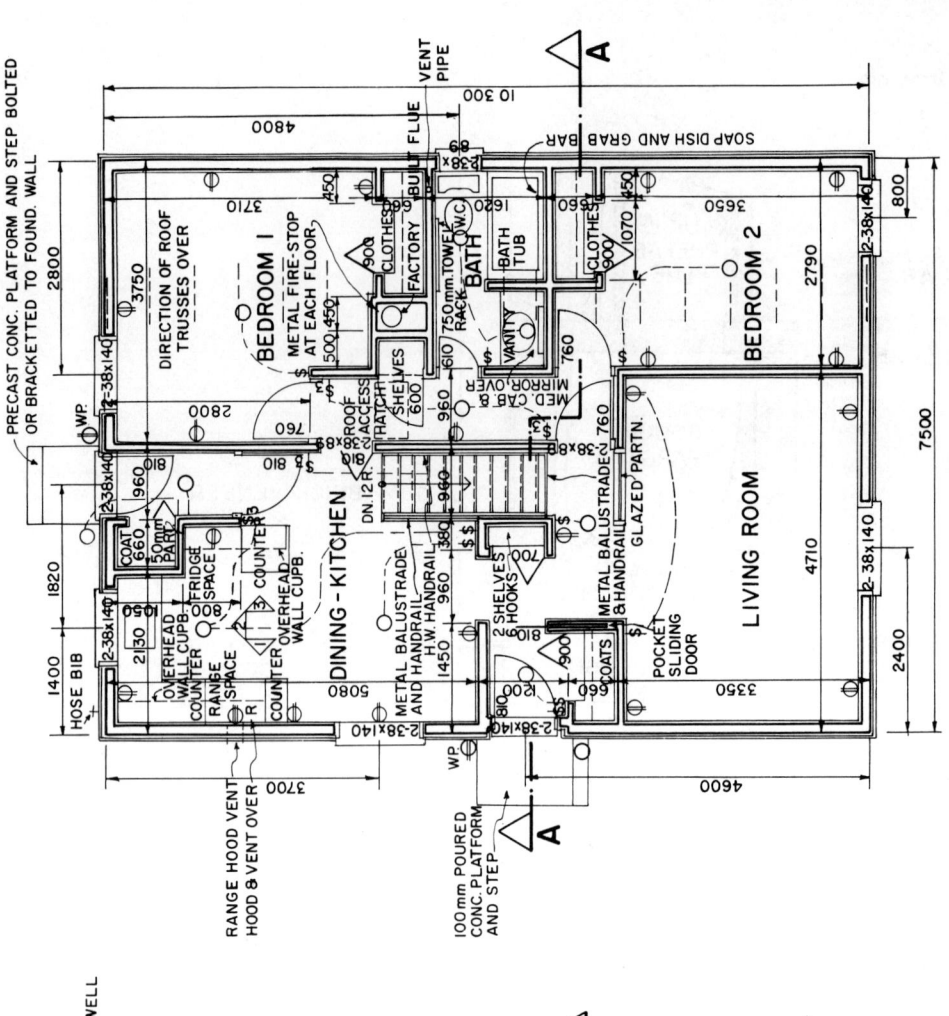

FLOOR PLAN

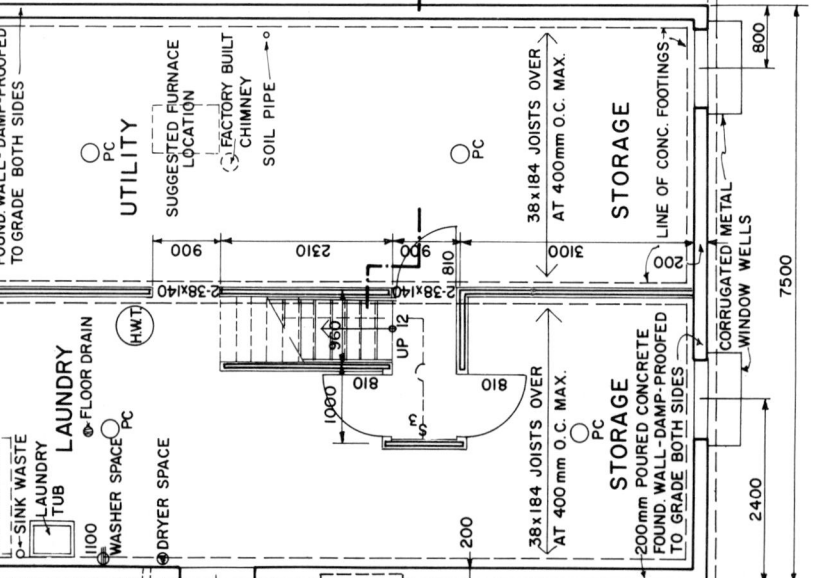

BASEMENT PLAN

WOOD FRAMED BUNGALOW

MET. 3

(Page 1 of 3)

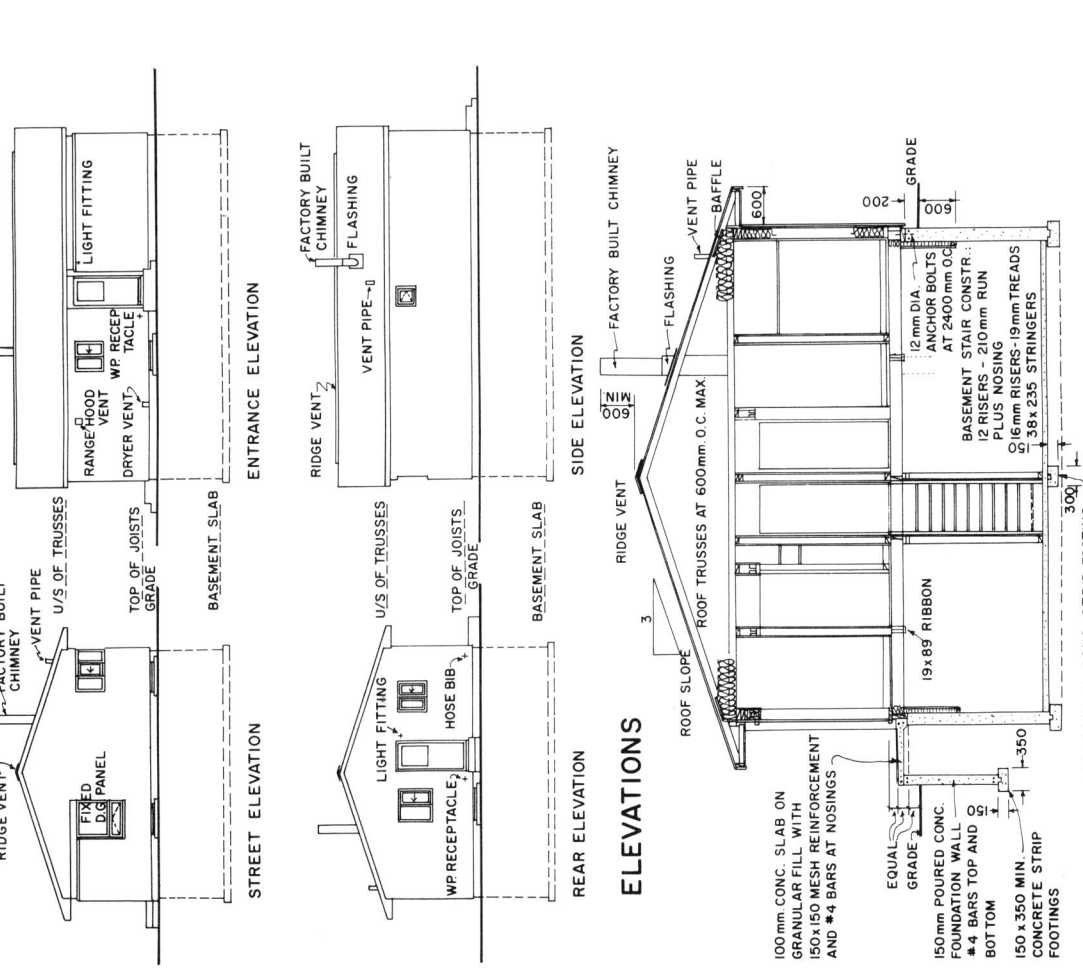

WOOD FRAMED BUNGALOW (Page 2 of 3)

MET. 3

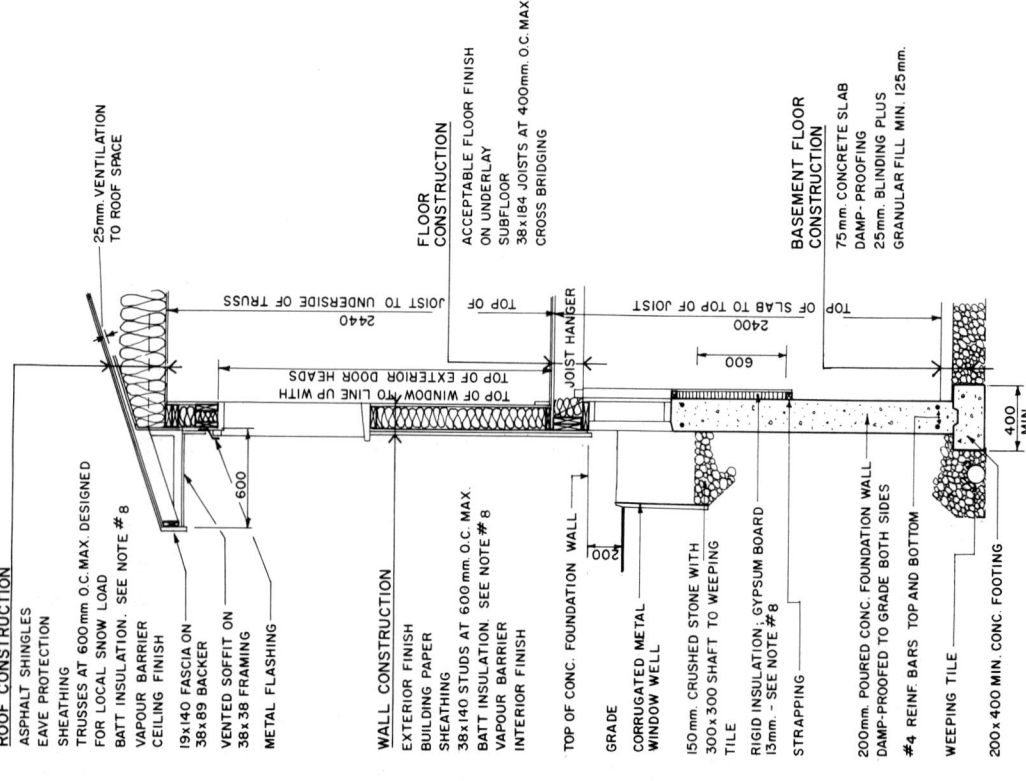

DETAIL WALL SECTION
WOOD FRAMED BUNGALOW

(Page 3 of 3)

MET. 3

GENERAL NOTES:

1. ALL FOOTINGS AND FOUNDATIONS SHALL REST ON UNDISTURBED SOIL OR ROCK. FOOTINGS AND FOUNDATIONS MAY HAVE TO BE INCREASED TO SUIT EXISTING LOCAL SOIL CONDITIONS.

2. DOUBLE JOISTS REQUIRED UNDER PARTITIONS RUNNING PARALLEL TO THE FLOOR JOISTS MAY BE BLOCKED APART 200mm. AT 1200mm.O.C. TO PERMIT PASSAGE OF HEATING DUCTS ETC.

3. THE POSITION AND NUMBER OF ELECTRICAL OUTLETS MAY BE RELOCATED BUT MUST IN ALL CASES MEET THE MINIMUM REQUIREMENTS OF THE RESIDENTIAL STANDARDS AND COMPLY WITH LOCAL CODES.

4. PROVIDE A ROD AND AT LEAST ONE SHELF IN ALL COAT AND CLOTHES CLOSETS.

5. ALL FLOOR JOISTS ARE DESIGNED TO NO. 1 GRADE SPRUCE. ANY CHANGE OR SUBSTITUTION MUST BE APPROVED BY AUTHORITY HAVING JURISDICTION.

6. ALL CONSTRUCTION MUST COMPLY WITH THE RESIDENTIAL STANDARDS CURRENT AT THE TIME OF CONSTRUCTION AND ANY APPLICABLE CODES.

7. THESE DRAWINGS ARE INTENDED AS GUIDES ONLY. THE FINAL CHOICE AND USE OF MATERIALS, AND METHODS AND DETAILS OF CONSTRUCTION, IS THE RESPONSIBILITY OF THE USER.

8. TOTAL R VALUES MUST COMPLY WITH CURRENT C.M.H.C. STANDARDS AS APPLICABLE TO LOCATION. 38x89 STUDS AT 400mm. O.C. MAY BE USED FOR EXTERNAL WALLS IN LIEU OF 38x140 SHOWN IF REQUIRED R VALUES CAN BE ACHIEVED.

9. WINDOW SIZES MUST COMPLY WITH CURRENT C.M.H.C. STANDARDS.

N.B. ALL DIMENSIONS ARE GIVEN IN MILLIMETRES UNLESS OTHERWISE NOTED.

ROOF CONSTRUCTION
ASPHALT SHINGLES
EAVE PROTECTION
SHEATHING
TRUSSES AT 600mm. O.C. MAX. DESIGNED FOR LOCAL SNOW LOAD
BATT INSULATION. SEE NOTE #8
VAPOUR BARRIER
CEILING FINISH
19x140 FASCIA ON 38x89 BACKER
VENTED SOFFIT ON 38x38 FRAMING
METAL FLASHING

WALL CONSTRUCTION
EXTERIOR FINISH
BUILDING PAPER
SHEATHING
38x140 STUDS AT 600 mm. O.C. MAX
BATT INSULATION. SEE NOTE #8
VAPOUR BARRIER
INTERIOR FINISH

TOP OF CONC. FOUNDATION WALL
GRADE
CORRUGATED METAL WINDOW WELL
150mm. CRUSHED STONE WITH 300x300 SHAFT TO WEEPING TILE
RIGID INSULATION; GYPSUM BOARD 13mm. - SEE NOTE #8
STRAPPING
200mm. POURED CONC. FOUNDATION WALL DAMP-PROOFED TO GRADE BOTH SIDES
#4 REINF BARS TOP AND BOTTOM
WEEPING TILE
200x400 MIN. CONC. FOOTING

FLOOR CONSTRUCTION
ACCEPTABLE FLOOR FINISH ON UNDERLAY
SUBFLOOR
38x184 JOISTS AT 400mm. O.C. MAX CROSS BRIDGING

BASEMENT FLOOR CONSTRUCTION
75mm. CONCRETE SLAB
DAMP-PROOFING
25mm. BLINDING PLUS GRANULAR FILL MIN. 125mm.

25mm VENTILATION TO ROOF SPACE
2440 JOIST TO UNDERSIDE OF TRUSS
TOP OF WINDOW TO LINE UP WITH TOP OF EXTERIOR DOOR HEADS
600
JOIST HANGER
2400 TOP OF SLAB TO TOP OF JOIST
600
400 MIN

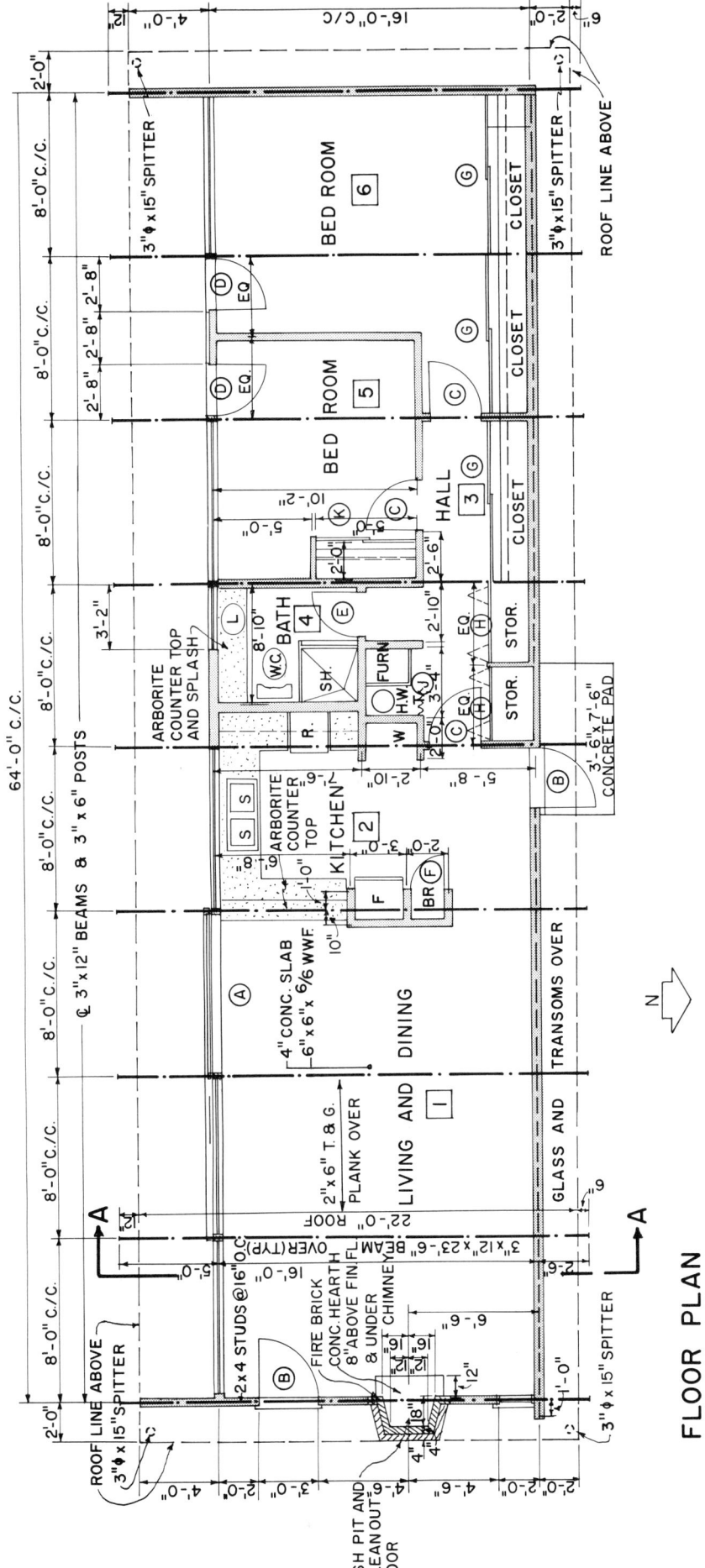

FLOOR PLAN

POST AND BEAM HOUSE

IMP. 4

(Page 1 of 2)

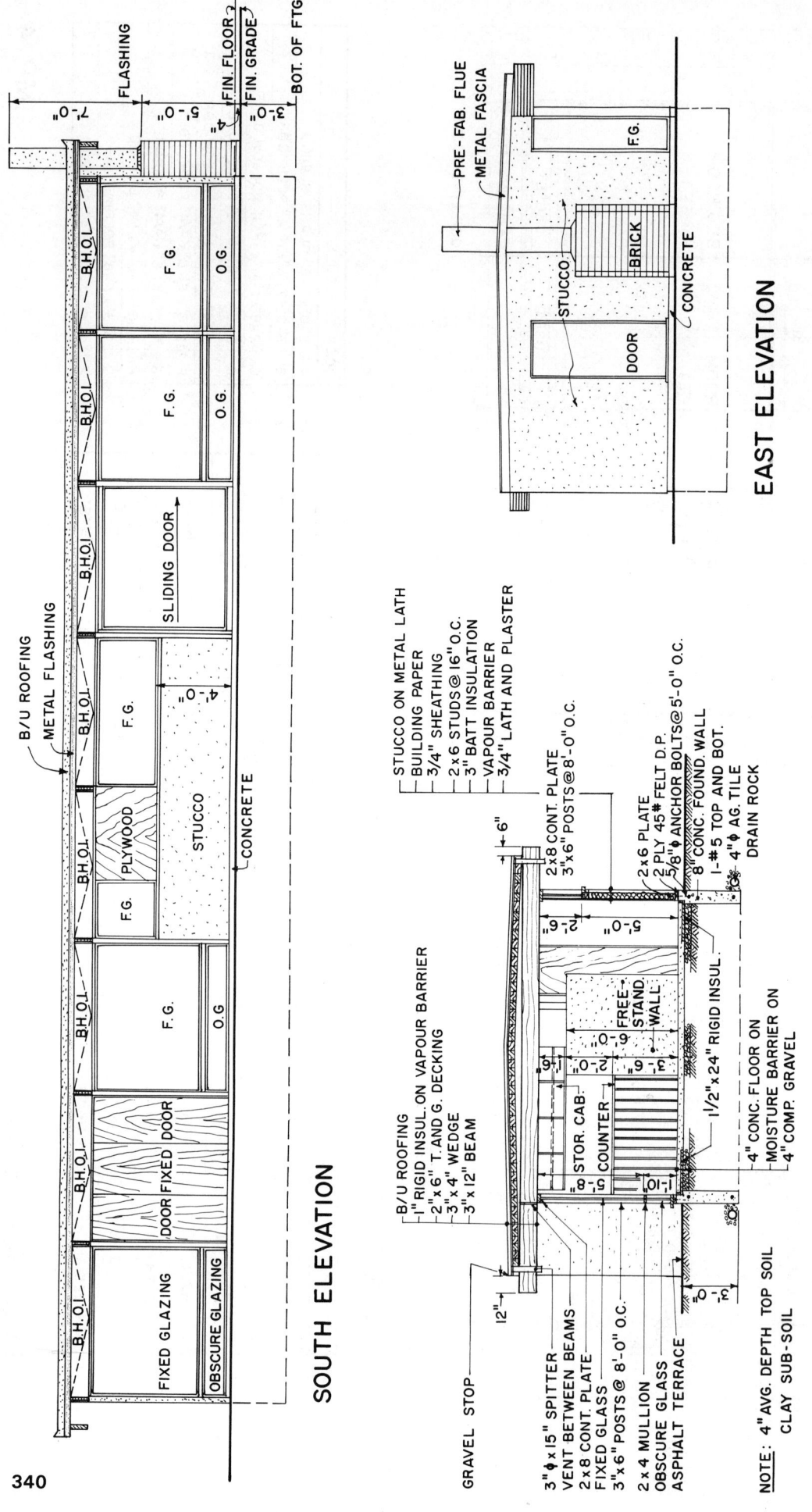

POST AND BEAM HOUSE — IMP. 4 (Page 2 of 2)

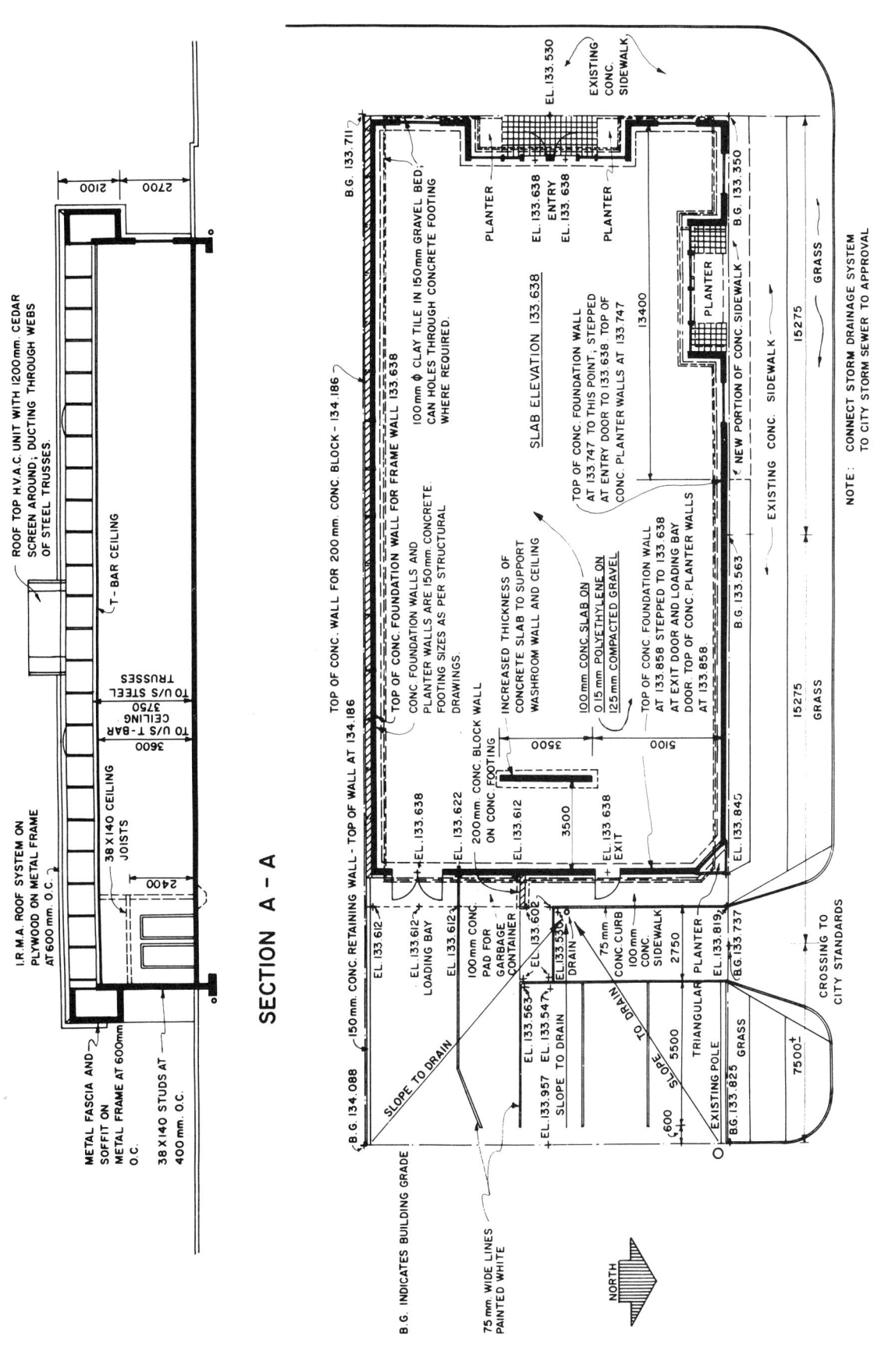

BUILDING GRADE PLAN - FOUNDATION AND SITE DRAINAGE PLAN
RETAIL STORE
(Page 1 of 8)

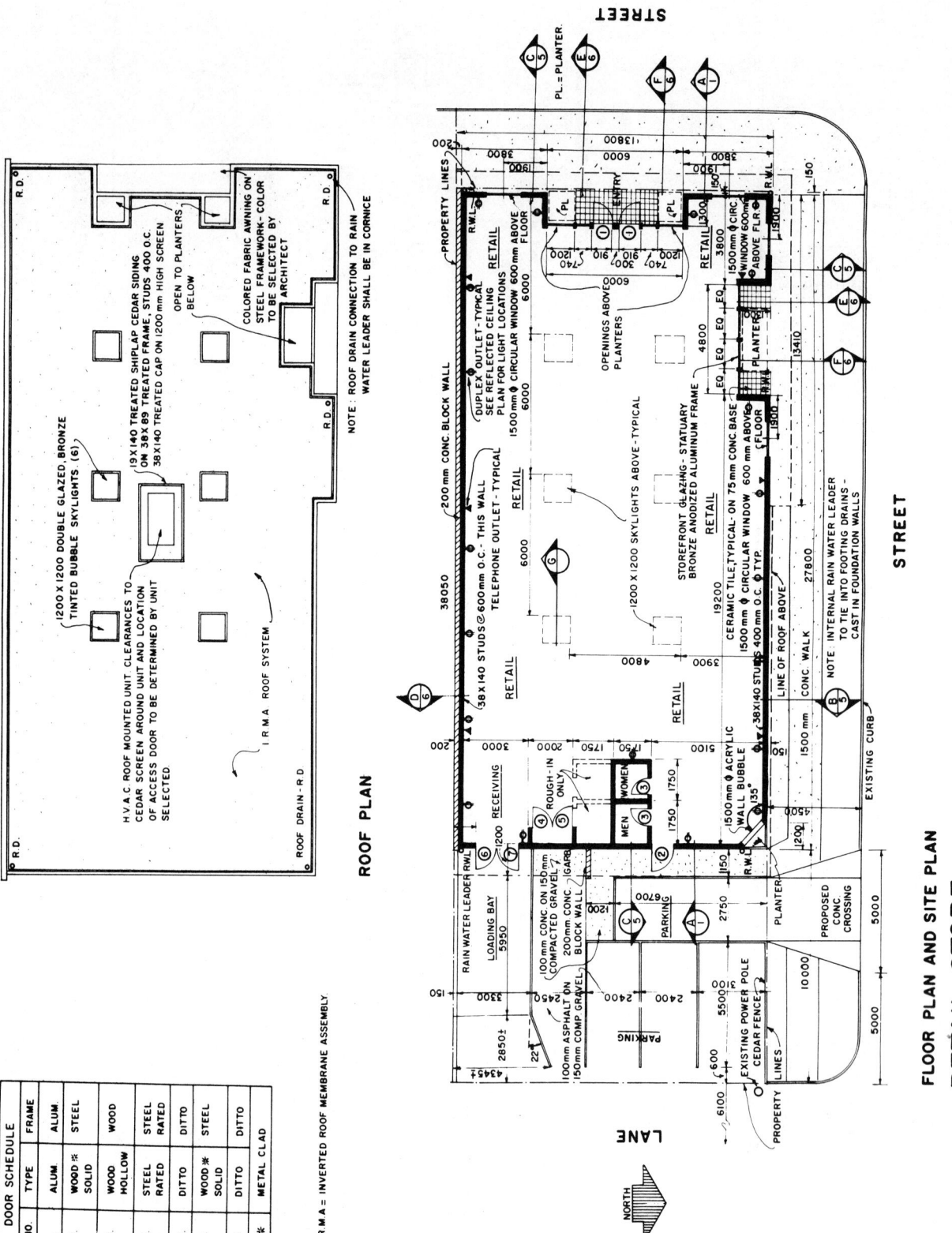

RETAIL STORE

SOUTH ELEVATION

- CANTILEVERED FROM 11th TO 22nd BLOCK COURSE
- BOND BEAM
- CONCRETE BLOCK WALL SEALED AND PAINTED
- CANTILEVERED BLOCK WALL SEE STRUCTURAL DRAWINGS - FROM 15th TO 22nd BLOCK COURSE
- BOND BEAM
- PAINTED CONCRETE BLOCK WALL
- CEDAR FENCE
- 1200
- 2400
- 1200

WEST ELEVATION

- METAL FASCIA
- 1500 ⌀ CIRCULAR WINDOW - TYPICAL
- COLORED FABRIC AWNING
- STOREFRONT GLAZING
- 1200 X 1200 SKYLIGHTS TYPICAL
- CONC. BLOCK WALL
- CEDAR SIDING
- 1350 TYP

EAST ELEVATION

- CONC. BLOCK WALL
- STUCCO
- PAINTED DOORS
- METAL FASCIA
- CEDAR SIDING
- 1500 ⌀ WALL BUBBLE

NORTH ELEVATION

- PAINTED CONC. BLOCK WALL
- CEDAR FENCE
- METAL FASCIA
- 1500 ⌀ WALL BUBBLE
- CEDAR SIDING
- ROOFTOP UNIT ENCLOSURE - TYPICAL
- 1500 ⌀ CIRCULAR WINDOW
- COLORED FABRIC AWNING
- STOREFRONT GLAZING
- SIGNAGE AS PER CITY SIGN BYLAW - TYPICAL
- 1200

MET. 4

(Page 3 of 8)

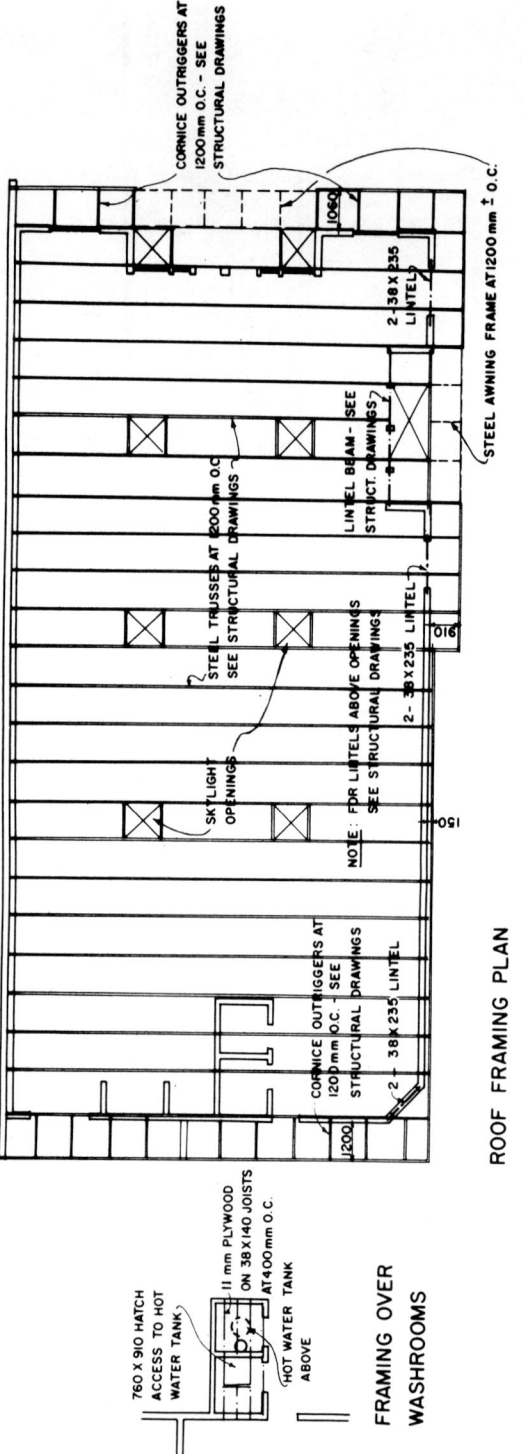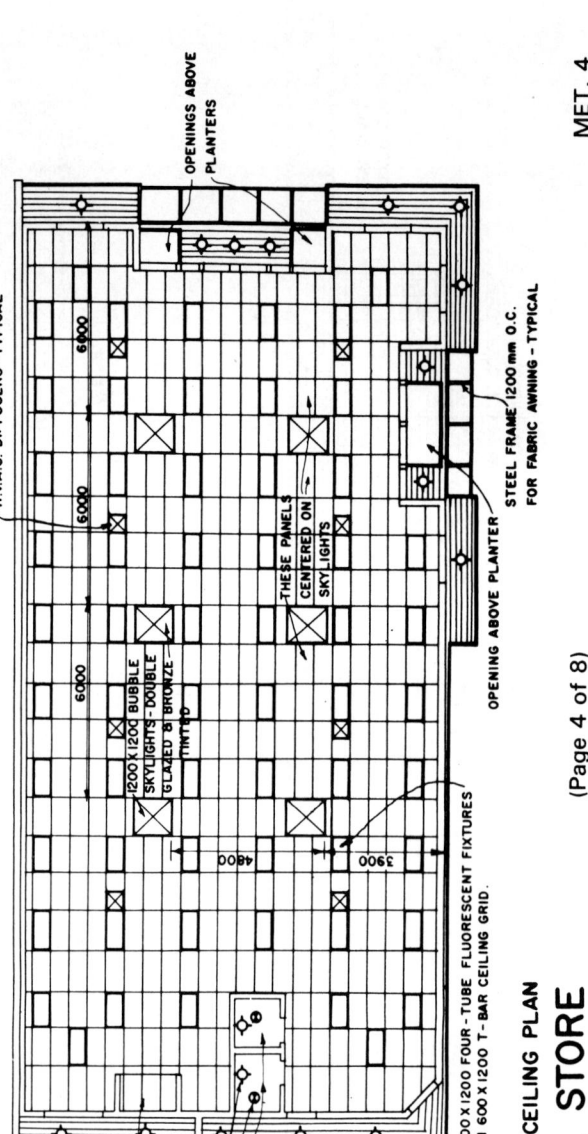

RETAIL STORE

PART FLOOR PLAN

SECTION B - FASCIA

SECTION C - ROOF CORNICES

MET. 4

(Page 5 of 8)

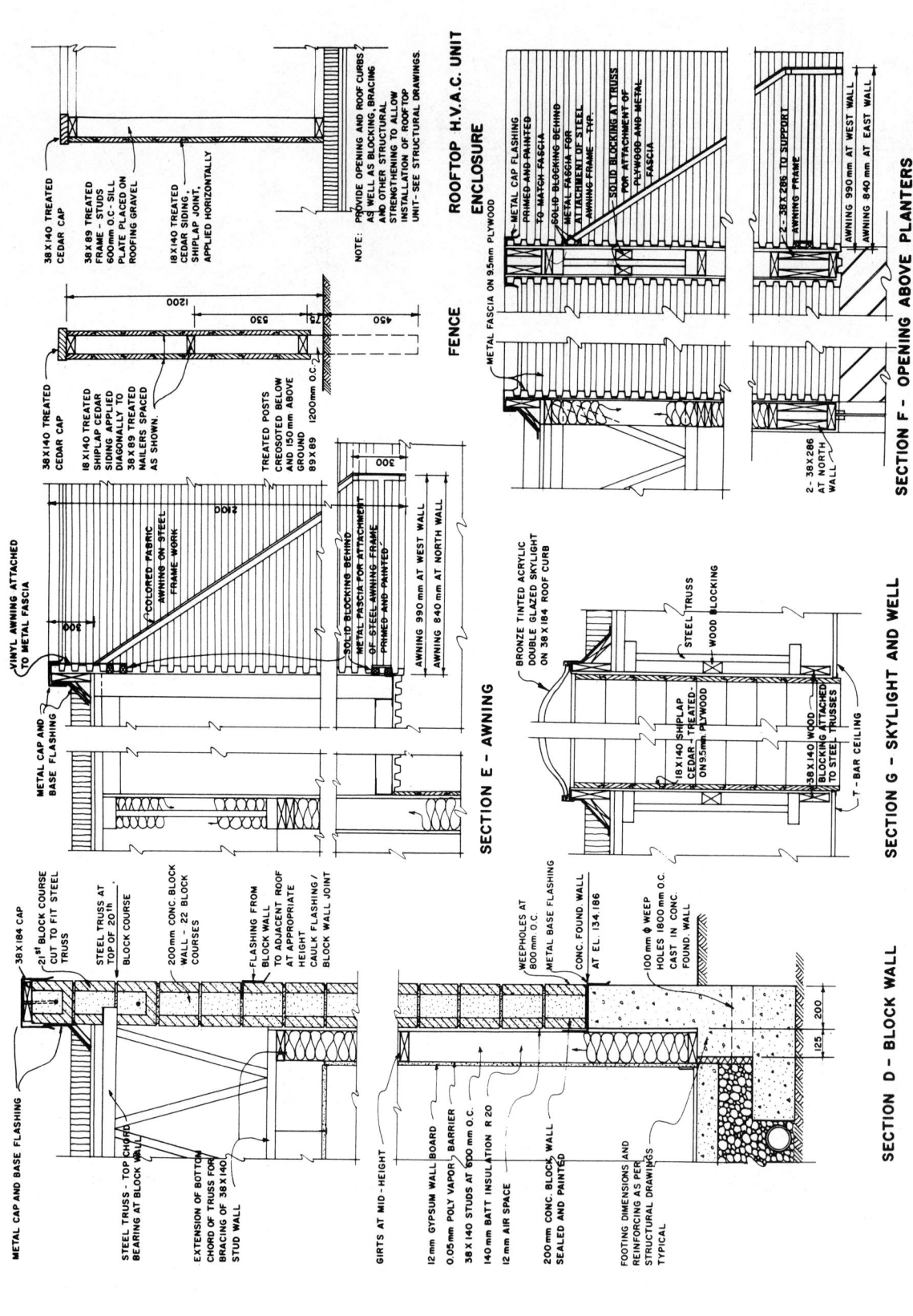

RETAIL STORE

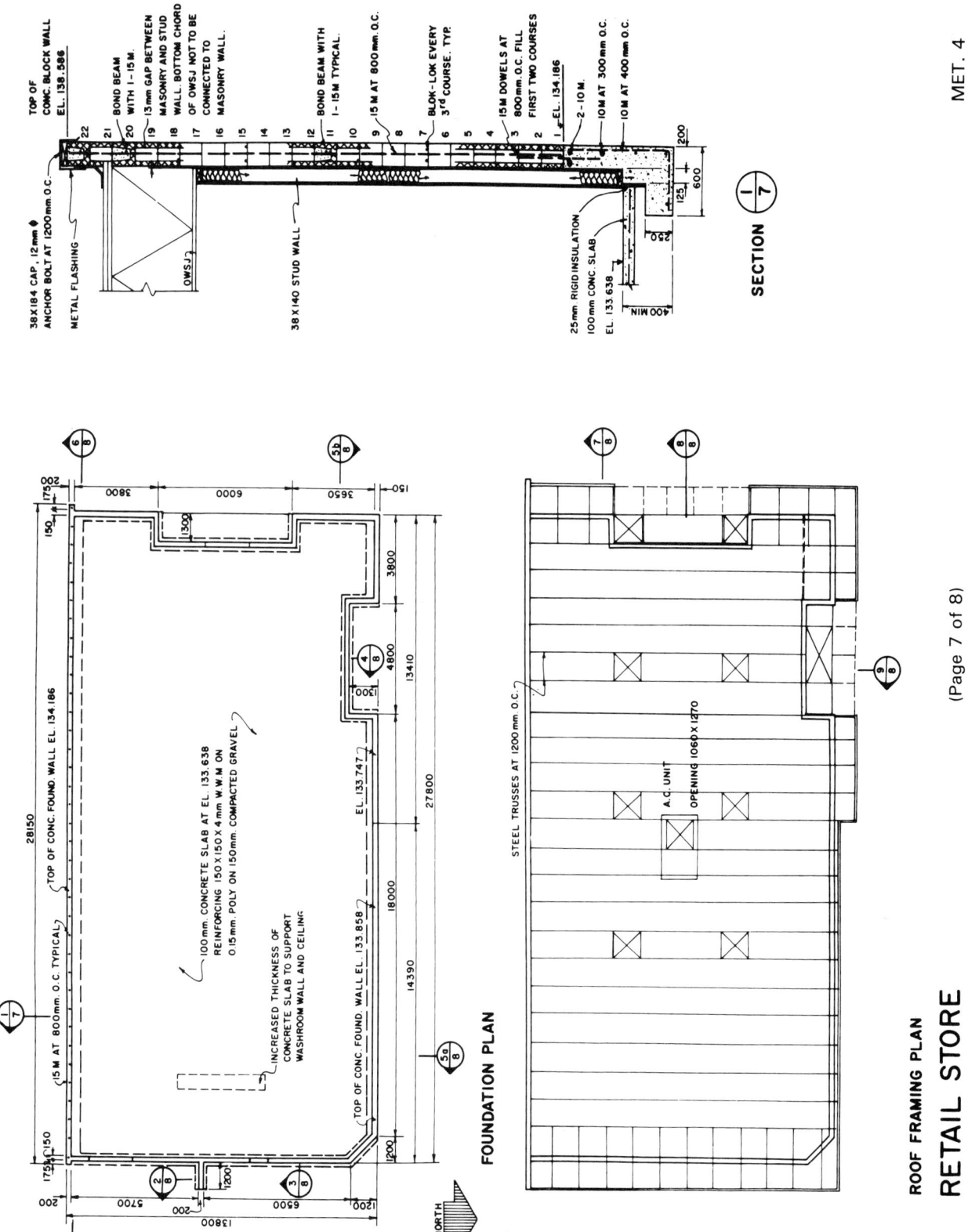

ROOF FRAMING PLAN
RETAIL STORE

(Page 7 of 8)

MET. 4

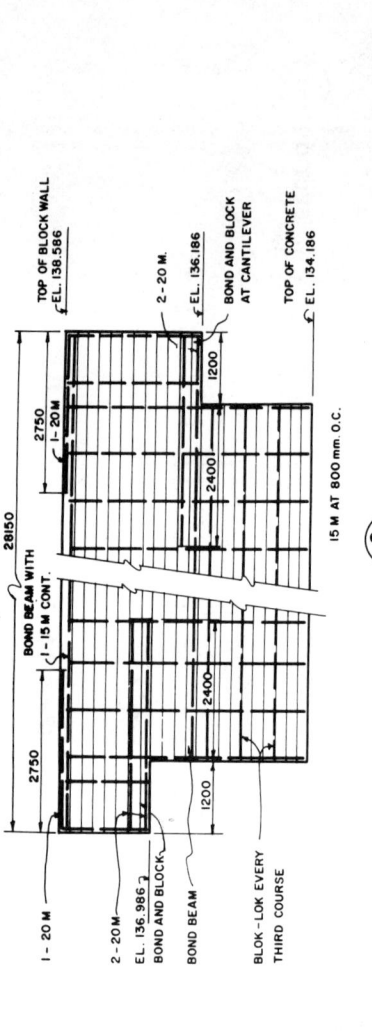
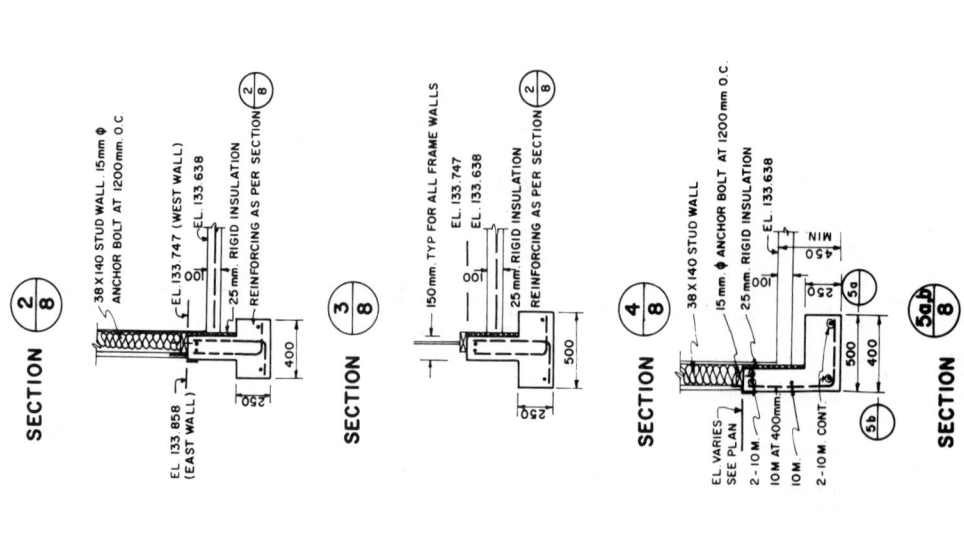

RETAIL STORE

MET. 4

INDEX

Abbreviations, 305
  in descriptions, 306
  in drawings, 305
  in general, 65
  in materials, 306
  in measurement, 306
  in particular, 305
  in specifications, 306
Abilities, estimator's, 15
Acceptance, materials, 102
Acoustic tile, 52
Acre, 76
Activities in construction, 9
Actualities, cost, 103
Acute, definition, 96
Addenda, 23, 141
Addition, 80
Adrian, James, 317
Algebra:
  association, 88
  commutation, 88
  distribution, 88
  equations, linear, 88
  equations, quadratic, 88
  in general, 88
Allowances:
  cash, 40
  contingency, 117
  slope, 161–62, 177
Ampere, 72
Analysis:
  definition, 45, 101
  of measurement, 45
Angle of repose, 161
Applications:
  in general, 5, 147
  in measurement, 159
  in pricing, 219
Applications, computers, 261, 265
Approximate:
  estimates, 67
  quantities, 58
Arc, 96
Architects, 328
Area method of measurement, 216
Arithmetic, 80
Asphalt, 52
Assembly, formwork, 151
Assignments:
  on bibliography, 42
  on bidding and control, 146

  on calculation, 100
  on computer applications, 304
  on construction costs, 146
  on contractors' practices, 326
  on designers' practices, 326
  on elements of pricing, 146
  on estimating methods, 100
  on estimating procedure, 42
  on estimating technique, 100
  in general, 5
  grading of, 7
  on materials methodology, 303
  on measurement by contractors, 304
  on measurement by designers, 304
  on measurement systems, 100
  on mensuration, 100
  on pricing selected elements, 304
  on pricing selected trades, 304
  on pricing techniques, 146
  purpose of, 6
  on style in estimating, 99
  on types of estimates, 42
  on women in building, 42
  on working conditions, 42
Associations, 319
ASTM, 65
Attitudes toward estimating, 25
Authors, 317
Automation Bid Center, 284
Averages, 84
Award of contracts, 134

Backfill, 148
Barrie, Donald, 317
Base, numerical, 72
Bathurst, Peter, 316
Bemben & Kuzych, 328
Benefits, cost, 137
Bidding:
  bonds, 141
  depository, 134
  elements of, 134
  in general, 13, 15
  instructions for, 26, 134
  invitations to, 134
  number of variables, 135
  pricing for, 131
  procedures in, 133–34
  rejection of, 135
Bid Forms, 26, 32, 135
Bit, 262

Blockings, 154, 165
Blocks, 52
Board foot, 76, 193
Bonds:
  bid, 141
  labor and material, 141
  maintenance, 141
  performance, 141
Braces:
  in carpentry, 155
  in formwork, 151
Bricks, 52
Broadscope titles, 309
Budget:
  figures, 131
  realistic, 23
Building:
  characteristics, 107
  construction, 3, 106–7
  elements, 208
  high-rise, 107
  low-rise, 108
  paper, 154, 165
  shape, 107
  size, 107
  type, 107
Bushel, 76
Byles & Weston, 328
Byte, 262

CADD program, 291
Calculations:
  addition, 80
  arithmetic, 80
  averaging, 84
  casting out nines, 82
  definition, 79
  division, 81
  duodecimal, 82
  in general, 14, 79
  machines, 92
  median, 85
  mental, 90
  mode, 85
  multiplication, 81
  percentages, 83
  perimeters, 85–86
  preliminary, 64, 167
  side, 174
  signs, 92
  square roots, 83

Calculations (cont.)
  squaring, 83
  subtraction, 81
  symbols, 92
  techniques, 91–92
Calculators, 92
Canada, 4, 71, 320
Canada Mortgage Corporation, 328
Canadian Institute of Quantity Surveyors, 47, 207, 246
Candela, 72
Cannon, Kenneth, 316
Carpentry:
  blockings, 154, 165
  brace, 155
  building paper, 154, 165
  collar beam, 155
  decking, 154, 165
  finish, 152, 203
  framing, general, 154, 165
  framing, roof, 155
  framing, wall, 155
  header, 155
  insulation, 154, 165
  joist, 155
  labor items, 154, 165
  measurement of, 165
  plate, sill, 155
  plate, sole, 155
  plate, top, 155
  pricing of, 235 et seq
  rafter, 155
  ridge, 155
  rough, 152
  sheathing, 154, 165
  stringer, 155
  stud, 155
  subfloor, 155
  trusses, 154, 165
Case studies, 25, 270 et seq
Cash allowances, 40
Cast-in-place concrete, 150
Casting out nines, 82
CGSB, 65
Chain, measuring, 76
Changes in contracts, 141
Characteristics, building, 107
Checklists, 26, 30, 61, 67, 323
Ching, Francis, 316
Chord, 96
CIQS, 47, 159, 207
Circle, 76, 96
Circumference, 96
Clarity of thought, 15
Classification:
  of construction, 2, 3
  of elements, 68
  of estimates, 19
  of work, 23, 64
Clay, 162
Clearing, 148
Clough, Richard, 315
CMAC Computer Systems, 270
CMHC, 315, 317
Coding of costs, 138
Collar beam, 155
Collier, Keith, 315, 317
Commands, computer, 263
Commercial construction, 3
Communication, 14
Competence of designers, 21
Components of cost, 122
Computation of prices, 121
Computers:
  applications, 261, 265
  bit, 262
  byte, 262
  case studies, 270 et seq
  cadd, 291
  commands, 263
  cpu, 261
  digitizer probe, 285
  discs, floppy, 264
  discs, hard, 264
  economies, 269
  equipment, 268
  fields, 265
  firmware, 264
  hardware, 262
  input, 263
  liveware, 264
  mainframe, 262
  memory, 264
  micro, 263
  modem, 263
  mouse, 263
  operating systems, 270
  output, 263
  peripherals, 263
  pocket, 263
  principles, 261
  programming, 264
  program examples, 270, 278, 285, 292, 296
  program selection, 267
  scrolling, 265
  software, 264
  spreadsheets, 265
  terminology, 261
  touchscreen, 263
  word processing, 266
Concrete:
  cast-in-place, 150
  consolidation, 151
  curing, 152
  deductions from, 164
  finishing, 151
  gravel under slab, 171
  hydration, 151
  ingredients, 150
  measurement of, 164
  mixing, 151
  placing, 151
  pricing of, 230
  proportioning, 151
  screeding, 151, 171
  vapor barrier, 171
Cone, 96
Consolidation of concrete, 151
Construction:
  activities, 9
  building, 3, 106–7
  classifications of, 2–3
  commercial, 3
  contracts, 108
  disciplines in, 3–4
  finishes, 107
  heavy, 2
  industry, 106
  knowledge of, 1
  market, 117
  nonresidential, 3
  organization, 106
  principles of, 14
  residential, 3
  simplicity of, 48
  standardization of, 48
Consultants, 22
Containment of cost, 137
Contents of contracts:
  addenda, 23, 141
  bonds, 141
  changes, 141
  disputes, 141
  documents, 15, 141
  drawings, 141
  insurance, 142
  interpretation, 14, 142
  law, 142
  parties, 142–43
  payments, 142
  progress, 142
  specifications, 142
  time, 142
  work, 142
Contingency allowances, 117
Contractors:
  measurement by, 159
  pricing by, 219
Contracts:
  award of, 134
  construction, 108
  contents of, 140
  discounting, 119–20
  elements of, 143
  in general, 13
  modification of, 140
  privity in, 145
  standard forms of, 140
  types of, 145
Control of cost, 15, 135, 138
Conversion:
  of board feet, 193
  of data, 16
  factors, 166–67
Cosecant, 90
Cosine, 90
Cost:
  accounting, for, 13, 138
  analysis, 137
  benefits, 137
  coding, 138
  components of, 122
  containment, 137
  control, 15, 135, 138
  data, 104
  definition of, 101–2
  escalation of, 117
  estimating, 138
  planning, 138
  reports, 104
  in use, 138
  of work, 138
Cost data:
  actualities, 103
  factual, 112
  probabilities, 103
  productivity, 114
  sources of, 103
Costs:
  components of, 113, 118, 122
  of equipment, 113, 118, 124, 126
  factors of, 102
  in general, 13
  of labor, 113, 118, 123
  of materials, 112, 118, 122
  operating, 113, 126
  of overhead, 113, 119, 127
  owning, 113, 126
  summary, 58
Cotangent, 90
CPU, 261
Crane, tower, 110, 231
Creativity in estimating, 15
Crew sizes, 226, 231–32
Cube, 96
Curing concrete, 152

Dagostino, Frank, 317
Data:
  books of cost, 104
  difficulties with, 104, 117
  on equipment, 113, 116
  establishment of, 112
  factual, 112
  on labor, 113–14
  on materials, 112, 114
  on overhead cost, 113
  productivity, 114–16
  recording of, 62
  risk, 116–17
  sources of cost, 102
Day, 76
Deatherage, George, 315
Decimal fractions, 80
Decking, 154, 165
Deductions:
  from concrete, 164
  from drywall, 165
Definitions:
  acute, 96
  angle of repose, 161
  arc, 96
  backfill, 148
  from building elements, 209, et seq
  calculation, 79
  carpentry, finish, 152
  carpentry, rough, 152
  chord, 96
  circle, 96
  circumference, 96
  clearing, 148
  cone, 96
  consolidation, 151
  cost, 101–2
  cost accounting, 138
  cost analysis, 137
  cost benefits, 137
  cost coding, 138
  cost containment, 137
  cost control, 138
  cost escalation, 117
  cost estimating, 138
  cost in use, 138
  cost of work, 138
  cost planning, 138
  cost process, 138
  cube, 96
  curing, 152
  dewatering, 148

Index

diameter, 96
discounting, 119
disposal, 148
drywall, 156
element, 54
ellipse, 97
equilateral, 97
equipment, 110
excavating, 148
from finish carpentry, 204
finishing, 107, 148, 151, 156
from formwork, 164
fringe benefits, 113
grading, 148
gypsum, 156
hexagon, 97
isosceles, 97
from masonry, 203
from membrane roofing, 204
method, 45
mixing concrete, 151
mobilization, 111, 126
normal, 97
octagon, 97
from painting, 205
parallelogram, 98
payroll burden, 113
placing, 151
plant, 110
polygon, 98
price, 101
prism, 98
privity, 145
profit, 119
protection, 148
pyramid, 98
quadrilateral, 98
radius, 98
rate, labor, 110
rate, wage, 110
rectangle, 98
from resilient flooring, 205
right angle, 98
scalene, 98
screeding, 151
sector, 98
segment, 98
shrinking, 149
sphere, 98
square, 98
stripping, 148
swelling, 149
system, 45
technique, 45
tools, 110
trapezium, 98
trenching, 148
triangle, 99
trim, 157
trimming, 148
unit price, 109
unit rate, 109
value, 101
work, item of, 109
work, unit of, 109
Degrees, 73, 76
Delay, 117
Delivery, materials, 102
Depositing bids, 134
Description:
 of elements, 69
 of items, 23
 of work, 65
Design:
 formwork, 150
 concrete, 150
Designers:
 measurement by, 207, 245
 primary, 22
Detail:
 amount, 21
 quality, 21
 type, 20
Dewatering, 148
Diameter, 96
Digitizer probe, 285
Dimensions:
 derived, 65
 direct, 65
 drywall boards, 156
 extension of, 23
 of items, 23
 in measurement, 65

reflected, 49, 177
side-noted, 65
Disciplines in construction, 3-4
Discounting:
 cash, 119
 contractual issues, 119
 trade, 119
 volume, 119
Discs:
 floppy, 264
 hard, 264
Disposal of earth, 148
Disputes in contracts, 141
Distribution of time, 115
Division:
 dividend, 81
 divisor, 81
 process, 81
 remainder, 81
Division, masterformat, 309
Documentation, 14
Documents:
 contracts, 15, 141
 deposits for, 50
Dram, 76
Drawings:
 contract, 141
 detail, 20-21
 project, 5
Drywall:
 configurations, 156
 deductions from, 165
 in general, 156
 gypsum, 156
 measurement of, 165
 pricing of, 241
 trim, 157
Duodecimal, 80, 82
Duties, estimator's, 15

Earthworks:
 angle of repose, 161
 backfill, 148, 160
 clay, soft, 162
 clay, stiff, 162, 221
 clearing, 148, 160
 dewatering, 148, 160
 disposal, 148, 160
 excavating, 148, 160
 excavation, basement, 161
 excavation, bulk, 161
 finishing, 148, 160
 grading, 148
 gravel, 221
 grids in, 162-63
 hardpan, 162, 221
 loam, 162, 221
 materials encountered, 149, 220
 measurement of, 160
 pricing of, 220 et seq
 protection, 148, 160
 reduced level digging, 161
 repose, angle of, 161
 rock, 161, 221
 sand, running, 162, 220
 shoring, 149
 shrinkage, 149, 161
 slope allowance, 161-62, 177
 stripping, 148, 160
 swelling, 161
 topsoil, 162, 221
 trenching, 148, 160
 trimming, 148
 work spaces, 162
Economies, computer, 269
Economists, building, 23
Efficiency of equipment, 220
Elemental analysis:
 description, 57
 process, 245
 summary sheets, 254
Elements of buildings:
 cladding, exterior, 210, 248
 contingencies, 214, 250
 definition of, 54
 doors, 211, 248
 equipment, 212, 249
 finishes, interior, 212, 249
 fittings, 212, 249
 measurement of, 57, 208
 overhead costs, 214, 250
 partitions, interior, 211, 248
 profit, 214, 250

services, 213
site development, 213
specifications for, 246
stairs, 211, 249
structure, 209, 246
substructure, 209, 246
vertical movement, 211, 249
Elements of contract, 143
Ellipse, 97
Engineering:
 planning, 139
 scheduling, 139
 value, 140
England, 75
Entrances, measurement, 204
Equations, algebraic, 88
Equilateral, 97
Equipment, 110
 backhoe, 221 et seq
 buckets, 232
 buggies, 231
 chutes, 231
 compactor, 125
 compressor, 111
 for concrete, 231
 conveyors, 231
 crane, tower, 110, 231
 fastener, powder, 111
 in general, 124
 hoists, 231
 operating costs, 113
 owning costs, 113
 pumps, 231
 vibrators, 231
Equipment, computer, 268
Erection, formwork, 150, 229
Escalation of cost, 117
Estimates:
 approximate, 19-20, 22, 67
 area method, 55
 base unit, 54
 classification of, 19
 contents of, 21
 criteria for, 22
 detailed, 19-20, 23
 elements, 22
 factor analysis, 55
 in general, 19
 identification, 62
 specific parameter, 55
 summarization of, 22, 32
 types, 54-55
 volume method, 55
Estimating:
 attitudes toward, 25
 components of, 10
 of cost, 138
 definition of, 10
 importance of, 9
 objectives of, 1
 principles of, 10
 process, 10, 24
 risk in, 10
Estimators:
 abilities, 15
 contractor's, 51
 designer's, 51
 duties, 15
 knowledge, 13
 novices, 25, 133
 skills, 14
Examples—general:
 carpentry work, 189, 194
 concrete work, 172, 175, 183, 191
 drainage work, 170, 180
 element, measurement, 215
 element, pricing, 245-56
 equipment costs, 127-28
 excavation work, 169, 179
 finishes, 198
 formwork, 173, 176, 183
 interest calculation, 125
 masonry work, 33 et seq
 measurement, 166
 miscellaneous work, 186
 pricing, 219
 profit margins, 117
 recapitulation, 38-40
 trades, measurement, 167
 trades, pricing, 219
Examples—imperial:
 carpentry, 236-37
 formwork, 227, 229

Examples—imperial (cont.)
  gypsum wallboard, 241-42
  perimeters, 87
  pricing excavation, 121
  pricing paintwork, 123
  recording data, 66-67
Examples—metric:
  areas & volumes, 56
  carpentry, 237
  concrete, 223
  equipment, 223
  excavation, 221
  industrial, 71
  perimeters, 87
  plywood walls, 52
  pricing brick veneer, 122
  pricing roofing, 121
  recording data, 66-67
Excavation:
  backfill, 148
  bulk, 148
  clearing, 148
  dewatering, 148
  disposal, 148
  finishing, 148
  grading, 148
  mass, 148
  materials of, 149
  preparation for, 149
  process, 148
  shrinking, 149
  stripping, 148
  swelling, 149
  trenching, 148
  trimming, 148
Extension:
  in general, 23
  in particular, 67
Extra-over technique, 48, 177

Fabrication, formwork, 150
Factors, productivity:
  carpentry, rough, 235-36, 238
  concrete placing, 231
  drywall, gypsum, 242
  earthworks, 220
  equipment, 116, 223
  formwork, concrete, 226-27, 229
  general, 114
  labor, 114
  multiplication, 81
  specific, 115
Factors of cost:
  building type, 107
  contract, 108
  direct, 102
  indirect, 103
  location, 107
  others, 108
  physical realities, 107
Factual cost data, 112
Falsework in formwork, 150
Fasteners, powder, 111
Fathoms, 76
Feet, 76
Ferry, Douglas, 54, 316
Fields, spreadsheet, 265
Finish carpentry, 152, 203
Finishing:
  of concrete, 151
  of drywall, 156
  of grades, 148
  in construction, 107
Firmware, 264
Flowchart, cost, 269
Foot board conversion, 76, 193
Format for estimates, 62, 68
Forms:
  bidding, 26
  elemental analysis, 254
  general estimate, 36-37
  measurement, 27
  pricing, 27
  quantity take-off, 34
  recapitulation, 35, 39
Formulae, areas/volumes, 96-98
Formwork:
  accessories, 152
  assembly, 151
  braces, 151
  deductions from, 164
  design, 150
  erection, 150
  fabrication, 150
  falsework, 150
  hydrostatic pressure, 150
  measurement, 163
  narrow widths, 164
  removal, 150
  repair, 150
  re-use, 150, 226
  stability, 150
  studs, 151
  ties, 151-52
  wales, 151
Foxon, E. B., 278
Framing, carpentry, 237
Fringe benefits, 113
Functions:
  algebraic, 88
  arithmetic, 80
  geometric, 89
  management, 9
  trigonometric, 89
Furlongs, 76

Gallons, 76
Gender, male/female, 16
Geometry, 89
Gill, 76
Glazing, measurement, 205
Godel, Jules, 316-17
Goldsmith, Immanuel, 315
Gooch, Kenneth, 316
Grading:
  of assignments, 7
  of surfaces, 148
Grains, 76
Gravel under slabs, 171
Gross floor area, 254-55
Guestimating, 10
Gutenberg, 267
Gypsum:
  composition, 156
  dimensions, 156
  drywall, 156
  fastenings, 156
  finishing, 156
  taping, 156-57
  trim, 157
  types, 156

Hanscomb, Roy, 316
Hardie, G. M., 118, 317
Hardpan, 162
Hardware, 262
Header, 155
Heavy construction, 2
Helyar, Frank, 54, 317
Hexagon, 97
Hoists, 231
Hornbostel, Caleb, 316-17
Hudson, Alfred, 315
Hundredweight, 75
Hundredweights, 76
Hunt, William, 317
Huntington, Whitney, 316
Hydration of cement, 151
Hypoteneuse, 89

Identification of estimates, 62
Imperial system:
  development, 75
  tables, 76
Inch, 76
Independence, 15
Industry:
  construction, 106
  economic force, 107
  fragmentation, 106
  organization, 106
Ingredients of concrete, 150
Input, 263
Instructions to bidders, 26, 134
Insulation, 154, 165
Insurance in contracts, 142
Interest on money, 125
Interpretation of contracts, 14, 142
Investment of money:
  recovery of, 125
  return on, 125
Invitations to bid, 134
Isosceles, 97
Items of work:
  definition, 45
  description, 23
  dimensions, 23

Jabine, William, 315
Joist, 155

Kaypro computer, 268
Kharbanda, Om Prakash, 316
Kelvin, 72
Knot, 76
Knowledge:
  categories of, 13
  of construction, 1

Labor:
  costs, 113
  items, 154, 165
  productivity, 220 et seq
  unrest, 117
Landsdowne, David, 316
Law of contract, 142
League, 76
Lewis, Jack, 317
Linear:
  equations, 88
  measurement, 160
Liter, 73
Liveware, 264
Loam, 162, 221
Lumber, 53

Machines, calculating, 92
Maguire, Byron, 317
Mainframe, 262
Management Data Base Systems, 295
Man-hours, 53
Market, construction, 117
Masonry work, 31, 203
Masterformat:
  Broadscope, 309
  Division, 309
  Narrowscope, 309
  Section, 309
  Uniform System, 309
Masterformat Sections:
  general scope App B, 309
  02200 Earthworks, 148
  03100 Concrete Formwork, 149
  03300 Concrete in Situ, 150
  04200 Unit Masonry, 203
  05100 Metal Framing, 203
  06100 Rough Carpentry, 152
  06200 Finish Carpentry, 203
  07500 Membrane Roofing, 204
  07600 Metal Flashing, 204
  08400 Entrances, 204
  08800 Glass & Glazing, 205
  09250 Gypsum Drywall, 156
  09650 Resilient Floors, 205
  09900 Painting, 205
  selected trades, 148
Materials:
  acceptance of, 102
  acoustic tiles, 53
  asphalt, 52
  blocks, 52
  bricks, 52
  clay, soft, 162
  clay, stiff, 162
  concrete, 150, 230
  delivery of, 102
  drywall, gypsum, 53, 241
  earthworks, 220
  in excavation, 221
  factors of cost, 102
  formwork, concrete, 225
  framing, carpentry, 237
  in general, 14
  hardpan, 162
  installation of, 102
  loam, 162, 221
  lumber, 53
  production of, 102
  rock, 162
  roofing felt, 52
  sand, running, 162
  topsoil, 162, 221
McNulty, Alfred, 317
Means, Robert Snow, 316
Measurement:
  accuracy in, 46
  angular, 76

Index

area methods, 55
  of areas, 216
  of carpentry, finish, 203
  of concrete work, 164
  confidence in, 47
  contractors, 159
  cubical, 160
  development of, 72, 75
  of earthworks, 160
  economy of, 46
  elemental, 57
  of elements, 54
  of entrances, 204
  extra-over technique, 48
  of fabrication, metal, 203
  of flashing, metal, 204
  flexbility in, 48
  of flooring, resilient, 205
  format, 62
  of formwork, 163
  of framing, metal, 203
  in general, 5, 15-16, 45
  of glass, glazing, 205
  gross in place, 53
  of gypsum drywall, 165
  history of, 72, 75
  imperial system, 75
  linear, 160
  of masonry work, 203
  materials take-off, 52
  methods of, 51
  metric system, 71
  net method, 52
  numerical, 160
  overall, 47
  of painting, 205
  of parameters, 54
  of quantities, 21
  repetition of, 49
  of roofing, membrane, 204
  of rough carpentry, 165
  separate, 160, 202
  simplicity of, 47
  standards of, 46
  style in, 49
  superficial, 160
  systems of, 71, 75
  take-off, 23, 58
  techniques of, 61, 202 et seq
  of voids, 47
  volume method, 55
  of volumes, 216
  of wants, 47
Measures:
  area, 76
  dry, 76
  English, 76
  linear, 76
  liquid, 76
  mariner's, 76
  metric, 72-73
  miscellaneous, 76
  surveyor's, 76
  time, 76
  volume, 76
Median, 85
Meier, Hans, 318
Membrane roofing, measurement, 204
Memory, 264
Mensuration, 15
  areas, 95
  capacities, 96
  definition, 95
  in general, 95
  lengths, 95
  volumes, 96
Metal work, measurement, 203-4
Meter, 72
Method, definition of, 45
Methods:
  approximate quantities, 58
  for contractors, 51, 159
  of construction, 14
  for designers, 53, 209
  elemental analysis, 54, 57
  of measurement, 51
  parameter estimates, 54-55
  of pricing elements, 245
  of pricing trades, 219
Metric system:
  components, 72
  development, 72
  elements, 72

prefixes, 73
  relationships, 73
  symbols, 74
Mexico, 4
Microcomputer, 263
Microlight Computer Systems, 291
Microline printer, 268
Mile:
  nautical, 76
  statute, 76
Minute, 76
Mixing concrete, 151
Mobilization, 111, 126
Mode, 85
Modem, 263
Modification of contracts, 140
Money:
  interest on, 125
  investment of, 125
Month, 76
Multiplication:
  factors, 81
  process, 81
  product, 81
Mouse, 263

Narrowscope titles, 309
Negotiation, 15-16
Neil, James, 316
Nonresidential construction, 3
Normal, 97
Numbers:
  bases, 80
  cardinal, 80
  conventions with, 80
  fractions of, 80
  integers, 80
  ordinal, 80
  types of, 80
  whole, 80

Objectives:
  of the book, 1
  of the case study, 26
  of estimating, 1
O'Brien, James, 317
Octagon, 97
Operating costs, 113, 126
Operating Systems, 270
Operator costs, 126
Organization of construction, 106
Ounce, 76
Output, 263
Overall, 53
Overhead:
  direct cost, 113, 127
  indirect cost, 113, 130
Owning costs, 113, 126

Painting, measurement, 205
Parado, 10
Parallelogram, 98
Parameters:
  area method, 216
  definition of, 54
  volume method, 216
Parties to contract, 142-43
Payments, contract, 142
Payroll burden, 113
Peck, 76
Percentages, 83
Perches, 76
Perimeter adjustment, 87
Peripherals, 263
Peurifoy, Robert, 317
Pi, 96-97, 99
Pint, 76
Placing concrete, 151
Planning, 139
Plant, 110
Pocket computer, 263
Polygon, 98
Pounds, 76
Preliminary calculations, 64
Prices:
  analysis, 58, 252
  application of, 130
  final, 131
  unit, 22, 109
Pricing:
  carpentry, rough, 235
  concrete, 230

drywall, gypsum, 241
earthworks, 220
elements of, 109
formwork, concrete, 225
in general, 5, 16, 22-23, 101
of elements, 250
process, 131
techniques of, 121, 131
terminology of, 109
Principles of computers, 261
Principles of construction, 14
Prioritization, 11
Prism, 98
Privity of contract, 145
Probabilities, cost, 103
Problem solving, 15
Process:
  of bidding, 133
  of cost control, 138
  of pricing, 131
Production, materials, 102
Productivity:
  data, 114
  factors, general, 14, 114
  factors, specific, 115
  of equipment, 114, 223
  of labor, 23, 114, 226, 231, 235, 241
  of materials, 114, 229
Professional development, 16
Profit margins, 117
Program development, 57
Program examples, 270, 278, 285, 292, 296
Programming, 264
Program selection, 267
Progress of work, 142
Project drawings, 327
Proportioning concrete, 151
Protection of work, 142, 148
Pumps, concrete, 231
Punctuality, 15
Purpose of assignments, 6
Pyramid, 98
Pythagoras theorem, 89, 174

Quadratic equations, 88
Quadrilateral, 98
Quantities, approximate, 59
Quantity of work, 130
Quantity sheet, 34
Quart, 76

Radius, 98
Rafter, 155
Rate:
  labor, 110
  unit, 110
Recapitulation sheet, 38-39
Rectangle, 98
Reflected dimensions, 49, 177
Regulations, 14
Reiner, Laurence, 317
Rejection of bids, 135
Relationships in contract, 143-44
Removal of formwork, 150
Residential construction, 3
Resilient flooring, measurement, 205
Richardsons, 316
Ridge, 155
Right-angle, 76, 98
Risk:
  capital, 117
  in construction, 9
  determination of, 116-17
  in estimating, 10
  in time & pressure, 11
Rock, 162
Rods, 76
Roods, 76
Roof framing, 155
Roofing felt, 52
Rosen, Harold, 318
Rough carpentry, 152

Sand, 162
Sarriel, Edward, 316
Scalene, 98
Schedules, finish, 165
Scheduling, 139
Screeding, 151, 171
Scrolling, 265
Secant, 90
Seconds, 76

Section:
  in specifications, 45
  in survey, 76
Section, masterformat, 309
Sections, 76
Sector, 98
Seeley, Ivor, 317
Segment, 98
Shapes:
  of buildings, 107
  circular, 95
  conical, 96
  cubical, 96
  cylindrical, 96
  irregular, 95-96
  prismoidal, 96
  pyramidal, 96
  quadrilateral, 95
  spherical, 96
  triangular, 95
Sheathing, 154, 165
Shoring, temporary, 149
Shrinking of earth, 149, 161
Side-noting, 65
Signs, calculation, 92
Sill plate, 155
Simplicity of construction, 48
Sine, 91
SI system, 72
Site visit, 16
Sizes:
  of buildings, 107
  of materials, 52
Skills, estimator's, 14
Slope allowances, 161-62, 177
Smith, Ronald, 316
Software, 264
Sole plate, 155
Sources of cost data:
  bonding companies, 103
  company records, 103
  consultants, 104
  government, 105
  insurance companies, 103
  reports, 104
  subcontractors, 103
  supply houses, 103
  labor unions, 103
  precautions with, 105
  press, 105
Specifications:
  and contracts, 142
  data, 21
  outline, 29, 58, 246
  spec notes, 23
Sphere, 98
Square, 98
Square roots, 83
Squaring, 83
Stability of formwork, 150
Stadia, 75
Standard forms of contract, 140
Standardization of construction, 48
Standards in measurement, 46
Stones, 76
Stringer, 155
Stripping of topsoil, 148, 160
Stud, 155
Subfloor, 155
Subtraction:
  minuend, 81
  process, 81
  remainder, 81
  subtrahend, 81
Summaries:
  of cost, 58
  elemental, 254, 256
  estimate, 22-23, 38-39
Summarization, 22
Surveyors, quantity, 22
Swelling of earth, 149
Symbols, calculation, 92
Synthesis, 101
Systems of measurement:
  definition, 45
  imperial, 75
  metric, 71

Take-off, 23, 58
Tangent, 90
Taping drywall, 156
Taxes:
  effect on cost, 23, 108
  indirect cost, 103

Technique:
  definition of, 45
  of measurement, 61, 67, 202, 207
  of pricing, 121, 131, 219
  of summarizing, 24
Temperature, 73
Terminology:
  actualities, 103
  cost, 102
  elemental, 54
  equipment, 110
  estimates, approximate, 19-20
  estimates, detailed, 19-20
  extension, 67
  extra-over technique, 48
  in gender, 16
  gross in place, 53
  labor rate, 110
  man-hours, 53
  method, 45
  mobilization, 111
  net in place, 52
  overall measurement, 47
  parameters, 54
  perimeter adjustment, 87
  probabilities, 104
  plant, 110
  price, 102
  reflected dimensions, 49
  system, 45
  technique, 45
  tools, 111
  unit price, 109
  unit rate, 109
  unit symbols, 67
  utilization hours, 53
  value, 102
  voids, 47
  wage rate, 110
  wants, 47
  work, definition, 52
  work, items of, 45
  work, units of, 45
Terminology, computers, 261
Ties in formwork, 151
Time:
  in contracts, 142
  cycle, 220
  definition of, 142
  distribution of, 115
  fixed, 220
  measurement of, 52, 74-75
  nonproductive, 115
  productive, 115
  semiproductive, 115
  variable, 220
Ton, 76
Tools, 110
Topsoil, 162, 221
Touchscreen, 263
Township, 76
Trades, selected:
  concrete, cast-in-place, 150
  drywall, gypsum, 156
  earthworks, 148
  entrances, 204
  finish carpentry, 203
  formwork for concrete, 149
  glass and glazing, 205
  membrane roofing, 204
  metal fabrications, 203
  metal framing, 203
  painting, 205
  resilient flooring, 205
  rough carpentry, 152
  sheet metal flashing, 204
  unit masonry, 203
Trapezium, 98
Trenching, 148, 160
Triangle, 99
Trigonometry:
  in general, 89
  hypoteneuse, 89
  side, adjacent, 89
  side, opposite, 89
Trimming:
  of carpentry, 153
  of drywall, 157
  of excavation, 148
Trusses, 154, 165
20/80 rule, 11
Types:
  of building, 107
  of contract, 145

Uniform system, 309
Unit:
  masonry, measurement, 203
  prices, 121, 220
  symbols, 67
  of work, 45
United States, 4, 71, 75, 320
Units of measurement:
  acres, 76
  ampere, 72
  board foot, 76
  bushels, 76
  candela, 72
  chains, 76
  circle, 76
  days, 76
  degrees, 73, 76
  drams, 76
  fathoms, 76
  feet, 76
  furlongs, 76
  gallons, 76
  gills, 76
  grains, 76
  hundredweights, 76
  inches, 76
  kelvin, 72
  kilogram, 72
  knot, 76
  league, 76
  liter, 73
  meter, 72
  miles, 76
  minutes, 76
  mole, 72
  months, 76
  nautical mile, 76
  ounces, 76
  pecks, 76
  perches, 76
  pints, 76
  pounds, 76
  quarts, 76
  right angle, 76
  rods, 76
  roods, 76
  seconds, 76
  stones, 76
  temperature, 73
  tons, 76
  township, 76
  weeks, 76
  yards, 76
  years, 76
Utilization-hours, 53

Value engineering, 139-40
Vance, Mary, 317
Vapor barriers, 171
Variables in bidding, 135
Vibrators, concrete, 231
Voids, 47
Volume measurement, 216

Wales, 151
Walker, Frank, 316-17
Walker, N., 315
Wall framing, 155
Wants, 47
Ward, Sol, 316
Wass, Alonzo, 316
Waste factor, 53
Watson, Donald, 317-18
Weather conditions, 117
Weeks, 76
Wine gallon, 75
Women in construction, 16
Word processing, 266
Work:
  architectural, 4
  cost of, 138
  definition of, 142
  electrical, 4, 53
  items of, 45, 109
  mechanical, 4, 53
  spaces, 162
  structural, 4
  units of, 45, 109

Yard, 76
Year, 76

Zehner, J. R., 316